WARFARE ETHICS IN COMPARATIVE PERSPECTIVE

This volume explores East Asian intellectual traditions and their influence on contemporary discussions of the ethics of war and peace.

Through cross-cultural comparison and dialogue between East and West, this work charts a new trajectory in the development of applied ethics. A sequel to the volume *Chinese Just War Ethics*, it expands the range of the earlier work and includes attention to Japan and other Eastern and Western traditions for contrastive reflection and engages with the full range of Chinese intellectual traditions for comparative analysis. The book scrutinizes pioneering works such as *the Mengzi, the Han Feizi*, and *the Seven Military Classics*, investigating their influence in subsequent times. It also engages with new texts and thinkers such as *the Four Books of the Yellow Emperor*, Zeng Guofan, Chiang Kai-shek, and Mao Zedong, along with examining recent writings of the scholars of the People's Liberation Army. The final section of the book identifies and discusses some emerging issues in the comparative study of military ethics, just war and peace that derive from the preceding sections. The volume editors then offer some concluding remarks at the end of the book.

This book will be of much interest to students of the ethics of war and peace, just war theory, military ethics, Asian studies and International Relations in general.

The late **Sumner B. Twiss** was Distinguished Professor Emeritus of Human Rights, Ethics, and Religion at Florida State University, USA, and author of many books.

Ping-cheung Lo is Dean and Rebecca Stephan Professor of Chinese Studies at Fuller Theological Seminary, USA, and author or editor of over a dozen books, in both Chinese and English.

Benedict S. B. Chan is the Director of the Centre for Applied Ethics, an Associate Dean in the Faculty of Arts and an Associate Professor in the Department of Religion and Philosophy at Hong Kong Baptist University.

WAR, CONFLICT AND ETHICS

Series Editors: Michael L. Gross *University of Haifa* and James Pattison *University of Manchester*

Ethical judgments are relevant to all phases of protracted violent conflict and inter-state war. Before, during, and after the tumult, martial forces are guided, in part, by their sense of morality for assessing whether an action is (morally) right or wrong, an event has good and/or bad consequences, and an individual (or group) is inherently virtuous or evil. This new book series focuses on the morality of decisions by military and political leaders to engage in violence and the normative underpinnings of military strategy and tactics in the prosecution of the war.

Moral Injury and Soldiers in Conflict
Political Practices and Public Perceptions
Tine Molendijk

The Empathetic Soldier
Kevin R. Cutright

Law, Ethics and Emerging Military Technologies
Confronting Disruptive Innovation
George Lucas

Ethics at War
How Should Military Personnel Make Ethical Decisions?
Deane-Peter Baker, Rufus Black, Roger Herbert and Iain King

Military Necessity and Just War Statecraft
The Principle of National Security Stewardship
Edited by Eric Patterson and Marc LiVecche

Warfare Ethics in Comparative Perspective
China and the West
Edited by Sumner B. Twiss, Ping-cheung Lo, and Benedict S. B. Chan

For more information about this series, please visit: https://www.routledge.com/War-Conflict-and-Ethics/book-series/WCE

WARFARE ETHICS IN COMPARATIVE PERSPECTIVE

China and the West

Edited by Sumner B. Twiss, Ping-cheung Lo and Benedict S. B. Chan

LONDON AND NEW YORK

First published 2024
by Routledge
4 Park Square, Milton Park, Abingdon, Oxon OX14 4RN

and by Routledge
605 Third Avenue, New York, NY 10158

Routledge is an imprint of the Taylor & Francis Group, an informa business

© 2024 selection and editorial matter, Sumner B. Twiss, Ping-cheung Lo, and Benedict S. B. Chan; individual chapters, the contributors

British Library Cataloguing-in-Publication Data
A catalogue record for this book is available from the British Library

Library of Congress Cataloging-in-Publication Data
A catalog record has been requested for this book

ISBN: 978-1-032-37312-6 (hbk)
ISBN: 978-1-032-37311-9 (pbk)
ISBN: 978-1-003-33637-2 (ebk)

DOI: 10.4324/9781003336372

Typeset in Galliard
by Taylor & Francis Books

This work is dedicated to:

Centre for Applied Ethics
Hong Kong Baptist University

Center for the Advancement of Human Rights
Florida State University

CONTENTS

List of illustrations   x
Notes on citation styles and Romanization   xi
List of contributors   xii
Preface and Acknowledgments   xvi

1 Introduction   1
   Sumner B. Twiss, Ping-cheung Lo and Benedict S. B. Chan

PART I
Comparative Approaches or Visions   9

2 Theoretical vs. Practical Considerations in Doing
   Comparative Military Ethics: An Engaged View   11
   James Turner Johnson

3 Clausewitz vs. Sunzi: Comparing Western and Chinese
   Ways of War and their Ethics   30
   C. Anthony Pfaff

4 East/West Just War Dialogues: Reflections on the Larger
   Implications   47
   Martin L. Cook

PART II
Chinese Thinking about War and Peace   63

5 Just Cause in Mengzi and Gratian: Similar Ideas, Different
   Receptions and Legacies   65
   Ping-cheung Lo

6 Seven Military Classics: Martial Victory through Good
   Governance   91
   Yvonne Chiu

7 Normativity of War and Peace: Thoughts from the
   Han Feizi   113
   Eirik Lang Harris

8 War and Peace According to Huang-Lao Philosophy: Based
   on the *Huangdi sijing*   126
   Ellen Y. Zhang

9 Zeng Guofan's Military Ethics   157
   Jonathan Chan

10 Mao Zedong's Ethics of War (1927–49)?   170
   Sumner B. Twiss

11 Chiang Kai-shek's Military Ethics: An Analysis of His
   Wartime Rhetoric   200
   Sumner B. Twiss

12 A Survey of 21st-Century PLA Scholarship on the Role of
   Military Ethics in Warfare   223
   Mark Metcalf

13 Moral Warfare: Weaponizing Ethics to Weaken, Divide, and
   Smash the Enemy   239
   Mark Metcalf

PART III
New Comparative Horizons on Just War and Peace   255

14 Adjusting Authority: Legitimacy and War in Muslim and
   Christian Traditions   257
   John Kelsay

15 The Right of Self-Defense and the Organic Unity of
   Human Rights   278
   David Little

16 Confucianism, Kant, and the Pacifist Tradition in the
Constitution of Japan 296
Benedict S. B. Chan

17 Conclusion 314
Benedict S. B. Chan and Ping-cheung Lo

Index 317

ILLUSTRATIONS

Table

5.1 The usage of the word 征 (zheng) in the Mengzi and its
English translations   71

NOTES ON CITATION STYLES AND ROMANIZATION

All citation references to a Chinese text are first to the page number of an English translation, and then to standard chapter and /or section numbers.

Romanization of Chinese names and terms in the text mainly follows Beijing's pinyin system. However, as some names and terms are more common in Wade-Giles or other systems, they are not converted to the pinyin system. Author names and book titles are also not affected.

CONTRIBUTORS

Benedict S. B. Chan is the Director of the Centre for Applied Ethics, an Associate Dean in the Faculty of Arts and an Associate Professor in the Department of Religion and Philosophy at Hong Kong Baptist University. He obtained his Ph.D. in Philosophy from the University of Maryland, College Park. His Areas of Research Specialization (AOS) include Applied Ethics & Moral Philosophy, Social & Political Philosophy, and Comparative Philosophy (Chinese & Western). He has published in academic journals such as *Dao: A Journal of Comparative Philosophy, Global Policy, International Journal of Chinese Comparative Philosophy of Medicine, Journal of Bioethical Inquiry*, and *Philosophia*. Website: http://bchan.info.

Jonathan Chan was a faculty member of Religion and Philosophy Department of Hong Kong Baptist University. He retired from the university in the summer of 2019 and was then appointed as Professor of Centre for Global Studies and Director of Integrative Thinking Programme of Shantou University in mainland China in 2019–21. He co-founded *Critical Thinking Web* with Professor Joe Lau in 2004 (website address: http://philosophy.hku.hk/think). He has published numerous articles in journals including *Journal of Religious Ethics, Journal of Military Ethics*, and various significant Chinese journals such as *Zhe xue yan jiu* (*Philosophical Investigation*) and *Zhe xue yu wen hua* (*Philosophy and Culture*), and also in book series such as *Philosophy and Medicine, Annals of Bioethics, Value Inquiry Book Series, Advances in Global Bioethics*, and *Philosophical Studies in Contemporary Culture*. Chan's research areas include the methodology of critical thinking, philosophical logic, moral and political philosophy, bioethics, environmental ethics, military ethics, and Chinese philosophy. At present he is working on a project which aims to provide a philosophical analysis of the Four Books of Confucianism.

Yvonne Chiu is a professor of strategy, policy, and warfare at the U.S. Naval War College and the author of *Conspiring with the Enemy: The Ethic of Cooperation in Warfare* (Columbia University press, 2019), which won the International Studies Association–International Ethics Book Award 2021 and North American Society for Social Philosophy Book Award 2020. She writes on just war theory, military and foreign affairs, and authoritarianism. She has been a National Fellow at the Hoover Institution, a Member of the Institute for Advanced Study, a postdoctoral fellow at the Political Theory Project (Brown University) and an Assistant Professor of Politics at the University of Hong Kong.

Martin L. Cook retired from the U.S. Naval War College in 2017 as the Admiral Stockdale Professor of Professional Military Ethics. He now holds that title as an Emeritus Professor. Upon retirement, Dr. Cook volunteered to teach at Ashesi University in Ghana and lived and worked in Ghana for one semester in Fall 2016. Dr. Cook was previously Professor and Deputy Department Head in the Philosophy Department at the U.S. Air Force Academy, and the Elihu Root Professor of Military Studies and Professor of Ethics at the U.S. Army War College. In his earlier career, he was tenured at Santa Clara University, and taught at Gustavus Adolphus College, The College of William and Mary, and St. John's College in Santa Fe. He was selected for the Fulbright Program Specialist roster for 2018–2021. Dr. Cook received his Ph.D. and M.A. from The University of Chicago and his B.A. from the University of Illinois. He recently stepped down from serving for seven years as the co-editor of *The Journal of Military Ethics*. He has lectured on topics of ethics, military ethics and international affairs to audiences in Hong Kong, Singapore, Australia, Germany, Serbia, Greece, Colombia, Ecuador, Sweden, France, Norway, Ghana, the Vatican and the U.K. He now resides in Colorado Springs and continues to lecture and consult in the U.S. and internationally.

Eirik Lang Harris teaches in the philosophy department at Colorado State University. His research focuses on early Chinese conceptions of normativity in ethics and politics, with a particular interest in early Chinese political realists such as *Han Feizi*. He is the author of *The Shenzi Fragments: A Philosophical Analysis and Translation* (Columbia, 2016), co-editor of *Adventures in Chinese Realism: Classic Philosophy Applied to Contemporary Issues* (SUNY Press, 2022) and his partial translation of the *Han Feizi* appears in *Readings in Classical Chinese Philosophy*, 3rd ed. (Hackett, 2023). For more on his research and publications, see www.eirikharris.net.

James Turner Johnson is Distinguished Professor Emeritus of Religion at Rutgers University. His research and teaching have focused principally on the historical development and application of the Western and Islamic moral traditions related to war, peace, and the practice of statecraft. He is the author of eleven books, of which the most recent is *Sovereignty: Moral and Historical Perspectives* (Georgetown) and editor or co-editor of six more, of which the most recent (with Eric Patterson) is *The Ashgate Research Companion to Military Ethics* (Ashgate).

John Kelsay is Bristol Distinguished Professor of Religion and Ethics in the Department of Religion at Florida State University. His publications include *Arguing the Just War in Islam* (2007) and *Islam and War: A Study in Comparative Ethics* (1993), as well as numerous articles dealing with the ethics of war.

David Little retired in 2009 as Professor of the Practice in Religion, Ethnicity, and International Conflict at Harvard Divinity School, and was an Associate at the Weatherhead Center for International Affairs at Harvard University. He is now a fellow at the Berkley Center for Religion, Peace, and International Affairs at Georgetown University. From 1989 until 1999 he was Senior Scholar in Religion, Ethics and Human Rights at the United States Institute of Peace in Washington, D.C.

Ping-Cheung Lo is Dean and Rebecca Stephan Professor of Chinese Studies at Fuller Theological Seminary, USA. He is Professor Emeritus in the Department of Religion and Philosophy at Hong Kong Baptist University, where he has served for almost 31 years. He is the author of ten books, co-editor of nine books, and the author of about 80 published articles in the areas of Chinese–Western comparative warfare ethics, Chinese–Western comparative bioethics, and Confucian–Christian comparative ethics.

Mark Metcalf is a Visiting Lecturer of Global Commerce at the University of Virginia's McIntire School of Commerce. He is a retired naval cryptologist and a graduate of the United States Naval Academy. He has graduate degrees in Chinese Studies from the University of Arizona (MA) and the University of Cambridge (MPhil).

C. Anthony Pfaff (Colonel, U.S. Army, Ret.) is currently the research professor for Strategy, the Military Profession and Ethics at the U.S. Army War College's Strategic Studies Institute and a Senior Non-resident Fellow at the Atlantic Council and a Distinguished Research Fellow at the Institute for Philosophy and Public Policy at George Mason University. He has a bachelor's degree in Philosophy and Economics from Washington and Lee University; a master's degree in Philosophy from Stanford University, with a concentration in Philosophy of Science; a master's in National Resource Management from the Eisenhower School for National Security and Resource Strategy; and a Doctorate in Philosophy from Georgetown University. The views expressed in this chapter are his and do not necessarily reflect that of the United States Government.

The late **Sumner B. Twiss** was Distinguished Professor Emeritus of Human Rights, Ethics, and Religion at Florida State University and Professor

Emeritus of Religious Studies at Brown University. He was co-author and co-editor of ten books and author of over seventy articles in the areas of comparative religious ethics, intercultural human rights, comparative study of just war, global ethics, philosophy of religion, and biomedical ethics. He was the former co-editor of the *Annual of the Society of Christian Ethics* (1995–2001) and the *Journal of Religious Ethics* (2001–11), as well as former senior editor of the Georgetown book series *Advancing Human Rights* (2004–08). He was on the editorial boards of *Religious Studies Review* and *Journal of Religious Ethics*.

Ellen Y. Zhang is a professor in the Department of Philosophy and Religious Studies at the University of Macau and Research Fellow of the Centre for Applied Ethics at Hong Kong Baptist University. Her research interests include philosophy of religion, Chinese philosophy (Daoism and Buddhism), comparative philosophy and ethics. Zhang is editor-in-chief of the *International Journal of Chinese and Comparative Philosophy of Medicine* (HK) and member of the editorial board of the *Journal of Religious Ethics* (US) and *South China Quarterly* (Macau).

PREFACE AND ACKNOWLEDGMENTS

This volume is dedicated to the remembrance of Dr. Sumner B. "Barney" Twiss, the first editor of this volume, who passed away on May 22, 2023.

When I was a PhD student in religious ethics at Yale University, we all read *Comparative Religious Ethics: A New Method* by David Little and Sumner B. Twiss. Considering Barney's considerable cumulative contribution to the discipline of comparative religious ethics, I invited him to visit Hong Kong Baptist University in 2008 as a consultant in support of our goal to extend comparative ethics by including Chinese perspectives. At the end of his weeklong visit, we agreed to start a team research project called "Ethics of War and Peace in Chinese Thought," which would be coordinated and funded by the Centre for Applied Ethics of Hong Kong Baptist University, of which I was the Director. Barney was so interested in this research project that, to our surprise, he wanted to be a member of our research team rather than an adviser. Of course, he was most welcome. Accordingly, from 2010 to 2019 I saw and worked with Barney every year, either in conferences organized by our Centre for Applied Ethics or conferences in the U.S. and U.K. where we sometimes presented in the same panel. During this process, Barney also worked closely with my colleagues Jonathan K. L. Chan and Ellen Zhang. The first fruit of this collaborative research was *Chinese Just War Ethics: Origin, Development, and Dissent* (2015), which I and Barney coedited. This book was the first academic work on this topic in any language. We then pressed on with our collaborative work, which led to more conferences at Florida State University and the University of Virginia. In 2021, we began our editorial work on this volume and signed a contract with a publisher. We were in the final stages of manuscript submission in spring 2023, but then the shocking news came that Barney had passed away.

In recent years, most members of our research team and I have since moved on to other stages of academic life. The Centre for Applied Ethics of Hong Kong Baptist University has led the research field on Chinese–Western comparative warfare ethics since the 2010s and Barney was a key member of this academic development. Barney rightly saw that Chinese texts and traditions embodied sophisticated thinking about the ethics of war, which was not only analogous and comparable to Western thought about just wars, but also preceded the latter by many centuries. Barney explored this unfamiliar territory with us and was instrumental in helping us break new ground in the Sino–Western comparative warfare ethics literature. His efforts made us better at what we were trying to do and we would not have been able to accomplish our goals without his sound advice and active participation. His humble presence, humor, and scholarly excellence were deeply appreciated by me and my team members. We are very grateful that our mutually enriching collaboration lasted for more than a decade.

Barney was very friendly to me and to my colleagues. When I visited Florida State University in 2010, he drove me to St. George Island, and we toured the Florida Panhandle during our very enjoyable day off. My former colleague, Jonathan K. L. Chan, also received "VIP" hospitality when he visited Barney. During this period, Barney introduced me to James F. Childress, who was very willing to host two conferences focused on our research project. I met Barney in person for the last time during the second conference at the University of Virginia in 2019. His deteriorating health had prevented him from traveling alone again, but he was still collaborating with us on proofreading the Introduction to this volume during his last month on Earth. We understand from his family that this project was very dear to him; hence, we dedicate this volume to the loving memory of our esteemed colleague and dear friend.

Ping-cheung Lo
Fuller Theological Seminary, U.S.

1

INTRODUCTION

Sumner B. Twiss, Ping-cheung Lo and Benedict S. B. Chan

Overview of Volume

In 2015 P. C. Lo and Sumner Twiss published their co-edited volume *Chinese Just War Ethics: Origin, Development, and Dissent* (Routledge, 2015; paperback, 2016). It covered the major ancient Chinese intellectual traditions and texts on warfare ethics and compared these in varying degrees of depth with the Western just war tradition. With this sequel volume and with a third co-editor we extend beyond the range of our earlier work, especially temporally (into the 19th and 20th centuries, and even dipping in the 21st century). Although China and the West are still our lodestones for comparative purposes, we also include some attention to Japan and a few other traditions for contrastive reflection.

China has clearly become a major world power to be reckoned with, and this is particularly evident from its current military build-up (e.g., aircraft carrier and naval bases) and the way it flexes its economic muscle regionally and across the world. Particularly relevant here are its Belt and Road initiatives, territorial disputes in the South China Sea, peacekeeping forces and economic development in Africa, not to mention its apparent flouting of international legal norms and institutions, making for a great deal of uneasiness in the areas in which it operates. Such uneasiness is only reinforced by dark fears about China's military and territorial intentions as expressed by, for example, surrounding countries in East Asia (e.g., Japan, Vietnam, and the Philippines) and in the ongoing and increasing tensions between it and the United States.

As a continuation of the earlier volume, the present volume continues to engage with the full range of Chinese intellectual traditions for comparative analysis. From the period of the Warring States (425–221 BCE)

DOI: 10.4324/9781003336372-1

onward, Chinese intellectual discourse on warfare has been pluralistic. As in the earlier volume, we continue to explore the rich tradition of the Confucian, the Military Strategy, the Legalist, and the Daoist Schools. Chinese thought is diverse even today and we need to bear this in mind as we engage with China. As in our previous volume, we continue to probe deeper into the pioneering works such as the *Mengzi* (Mencius), the *Han Feizi*, and the *Seven Military Classics*. The book scrutinizes some subtle, hitherto neglected dimensions of these texts as well as investigating their influence in subsequent times. These three texts have influenced Chinese discourse for more than two thousand years and are still being studied today.

The book also breaks new ground in the following ways. In addition to extending our inquiry to developments in early modern and modern China and beyond, we engage with new texts and thinkers, such as the *Four Books of the Yellow Emperor*, Zeng Guofang, Chiang Kai-shek, and Mao Zedong. Many of these research topics are appearing in English for the first time. Moreover, unlike other works that look only at China's past, the previous volume already had some preliminary discussion of the views of the scholars of the People's Liberation Army of the People's Republic of China. The present volume steps up that preliminary effort and has two chapters examining recent, relevant writings of these PLA scholars.

Taken as an integral whole, then, this volume aims to chart a relatively new and timely trajectory in the development of the ethics of war and military ethics more broadly conceived—namely, cross-cultural comparison and dialogue between East and West on a subject of mutual concern. It does this by first introducing three comparative approaches (or visions) for the subject matter; then moving on to consider illustrative historical examples particularly in China (and secondarily related to Japan); and then concluding by discussing certain emerging topics in the ethics of war and peace in the contemporary period.

I Comparative Approaches or Visions

The three comparative approaches are in brief:

In Chapter 2, "Theoretical vs. Practical Considerations in Doing Comparative Military Ethics: An Engaged View," James Turner Johnson discusses adaptation and expansion of the interdisciplinary, dialogical, and historical vision associated with the *Journal of Military Ethics* that is broadly both academic and inter-professional in perspective. In Chapter 3, "Clausewitz vs. Sunzi: Comparing Western and Chinese Ways of War and their Ethics," C. Anthony Pfaff discusses comparing Western and Eastern "ways of war" - goals, means, ideals, restraints and the like - contrastingly symbolized and represented by Clausewitz and Sun Tzu among other thinkers. In Chapter 4, "East/West Just War Dialogues Reflections on the Larger Implications," Martin L. Cook discusses an international affairs approach comparing and contrasting the Westphalian-inflected (and

Western) international legal order with the apparent recent revival of competing non-Western cultural narratives and their concomitant particularistic visions for a new world order.

The point of limning these approaches in the book's first section is to highlight the dialogical complexity of cross-cultural military ethics and to prepare readers to appreciate better the nuances and references embodied in subsequent chapters. While we make no claim that these approaches are the only ones available - nor that they are mutually exclusive - we do think that they are helpful for understanding ongoing developments in East-West dialogue about warfare ethics.

I Chinese Thinking about War and Peace

A Pre-Modern

The second section of the book focuses on Chinese (and more broadly, East Asian) thinking about war and military ethics, again in dialogue with Western thinking about such matters. This section of the book is chronologically divided into two sub-sections - pre-modern and modern - which themselves are chronologically organized. The "pre-modern" chapters cover, respectively:

In Chapter 5, "Just Cause in Mengzi and Gratian: Similar Ideas, Different Receptions and Legacies," Ping-cheung Lo discusses just war thinking in Mengzi (Mencius) in contrast with Gratian's (pioneer figures of their traditions), thus initiating cross-cultural comparison between Confucian and Christian (Western) views. In Chapter 6, "*Seven Military Classics*: Martial Victory through Good Governance," Yvonne Chiu discusses the role of "good governance" in the Chinese military classics and Confucian just war thinking, especially with regard to *ad bellum* considerations. In Chapter 7, "Normativity of War and Peace—Thoughts from the *Han Feizi*," Eirik Lang Harris discusses "political realist" views of the legalist Han Feizi that reject considerations of morality in connection with warfare (only what benefits the state counts as important in what could be construed as "amoral" state consequentialism). In Chapter 8, "War and Peace According to Huang-Lao Philosophy: Based on the *Huangdi sijing*," Ellen Y. Zhang discusses syncretistic views of Huang-Lao philosophy, which combines aspects of legalism with those of classical Daoism (e.g., especially the *Laozi*'s restriction of warfare to self-defense of the state). In Chapter 9, "Zeng Guofan's Military Ethics," Jonathan K. L. Chan discusses Neo-Confucian views of Zeng Guofan, who commanded the Hunan Militia Army during the mid-19th-century Taiping rebellion and then had a hand in rebuilding Nanjing *post bellum*.

In addition to the fact that these chapters are crucially important for understanding properly the chapters in the second "modern" sub-section - particularly when cited - they also represent and inform positions that continue to be influential in Chinese thought and practice.

B. Modern

The more "modernist" sub-section of this part of the book includes chapters ranging across:

In Chapter 10, "Mao Zedong's Ethics of War (1927–1949)?", Sumner B. Twiss discusses Mao Zedong's practical theorizing (or at least rhetoric) in 1927–1949 about just war including *ad bellum, in bello,* and *post bellum* norms. In another chapter also from Twiss, Chapter 11: "Chiang Kai-shek's Military Ethics: An Analysis of His Wartime Rhetoric," Twiss discusses Chiang Kai-shek's military ethics (or at least rhetoric) about such norms during the Second Sino-Japanese War (World War II)—drawing on sources ranging from Confucianism and neo-Confucianism, Sun Tzu, Japanese Bushido, and the law of armed conflict as understood at the time. In Chapter 12 and Chapter 13, "A Survey of 21 Century PLA Scholarship on the Role of Military Ethics in Warfare" and "Moral Warfare: Weaponing Ethics to Weaken, Divide, and Smash the Enemy," Mark Metcalf discusses current military ethics of the People's Liberation Army, which focus largely on just cause and other *ad bellum* issues, while also at the same time "weaponizing" (so to speak) ethics itself as a tool of rhetorical warfare even during peacetime.

The chapters identify a number of competing options for comprehending the complexity as well as the backgrounds underlying and informing current discussions of military ethics in the Chinese context.

III New Comparative Horizons on War and Peace

The third and final section of the book identifies and discusses some emerging issues in the comparative study of just war and military ethics that derive from the preceding:

In Chapter 14, "Adjusting Authority: Legitimacy and War in Muslim and Christian Traditions," John Kelsay discusses "adjustments" in the *ad bellum* notion of legitimate authority either by "redefinition" (in cases of waning political power of a regime) or by "relocation" to other agents (in cases of revolution or rebellion against a regime) guided initially by Christian and Muslim examples and attempting (at least implicitly) to pose related questions for Confucian readjustments of legitimate authority. In Chapter 15, "The Right of Self-Defense and the Organic Unity of Human Rights," David Little discusses reconsideration of the priority, role, and significance of the collective right of self-defense with respect to just cause for going to war as well as re-deploying human rights thinking and practice more generally in warfare ethics. Lastly, in Chapter 16, "Confucianism, Kant, and the Pacifist Tradition in the Constitution of Japan," Benedict S. B. Chan discusses Japan's apparent turnabout from its former Samurai Bushido code to its somewhat surprising and controversial post-World War II adoption of a "pacifist" Constitution arguably influenced by both Western and Confucian sources. Chapter 16, along with the two others in this section, thereby concludes the volume on a somewhat positive and hopeful note about the viability and power of cross-cultural comparative military ethics. In this vein, we have a few further reflections about the volume.

Final Reflections

Realism and Pacifism

While most of the chapters in this book focus on military ethics and just war thinking, some of them do discuss ideas related to the two other important traditions in the ethics of war and peace: realism and pacifism. It is open for debate how we should understand realism and pacifism, and whether they are only two extreme ends in the traditional schema of realism, just war thinking, and pacifism. Nevertheless, at least it is academically valuable to discuss some ideas in realism and pacifism and how they can be related to some concepts in military ethics and just war thinking. For example, Eirik Harris and Ellen Zhang discuss in Chapter 4 and Chapter 5, respectively, how perspectives in Chinese Legalism and Huang-Lao thought can be related to realism. Although Legalism and Huang-Lao philosophy are closer to realism than pacifism, that observation does not mean (or imply) that these sources ignore the importance of peace. For example, Harris argues that drawing from Han Fei's commentary on the *Laozi*, we can see that "peace among rival states is possible, and, in many cases, preferable…That is to say, conflict and war is not inevitable in the international arena even if there are a number of roughly equally powerful states bordering one another." For her part, Zhang also argues that Huang-Lao thought is not a pacifist view while still speaking of peace in the negative sense, writing that "Huang-Lao's definition of peace is simply an absence of violence and war…In other words, its position on peace is *ipso facto* a perspective on violence or chaos."

Compared with Harris's and Zhang's chapters, Benedict Chan (Chapter 12) focuses more on pacifism, especially the Japanese pacifist tradition, one significant element of which is that it accepts the right of individual self-defense while denying the right of national defense. Chan thus argues that the Japanese pacifist tradition is not traditional or absolute pacifism, but rather a close relative of or even a pioneer in "contingent pacifism," a new understanding of pacifism within the just war tradition of thinking. Ping-cheung Lo, in his chapter on Mengzi (Mencius) and Gratian (SECTION II), contends that Menzi's legacy to subsequent Confucianism is ambiguous. The "pacifist bias" in subsequent Confucian political discourses can be traced back to him. As a just war theorist, Mengzi has a strong presumption against war whereas Gratian embraces a presumption against injustice. Some contemporary scholars who are on the side of Gratian characterize such a presumption against war position as "crypto-pacifist."

International Order, Human Rights, and Self-Defense

Some chapters consider that the ethics of warfare is imbedded in discussions of international law and order. Some also consider the ethics of warfare as an important topic from the perspective of the development of international institutions. Martin Cook in Chapter 2 argues that the "liberal" order created in the aftermath of the Second World War embedded in the institutions of United Nations (and the U.N. system more broadly) is now being challenged and potentially transformable by different traditions (such as Chinese, Islamic, and Russian) with different, particularistic visions of what the world order ought to look like. He even argues that current major conflicts are similar to "the last major historical transition in world order: the creation of the West-phalian system of sovereign states in 1648," and that the ensuing tensions are "analogous to forces building toward an earthquake." Although we cannot predict the exact subsequent landscape (as we cannot predict the landscape after an earthquake), we should expect great changes and serious conflicts in future international relations.

While Cook's chapter focuses on the future of international order, David Little's chapter (Chapter 11) focuses more attention on finding a moral foundation for international human rights considered as a major component of the current international order and thus for thinking about the ethics of war. Little argues that the right of self-defense is the moral foundation undergirding most (if not all) human rights and the current international order and how to construe properly the ethics of war.

In addition to comparing how Cook and Little think about warfare ethics and the international order, readers of this volume may also be interested in comparing Little's and Benedict Chan's perspectives on self-defense. For example, Little's chapter has a detailed discussion of Article 51 of the U.N., which claims that every state member of the U.N. has an inherent collective right to self-defense. In contrast to this perspective, Chan's focuses on the Japanese pacifist tradition and Article 9 of the Japanese Constitution, which accepts personal defense but not national defense. This makes the Japanese Constitution a unique case in international institutions, as it denies military action at the state-to-state level, while accepting collective self-defense at the international level. It is open for debate how this Japanese pacifist tradition may permit some kinds of self-defense, such as the alliance between the U.S. and Japan.

A Concluding Word

Finally, in reviewing the various chapters of this volume, the editors could not help but be struck by such recurrent features of Chinese military ethical thought, such as the perdurable influence of the *Seven Military Classics* (especially Sun Tzu); the distinctive role of good governance (and statecraft more generally) in *ad bellum* thinking and the role of morality in considering about legitimate authority and the right to go to war; the strong preference for avoidance of violence and the use of deceptive tactics in "outfoxing" the enemy; the defense, protection, and retention of civilization as a sufficient just cause; and the emergence of pacifism as a legitimate alternative to warfare, though there might be some ambiguity about how exactly to characterize this alternative.

Also, it should not be lost sight of that the chapters of this volume collectively bring Western and Eastern ideas of military and warfare ethics into constructive dialogue with each other, and further, that the intensive focus on China and its traditions provides a modicum of control over the parameters of such dialogue. All the authors are well-known scholars in their respective fields, lending confidence to the accuracy and adequacy of their findings. There are other possibly comparable works, but those collections are somewhat more diffuse and lack a sustained and concentrated focus that "models" what a constructive comparative military ethics dialogue could look like. For example, they try to incorporate myriad traditions, geographical areas, and discrete countries with little or no coordination about what their comparisons amount to. Our prospective volume, by contrast, does have a sustained and constrained focus, thereby serving as a coherent model for what a comparative cross-cultural dialogue can and should do: e.g., tracking the internal complexity of both Chinese and Western military ethics traditions in their own cultural contexts and also identifying important areas of ethical agreement and difference between the traditions when considered across the cultural divide. The book is intended for scholars and students alike and could be used as a primary or supplemental text in courses dealing with just war traditions and theory, military ethics, comparative ethics and philosophy, global ethics, Asian studies, moral problems, and international affairs. These courses could be at the intermediate (undergraduate) or advanced level (graduate) in colleges, universities, and military academies. All three editors have themselves taught such courses at varying levels.

Acknowledgments

The three editors would like to thank the following organizations and people for their generous support of this edited volume. We would like to express our gratitude to Mary Ho Yan Siu and Simon Sai Ming Wong from the Centre for Applied Ethics at Hong Kong Baptist University for their invaluable assistance in editing the manuscript and managing various administrative tasks. We also acknowledge the support of the Center for the Advancement of Human Rights at Florida State University (FSU) and to thank the faculty and students in the philosophy, religion, and ethics doctoral program in FSU's Department of Religion. Chapter 14 is an abridged version of the author's article, which was published under the same title in the *Journal of Law and Religion* 36, no. 3 (2021): 459–94. We are grateful to Cambridge University Press for granting us permission to reprint the article as a chapter in this book.

PART I

Comparative Approaches or Visions

2

THEORETICAL VS. PRACTICAL CONSIDERATIONS IN DOING COMPARATIVE MILITARY ETHICS

An Engaged View

James Turner Johnson

Introduction

Quincy Wright's *A Study of War*, originally published in 1942 during the Second World War (Wright 1942), stands as a landmark effort to identify how war is and has been conceived and practiced across the great variety of cultures of the world. Wright argued that wars across major cultural divides are typically fought with less restraint than wars between belligerents sharing a common culture: that is, conceptions of restraint in war do not cross cultural divides. Samuel P. Huntington more recently returned to this same theme under the rubric "the clash of civilizations" (Huntington 1993), though in addition to acknowledging the dangers of wars across major cultural lines, he suggested ameliorating this danger by efforts to identify commonalties through cross-cultural understanding and to build dialogical bridges.

A unique opportunity for pursuing this approach is provided by scholarship on the idea of just war in the ethical traditions of western and Chinese culture, the focus of this book. On the level of theory, comparative dialogue should aim at understanding both the ethical traditions being examined, uncovering both agreements and differences so as to maximize the former while managing the effects of the latter. But military ethics, a distinctive element of the present book, is not simply theory; it is a practical field having to do with the application of normative concerns to the choice of the use of military force and conduct in the use of that force. So, in examining military ethics, comparative dialogue aimed at bridging cultural divides must also examine this application in each of the cultural contexts.

DOI: 10.4324/9781003336372-3

This chapter begins with a brief examination of military ethics, how it should be conceived and what sorts of issues it addresses, then turns to identifying important areas of agreement and difference between the two culturally grounded ethical traditions of *bellum iustum* and *yizhan*, and lastly shifts to consider what can be said and what more needs to be said about the application of these ethical traditions to military practice, including training, discipline, rules of engagement, and integration of relevant international norms, especially including those bearing on the conduct of war.

What Is Meant by Military Ethics?

I come to the topic of military ethics with a certain background. Though I am not a soldier but an academic, the main focus of my scholarship has been western moral tradition on war (just war tradition), with a secondary and allied focus for much of the last three decades on the corresponding tradition in Islamic religion and culture, that of *jihad* of the sword. I have always been attentive to the interactions between military and non-military influences within both of these traditions and the practical application of the tradition to military affairs, including both the decision to engage in armed conflict and conduct in the use of arms within an overall purpose of serving the good of the political community. With this background, in 2002 I became a founding co-editor of the *Journal of Military Ethics*, serving in that capacity through 2009. The first issue (*Journal of Military Ethics* 1.1 [2002]: 2–3) included a short editorial by me with the title "Thinking Broadly About Military Ethics," aimed at introducing the journal to theorists and practitioners alike as aimed "to stimulate thinking about military ethics, [creating] a broad conversation on normative issues related to military force." Describing the then-current state of normative reflection and discourse on war I wrote,

> Normative issues related to military force are treated today in the distinctive contexts of law, professional military life, and a wide variety of academic disciplines. In some contexts the approach is regulative, in others descriptive; in some it is theoretical, in others practical; in some it is closely linked to the values of a particular culture, while in others the aim is to find common normative terms that rise above cultural differences.

I noted further that conversation across the various disciplinary, professional, and cultural lines has often been difficult and stated again the journal's aim to stimulate and facilitate such conversation, in order to "enrich and advance understanding of the broad spectrum of what is involved in military ethics."

This statement aimed to attract the attention of potential authors from all the arenas in which the ethics of war was being discussed, and I think my summary of the state of the discussion—or rather, distinct discussions—at that time was a fair one. But as the statement said plainly, it also aimed to define military ethics as treated within the pages of the journal as a conversation across disciplines and across cultures. While the journal has since succeeded in both these aims, work within the various specific disciplines has also continued, with the conversation in each one tightly discipline-specific. Should such work be considered "military ethics" or not? Some further specification was needed.

A bit over eight years after I wrote my initial broad definition of military ethics and after Baard Maeland and I, the founding co-editors of *JME*, had stepped down, the new co-editors, Henrik Syse and Martin Cook, sought to specify the field more tightly in their own inaugural editorial statement, "What Should We Mean by 'Military Ethics'?" in the first issue under their leadership (*Journal of Military Ethics* 9.2 [2010]: 119–22). They opened by noting "the great diversity of activities nominally gathered under that rubric [military ethics]." But then they went on to try to narrow the field by articulating their own "core understanding of what military ethics is and ought to be," stating as first and most important, "military ethics is a species of the genus 'professional ethics.'" They defined this field as meaning that it exists

> to be of service to professionals who are not themselves specialists in ethics but who have to carry out the tasks entrusted to the profession as honorably and correctly as possible. It is analogous to medical ethics or legal ethics in the sense that its core function is to assist those professions to think through the moral challenges and dilemmas inherent in their professional activity and … to enable and motivate them to act appropriately in the discharge of their professional obligations.

Syse and Cook followed this statement about military ethics as a species of professional ethics with three paragraphs in which they took specific aim at the burgeoning discussion of the ethics of war in the field of philosophy as a very different phenomenon from what they were describing. They wrote,

> There may be a place within the discipline of philosophy for … conceptual and intra-philosophical arguments, but unless they can be brought to bear on the professional activity of military personnel in some meaningful way, they are academic exercises of interest primarily to other academics within the same field.

In short, such work is not military ethics. Syse and Cook's immediate target here was the work of the philosophers—specifically the analytic philosophers —but how far it applies to work on the ethics of war in other disciplines is not clear. I share their rejection of the debate that had developed (and continues to develop) among the analytic philosophers: this debate has to do with philosophy, not military ethics. I also find little to learn from a good deal of recent religiously centered work ostensibly on just war, where the focus has increasingly been on attempting to cast the just war idea in such a way as to reject war entirely. As for historical work of the sort having to do with understanding the relevant ethical traditions, the focus of much of my work and that of the present volume, later in their discussion of the characteristics of military ethics Syse and Cook made a place for such historical work as one of the defining features of military ethics.

Among these defining features, as Syse and Cook identified them, the first is relevance to the military context and the needs of military practitioners.

They stated as their second defining feature of military ethics that the law of armed conflict (LOAC) is a "fundamental component" of it. They amplified this in two ways: first, LOAC is the body of law unique to the military profession; second, since "the law necessarily lags behind and requires interpretation so as to remain relevant, ... explorations to the limits of current legal guidance, and proposals for modification of law ... make a practical contribution to the body of military ethics."

Then, returning to the earlier discussion, they identified as their third defining feature for military ethics that "historical contributions ... to critical thinking about war and the military profession are an essential piece of a comprehensive understanding of professional military ethics." As with the case of philosophy, though, "historical analysis as a purely historical exercise is not a contribution to professional military ethics," though they do not specify further what this might mean. The modern field of military ethics is rooted in history and historical work, and I have made a point of noting in my own writing that both the just war and *jihad* traditions have, in their development, drawn on sources from various perspectives that have importantly included actual military practice. The publication of Lo and Twiss's *Chinese Just War Ethics* (Lo and Twiss 2015) shows the same to be the case with the Chinese traditions on ethics and war. To be fair to Syse and Cook, their critical language directed at the analytic philosophers—and to historical analysis as purely historical, whatever that may be—is focused on the systematic disconnect between such work and actual military practice. By contrast, historical and/or comparative work on the just war and *jihad* traditions, as well as the various Chinese traditions on the ethics of war, properly involves recognizing the dialogical presence in them of influences from the arena of the practice of war. This should be kept in mind going forward in the project of comparative study of the western tradition of *bellum iustum* and the Chinese tradition of *yizhan*: identifying and elaborating such influences should be a high priority.

Syse and Cook next turned to the field of religion, rejecting "confessionally specific beliefs" as a possible basis for military ethics but singling out as relevant "clarifications of what certain strands, traditions, or texts within an influential religious tradition have to say about military ethics." Happily, that includes my work, that of scholars like John Kelsay on *jihad*, as well as that of Lo and Twiss, the other contributors to their 2015 volume, and many of us contributors to the present volume. Again, this is something to keep in mind as further work on these traditions develops.

Syse and Cook's final point was that "there is some role for the hortatory in professional military ethics." What they meant by this includes "non-rational appeals that motivate" and "tales and examples of exemplary individuals and actions." But, they concluded, this may also backfire, and to choose appropriate examples and heroes "presupposes an antecedent grasp of excellence in military conduct and virtue." Their point here, I think, is not well enough developed to be entirely clear conceptually, though it brings to my own mind examples from all three of the historical traditions I have mentioned. Rather than being only an intellectual exercise, military ethics includes persuasion of those involved in the use of military force to act in certain ways. So far as any of our work involves this dimension of what Syse and Cook have defined as properly military ethics, I suggest we should be alert to tease out as specifically as possible how exemplary figures, events, or contexts shape military ethics and vice versa.

Before leaving Syse and Cook's discussion, I should note that both of them have military connections: Syse, though by profession an academic, is also a reserve military intelligence officer in the Royal Norwegian Defence Force, while Cook, when the above editorial was written, was on the faculty of the U.S. Naval War College, having previously been on the faculties of the Air Force Academy and the Army War College; he is now retired. Baard Maeland, who was co-editor with me, also had a professional military connection, serving as a chaplain in the RNDF and having taught at the Royal Norwegian Army War College. So all these men viewed military ethics in terms of their own professional service, and this testifies in another way to the importance of a relevant connection between scholarship on the ethics of war and the actual practice of war as experienced by military professionals.

I have spent a good deal of time on these two different characterizations of military ethics from the *Journal of Military Ethics*—much more on the latter more specific effort. It would be possible to discuss additional understandings of military ethics, including how the subject is defined in the various service academies and war colleges in the United States and their counterparts in other countries, but I suggest that the two examples I have sketched fairly represent the spectrum. That spectrum still exists, as acknowledged in the General Introduction to a book, *The Ashgate Research Companion to Military Ethics*, I recently co-edited with Eric Patterson, a serving Air National Guard lieutenant colonel as well as a professor of political science. We noted in that Introduction that the field of military ethics continues to be "broad and diverse" and described the purpose of the volume as to provide "a comprehensive and authoritative state-of-the-art review" of thinking from across these diverse perspectives on major issues in military ethics (Johnson and Patterson 2015, 1). If anyone needs a deeper immersion in the question what military ethics is, I recommend this book. But for the purposes of the present project, including both this book and the conferences that have fed into it, thinking in terms of the two descriptions I have described from the *Journal of Military Ethics* should suffice.

If I may sum up: military ethics is broad, diverse, and inherently interdisciplinary, and throughout it is tested fundamentally by the degree to which it has relevance for the conduct of war.

Comparing Traditions On Just War

The Western Tradition on Just War: Its Classic Form

To my mind any effort at establishing a comparative dialogue between traditions on war shaped in different historical and cultural frames requires as thick as possible an understanding of each and every tradition we seek to bring into this dialogue. This is not a maxim equally easy to honor in the case of all traditions, and in prior scholarship focused on comparing the ethics of war across cultures it has largely, if not entirely, been ignored. Some examples will illustrate this. An early comparative book, John Ferguson's *War and Peace in the World's Religions* (Ferguson 1977), focuses on the scriptures of the religions treated and does not attend at all to the post-scriptural development of the traditions on war of these religions; it also entirely omits addressing the broad Chinese tradition, which does not easily reduce to religion. James A. Aho's *Religious Mythology and the Art of War* (Aho 1981) similarly focuses on the scriptures for most of the religions it treats, and though it does provide a summary discussion of several early Confucian figures, it does not treat the western just war tradition as such at all. As for more recent volumes, the chapters in Torkel Brekke's *The Ethics of War in Asian Civilizations* (Brekke 2006) are very uneven in the depth with which they treat the particular civilizations covered, with some treated via their scriptures alone. The same unevenness as to depth characterizes the collection edited by Richard Sorabji and David Rodin, *The Ethics of War: Shared Problems in Different Traditions* (Sorabi and Rodin 2006), which also is limited in the number of traditions it includes. A later collection, *World Religions and the Norms of War* jointly edited by Vesselin Popovski, Gregory M. Reichberg, and Nicholas Turner (Popovski, Reichberg, and Turner 2009), considers both the scriptures and some later developments in the particular religious traditions it treats, though once more there is no discussion of Chinese tradition. (On comparative study of the ethics of war across cultures see further Johnson 2008.)

For the particular case of the western idea of just war, while the last fifty years have produced a large and growing body of writing on this subject, most of that work–including that of major contributors like Paul Ramsey, Michael Walzer, the United States Catholic bishops, Jean Bethke Elshtain, and others, as well as the whole body of work produced by recent analytical philosophers writing on the ethics of war–has had to do with contemporary reconstructions of the just war idea, not investigating the tradition of just war itself. Investigation into that tradition, by contrast, has been a major focus in my own work, and recently major sources in the western tradition on the ethics of war have been made available in the collection jointly edited by Gregory Reich-berg, Henrik Syse, and Endre Begby, *The Ethics of War: Classic and Contemporary Readings* (Reichberg, Syse, and Begby 2006). A narrower focus on key figures in the development of the historical just war tradition is provided by Daniel R. Brunsetter and Cian O'Driscoll's *Just War Thinkers* (Brunstetter and O'Driscoll 2018). Other work is currently under way. While there is a sense that one can never know all there is to be known, it is fair to say that a thick understanding now exists of the ethics of war in western cultural tradition.[1]

For the case of Chinese traditions a landmark is provided by P.C. Lo and Sumner B. Twiss's edited collection already mentioned above: *Chinese Just War Ethics: Origin, Development, and Dissent* (Lo and Twiss 2015). This book is unparalleled in its depth and breadth of coverage as a study of the various historical Chinese traditions on the ethics of war. In that it compares directly to my own on the historical origins and development of just war tradition in western culture.

I have noted above that the tradition on just war in western culture originated and developed historically with close interaction among inputs from different cultural sectors of the overall culture. While the roots of western just war thinking reach back into Roman, Greek, and Hebraic culture, and while the late classical Christian theologian Augustine provided pungent thoughts placing this pre-Christian idea within a Christian theological frame, the first comprehensive and systematic conception of just war in western culture did not appear until several centuries later as the result of the work of three generations of canonists beginning with Gratian's *Decretum* in the mid-twelfth century and continuing for roughly a hundred years in the writings of his successors, the Decretists and the Decretalists. This work, of which a broad swath of present-day writers on just war remains ignorant, provides the classic form of the western idea of just war, which was summarized and placed into a theological context by Thomas Aquinas in the late thirteenth century. That classic conception is best known today through Thomas's three requisites for a just war (*Summa theologiae* II/II, Q. 40, Art. 1): the authority of a prince (a temporal ruler with no temporal superior), a just cause (broadly, a response to injustice so as to vindicate justice; in Thomas's words, taken over from Augustine through Gratian, to restore that which has been wrongly taken and to punish evildoing), and a right intention (described as both the avoidance of a number of wrong intentions enumerated by Augustine and repeated by Gratian, and the overall aim of restoring a disordered peace).

Several further comments will clarify the meaning of this conception. First, Aquinas's three requisites corresponded directly to the then generally accepted conception of the three ends of politics, order, justice, and peace, drawn from Roman thought through Augustine. Second, "war" (Latin *bellum*) in this concept referred to collective use of armed force in the service of a *res publica*, a political community or commonwealth. There was a different word, *duellum* (the root of our word "duel"), for private quarrels using arms. But "war" *bellum*, might be just or unjust, depending on whether the three requisites were satisfied. Third, then, just war, *bellum iustum*, required the authority of a temporal ruler with no temporal superior, acting in the role of the one having ultimate responsibility for the good of the commonwealth and as the judge of last resort in the case of disputes within the commonwealth. Though the intellectuals who defined this classic conception of just war were all church-men (like all educated people of the day), either canonists or theologians, they explicitly reserved authorization of just war to temporal sovereigns, excluding the possibility of authorization by church authorities. Similarly, fourth, just cause was defined in terms of the vindication of justice, whose requirements in any given case the sovereign authority was to determine through reference to the natural law. Just war was not holy war, war waged on religious authority or in the cause of religion. Fifth, still on the matter of just cause, while Aqui-nas did not explicitly mention self-defense as a just cause, that the natural law allowed anyone justly to resort to self-defense while under attack had been settled before him by a prominent Decretalist, Pope Innocent IV, in the first part of the thirteenth century. But the use of arms in self defense, Innocent specified, could only be *incontinenti*, "on the spot." Once the moment of attack had passed, the case became a matter of determining justice and injus-tice in the use of arms, punishing the injustice and thus restoring justice; and this was a matter for government, not private action. Finally, that the end of a just war is peace was phrased by reference to further words taken from Augustine and by reflection on the interlocked relationship of the end of politics, whereby good order in the community was defined by peace, which in turn was established by the presence of justice; or correspondingly, peace was the result of the use of order to establish justice. The three goods of politics were thus reciprocally interlinked, and so were the three requirements for a *bellum iustum*.

Now, did this conception of just war meet the tests of a military ethic noted earlier? The answer to this question is provided by a closer look into the fact that the classic medieval conception of just war was the result of inputs from various sectors of the social, political, and religious order, including the military.

We should note two things about these various influences. First, while all those directly involved in debating and shaping this comprehensive and sys-tematic conception of just war were churchmen, who had received their edu-cations as a result of entering religious orders, at the time typically they were from the knightly class, that class in medieval culture whose members were the only ones possessing the *ius armae*, the right to carry arms; the functions of government also were lodged in the knightly class. When one joined the ranks of the church one put aside the responsibilities of a knight for temporal government and order, including the right to bear arms; yet knowledge of those responsibilities and that right remained, and relationships to family members and others exercising the functions of members of the knightly class also remained. This was exemplified notably in the case of Aquinas, whose father and brothers were prominent in their involvement in government and the use of arms. The churchmen who gave the idea of just war its classic form knew well the worlds of temporal government and combat; they were not writing in an elevated moral world remote from this reality.

A second thing to note about the enterprise that produced the first comprehensive and systematic understanding of just war is that it reflected and depended heavily on the recovery and interpretation of Roman law that was taking place at the same time the idea of just war was being defined and clarified. Particularly important here were the ideas of eternal law and natural law, both described as *ius*, that which is right, and the relationship of these kinds of law to the law of particular peoples (*ius gentium*) and to Roman law, which was a specific form of *lex lata*, law that is formally laid down or legislated. Within the Roman understanding of law all these forms of law were related to one another in a hierarchy, with each lower form defined by the higher. It was canonists during the twelfth and thirteenth centuries, particularly those connected with the University of Bologna, who were central to both the recovery and interpretation of Roman law and the definition of just war. The influence of Roman law appears particularly in the canonical conception of authority for a *bellum iustum*, where the function of such an authority is to determine the meaning of the natural law in each disputed case and shape and enforce the law of the *res publica* accordingly. Aquinas's conception and use of the idea of natural law and its relationship to divine law has generally been described as coming from his reading of Aristotle, but the pattern seen in his work is also that found in Roman law.

The idea of just war in its classic form also reflected earlier canonical decisions that related directly to the actual practice of war: canonical rulings from the tenth through the early thirteenth centuries defining noncombatant immunity by lists of classes of persons not normally involved in war and thus not to be attacked in the course of war; canonical rulings from the ninth and tenth centuries that imposed penance on warriors returning from battle who had fought with wrong intentions, as measured by the list given by Augustine in *Contra Faustum*; and canons from the Second Lateran Council in 1139 imposing restrictions on certain weapons. These canons became part of the *Corpus Juris Canonici* that remained in effect until early in the twentieth century and that established precedent for the protections of noncombatant immunity and weapons bans that are part of present-day international humanitarian law (also known as the law of armed conflict and the law of war). (For further discussion of these developments see Johnson 1975, Chapter I; Johnson 1981, Chapter V; and Johnson 2014, Chapter 1.)

These considerations show clearly that the classic idea of just war in western culture meets the tests of being a military ethic as described above. As for recent forms of the idea of just war, the work of Paul Ramsey, Michael Walzer, and Jean Bethke Elshtain all knowledgeably and intentionally engage military practice, while in my own work on present-day implications drawn from the classic conception of just war I have benefited from dialogue with military professionals and engaged military practice. The same, though, cannot be said of all recent work under the rubric of just war, including both religiously and philosophically rooted work.

The Chinese Traditions on *yizhan* (Righteous or Just War)

Let us now turn to the Chinese traditions. Among the traditions examined in Lo and Twiss's book not all equally engage the practice of war or the role of the use of military force in relation to government. In the present context I will focus on the traditions examined in two chapters in which reflection on the ethics of war clearly engages the military practice of the time, Lo's chapter *The Seven Military Classics*, or as he also calls them, the *Art of War* corpus (Chapter 2) and that of Twiss and Jonathan K. Chan on Wang Yang-ming (Chapter 7). The term *yizhan* employed in the historical works is usually translated "righteous war," but Lo observes it can also be read as "just war," with the same meaning as the present-day term *zhengyi zhanzhang*, translated "just war" (Lo and Twiss 2015, 34).

Characterizing *The Seven Military Classics* Lo writes,

> [The] first five treatises are the earliest Chinese treatises on military affairs. Hence, they cover a variety of military subject matters such as military organization, education and training, leadership and its required virtues, strategy and stratagem, tactics, geography, intelligence, psychology, economics, and logistics. Most of them also deal with statecraft; hence, their discussions of warfare are embedded within a larger perspective.
>
> *(Lo and Twiss 2015, 31)*

This list of topics covered certainly satisfies the test of relevancy for military ethics.

While the oldest of these treatises, *Master Sun's Art of War* (*Sunzi Bingfa*) was composed during the Spring and Autumn Period, 770–476 BCE, the second through fifth treatises all were written during the Warring States Period, 475–221 BCE. The sixth treatise was composed a bit later, in the period 220–100 BCE, and the seventh later still, during the Tang Dynasty, 618–916 CE. Lo comments of this last treatise, "it is more or less a summary of these six treatises in the form of a catechism for pedagogical purposes" (Lo and Twiss 2015, 31). Together these seven works occupy a preeminent place in Chinese historical reflection on war. Observing that they provided the basis for much of the extensive writing on military affairs during the Ming Period, Lo continues by noting, "this set of books is still being studied by military universities and academies in both the People's Republic of China (PRC) and Taiwan" (Lo and Twiss 2015, 31). He later qualifies this statement importantly, though, as we shall see.

In the United States and other Western countries only one of these early Chinese treatises has received significant attention, *The Art of War* by Master

Sun/Sun-tzu/Sunzi, which has been used frequently in military education and has been the subject of analysis by scholars interested in strategy, tactics, and other military affairs. The importance of the other treatises in this group within historical and contemporary China argues, though, that more attention should be given to them as a group.

Lo's examination of this group of treatises includes fairly frequent comparisons with the Western idea of just war, and it is convenient to take these as useful starting points for my own comparative remarks here.

Lo begins his analysis by observing "the tragic nature of warfare" as understood by these authors, writing, "Military violence is deeply deplorable; it is bad and tragic. It should be resorted to only very reluctantly when there are no better options ... ; that is, in contemporary terms, as a last resort" (Lo and Twiss 2015, 33). Commenting on this, Lo provides the first of his comparative judgments, using a reference to my work to establish the position found in just war tradition. Lo writes that by contrast to this view of *The Seven Military Classics* authors, "For Johnson, 'Force here is not evil in itself; it takes its moral character from who uses it, from the reasons used to justify it, and from the intention with which it is used'" (Lo and Twiss 2015, 35). Lo's language here is not quite correct: my point was not to state this view as a position I hold personally, but that this conception of the moral character of force properly represents the position taken in the classic form of Western just war thought. One does not in fact find the requirement that just war must be a last resort in the classic statement of western tradition on just war, though to be sure, it might have been a consideration of the sovereign as part of the exercise of government, and the requirement of last resort has become a prominent part of much recent just war thinking.

As to the woefulness of war, a similar judgment to that which Lo finds in several of the *Military Classics* can be found in Augustine's understanding of just war. Gregory Reichberg calls Augustine "the reluctant just war theorist" and writes, "Having to use force to keep the peace in this world is not something the wise or pious man does gladly, but merely out of necessity" (Reichberg, Syse and Begby 2006, 71). The reason, Augustine argues, is that maintaining the peace of this world, the earthly city (*civitas terrenae*) is for the Christian a moral obligation only so long as its existence serves God's purpose for it. When that purpose is fulfilled, the earthly city will be no more, yielding to the perfection of the heavenly city, the City of God (*civitas dei*). On this Reichberg cites Augustine in *City of God* XIX.7. But this way of thinking is not characteristic throughout Augustine's statements on just war; rather, he came to it only late in life. Earlier, in *Contra Faustum* XXII.74, a passage Augustine scholar Robert Markus regards as perhaps the single most important of Augustine's statements on just war, Augustine had conveyed a very different understanding of just war. That passage, but not the one from *City of God* XIX, appears in the cornerstone document of the medieval conception of just war, Gratian's *Decretum*, where it is cited in full. This passage begins with the following words; these are also cited in by Aquinas in his Question *On War*, cited earlier:

> What is to be blamed in war? Is it the death of some who are to die in any case, so that others may be forced to peaceful subjection? To reprove this is cowardice, not religion. What is rightly reproved in war are love of mischief, revengeful cruelty, fierce and implacable enmity, wild resistance, love of power, and such like. And it is generally to punish these things, when force is required to inflict punishment, that, in obedience to God and lawful authority, good men undertake wars … .
>
> (*Gratian,* Decretum, *Part II,* causa *23, canon 4; cited in Reichberg, Syse and Begby 2006, 112*)

This sentiment, and not the one reflected in the passage quoted by Reichberg from *City of God* XIX, expresses the perspective on just war from Augustine that was taken up in the tradition and became central in the idea of just war in its classic form. This leads me to think of the comparison between the classic form of *bellum iustum* and *yizhan* as depicted in *The Seven Military Classics* somewhat differently from Lo. On this let me first return to the correction of Lo's language I offered earlier. At times his language suggests that on just war I am making my own normative statement, so that he is comparing the historical Chinese conceptions to my own conception. Of course I do have my own conception of just war, and it relies heavily on the classic conception, but in the comparisons cited I am attempting to understand and represent the classic idea of just war, not my own normative understanding. The comparison that matters is not that between the authors of *The Seven Military Classics* and Johnson but between those Chinese sages and the medieval European thinkers who gave the western idea of just war its classic form.

As to the matter of last resort, in trying to resolve the tension between the classical conception of just war and recent conceptions in which last resort figures prominently, on this point I have offered the view that the classical conception of just war defines the morality of resort to war so as to establish the rightness of war in deontological terms, as duties to be obeyed when resorting to war, while the idea of last resort and other requirements offered in present-day conceptions of just war are prudential checks, requiring the responsible agent's use of prudential reasoning to modify the obligation defined by the deontological requirements of sovereign authority, just cause, and right intention. (On this see further Lo in Lo and Twiss 2015, 37–38.) This is not all I would say about the difference and relative priority of the classic requisites for just war and the contemporary additions, but I would note in favor of my doing so that Augustine himself in his correspondence with the Consul Marcellinus at the time of the Donatist controversy expressed the view that it is better to use force to punish evildoing in this life, so as to vindicate justice, than to allow God to do so in the next life, when greater punishments are available. This way of thinking does not seem to reflect any sort of requirement of last resort as understood in present-day just war discourse.

There is also the fact that the authors of *The Seven Military Classics* themselves made a point of distinguishing between righteous and unrighteous (or just and unjust) warfare, describing the former with terms such as "to chastise the unrighteous" and "to apply the punishment of rectification" (Lo and Twiss 2015, 38). It is suggestive to compare such language to that of Augustine in his correspondence with Marcellinus (Letter 139) and to that of Romans 13:4 on the moral obligation of the prince to punish evildoers (cited in Aquinas, *Summa theologiae* II/II. Q. 40, A. 1 and frequently cited in just war sources from the twelfth to the sixteenth centuries). My point is simply this: for the classic understanding of just war in the West, what is woeful and deplorable is injustice, and the idea of just war centers on the moral obligation of persons in governing authority to punish evildoing and seek to restore justice, specifying when they may morally employ armed force in seeking to carry out that obligation. This does not mandate that each and every instance of evildoing mandates a military reply, but whether such force is appropriate to a particular case is a matter of prudential judgment for the sovereign. Lo's analysis of the thinking found in *The Seven Military Classics* presents, as I read it, a picture not at all unlike this. Perhaps we need a closer look at this whole matter in the two traditions: it may be that the evils created by war are not the worst evils, or that where the use of arms is necessary to respond to severe unrighteousness or injustice the evils caused by war recede to a secondary place.

On the matter of moral conduct in the use of military force Lo cites a passage from one of the seven treatises, that by Master Sima:

> When you enter the offender's territory, do not do violence to his gods; do not hunt his wild animals; do not destroy earthworks; do not set fire to buildings; do not take the six domesticated animals, grains, or implements. When you see their very elderly or very young, return them without harming them. Even if you encounter adults, unless they engage you in combat, do not treat them as enemies. If an enemy has been wounded, provide medical attention and return him.
>
> *(Lo and Twiss 2015, 39)*

A page later Lo cites a passage from another author in this group, Taigong, laying out much the same restraints. Lo concludes that the ideas for proper war-conduct in *The Seven Military Classics* "are roughly analogous to the principle of discrimination or distinction" (Lo and Twiss 2015, 40). I suggest that the comparison would be less "roughly analogous" had Lo looked at the conception of noncombatant immunity as it was defined in medieval canon law, where the definition is not by means of a principle (which can, after all, be interpreted differently by different people) but rather by listings of specific classes of persons and property not to be directly attacked in war. These lists correspond quite directly to those given by Master Sima and Taigong.

One might usefully pursue this comparative analysis further, given time and space, but let me conclude these remarks on Lo's chapter by taking note of his section on contemporary Chinese moral discourse on war (Lo and Twiss 2015, 48–51). Here Lo surveys People's Republic of China publications dealing with the topic of just and unjust wars (*zhengyi zhanzheng yu feizhengyi zhanzheng*), including Mao Zedong's writings on this subject and three publications produced under the auspices of the People's Liberation Army: the *Military Science* volume of the *Encyclopedia of China*, a publication entitled *Strategic Science* prepared by the PLA's Strategy Research Unit of the Academy of Military Science, and a recent book, *What is Military Science?* Lo summarizes his judgment of the treatment of the just war idea in these works as follows: "[T]he current Chinese military establishment inherits the historical idea of just war with a twist. For one thing, only the *ius ad bellum* parts survive. For another, just cause is interpreted through the lens of communist ideology and practical need" (Lo and Twiss 2015, 49). A page later he approvingly quotes Andrew Scobell along the same lines:

> China's strategic behavior is influenced not just by a Realpolitik strand but also by a Confucian one. The combined effect is what I have dubbed a Chinese Cult of Defense, in which realist behavior dominates but is justified as defensive on the basis of a pacifist self-perception. ... The outcome of these two strategic cultures interacting is a China that assertively protects and aggressively promotes its own national interests, up to and including acts of war, but that rationalizes all military moves as purely defensive.
>
> *(Lo and Twiss 2015, 50)*

Thus while the tradition on the ethics of war established by *The Seven Military Classics* clearly brings moral reflection into engagement with military practice, this tradition of engagement, according to the judgment of Lo and Scobell, has not carried forward into contemporary Chinese military attitudes, policy, and practice. This contrasts sharply with the way in which various sorts of reflection on just war have made their way into western military thought, education, training, and practice. There is a significant gap between the cultures here on the subject of military ethics, or more specifically, conduct in war.

At the same time, a common feature is that while both traditions historically justified use of armed force to rectify injustice and evildoing, in both the contemporary Chinese "Cult of Defense" and in much recent just war thought self-defense has become the central justification for national resort to use of armed force.

Let me turn now to Twiss and Chan's chapter on Wang Yang-ming. Wang lived much later (1472–1529 CE) than the period when *The Seven Military Classics* were written and in a greatly different social and political context, that of the Ming imperial dynasty. Twiss and Chan write of Wang, "he is known as an outstanding Ming scholar, statesman, and general, deeply steeped in the

Confucian classics … as well as the military classics" (Lo and Twiss 2015, 153). This combination of features makes him especially interesting to be considered from the perspective of military ethics, as do elements in his career. Twiss and Chan concentrate on two particular episodes in Wang's career. The first was when, in his twenties, he offered a memorial to the Emperor responding to an imperial government invitation for recommendations addressing the situation on the frontier, which was menaced by military invasion. Wang's memorial on this occasion, Twiss and Chan comment, showed a marked influence from the military classics and especially *Sunzi Bingfa*, Master Sun's *Art of War* (Lo and Twiss 2015, 154–55). Wang here emphasized the need for broadly educated military leadership, both possessing strong military qualities and knowledgeable about conditions and the lives of the people on the frontier. He further stressed that troops assigned to the frontier should conduct themselves in a way that does not burden the local population, establishes order to the benefit of that population, and works with the population so as to increase opposition to the enemy. Military action against the enemy, argued Wang, should rest on local knowledge gained from interaction with the population as well as from scouting and spying on the enemy's dispositions. Such action should employ "tactics of ambush and surprise against the enemy's weaknesses like an unexpected flood of water" (Lo and Twiss 2015, 155). Wang's approach here offers striking parallels to that found in the U.S. Army and Marine Corps *Counterinsurgency Field Manual* of 2007 in its emphasis on fighting the enemy by working with the local population so as to improve living conditions and create support for the government troops along with opposition to the enemy and, so far as military action is concerned, fighting with flexible tactics aimed at the enemy's weaknesses. In this memorial Wang directly connects matters of ethics (care for the good of the population, expressed in various ways) and specific military activities in a way that accords well with the understanding of military ethics sketched at the beginning of this chapter.

The second episode in Wang's career highlighted by Twiss and Chan is when, later in life, Wang served as military governor of the territory of Ganzhou, where he was given the specific charge to suppress banditry and establish public order. In this capacity he "subdued the rebellion of the Prince of Ning who was not only preparing to undertake military action … to usurp the throne but also [was] acting tyrannically toward his own people" (Lo and Twiss 2015, 157). Twiss and Chan comment further,

> In the training of his troops at Ganzhou, Wang promulgated a set of rules for military conduct, some of which pertained to troop interaction with the civilian populace and others of which regulated social relations among the soldiers themselves.
>
> *(Lo and Twiss 2015, 164)*

They go on to summarize these rules, then observe,

> From these two sets of rules, then, one may fairly infer that Wang adheres to a *ius in bello* norm of respect for civilian infrastructure including property, social practices, and material and psychological well-being, not to mention an emphasis on moral qualities and policies essential to military control. The former norm comes close to acknowledging ... non-combatant immunity, although Wang does not explicitly state it here.
>
> *(Lo and Twiss 2015, 165)*

In both the classic western understanding of just war and in Chinese tradition the term "war" is used for collective military action as opposed to individual uses of arms, and "just war" (or "righteous war," in the Chinese traditions) is undertaken on behalf of the good of societal order by an authority responsible for that good. The enemy of that good may be either external or internal; the term "just war" (*bellum iustum* in the West, *yizhan* in China) is used for collective military action against both kinds of enemies. In the two episodes from Wang's career highlighted by Twiss and Chan the first has to do with external enemies menacing the well-being of the Chinese frontier areas, while the second has to do with internal policing. What is at stake in both cases is the collective use of armed force carrying out the responsibility for good government.

Wang's life and work exhibit a close connection between intellectual endeavor and practical activity in public service and provide a substantial basis for cross-cultural comparative dialogue on all aspects of this and on the scope of military ethics. His internalization of classical Confucian morality and specifically the conception of government and war conveyed in *The Seven Military Classics* shows up in his conduct as an official in the imperial bureaucracy, as a general, and as a territorial governor. The sorts of issues he confronted, as well as the morality he brought to bear in doing so, make for straightforward comparison with those found in European society during the period of consolidation and application of just Western just war tradition. They also, as I have already suggested, invite comparison with present-day military thinking on dealing with insurgency. There is no space here to pursue such comparisons, but it is important to note the suggestive possibilities that can be seen.

Conclusion

This chapter has addressed the cross-cultural comparative study of military ethics focusing on the traditions of just war as these came together in western and Chinese cultures and developed historically and as they are, or are not, present in contemporary military affairs. I have argued that military ethics is not simply an intellectual exercise of reflection on ethics and warfare but one that involves the integration of approaches and influences from a variety of disciplinary perspectives—especially those of religion, philosophy, law, political and military theory, and practical experience in government and military affairs —brought into engagement with the practical use of military force. To give substance to how to conceive military ethics I have summarized two editorial statements on this subject from the *Journal of Military Ethics*, highlighting its interdisciplinarity and its relevant engagement with problems in military practice. I then sought to examine this matter through a comparative study of the classical idea of just war in the West and in two examples from different periods of Chinese history, noting that in both these culturally different historical traditions the above tests of a military ethic are satisfied and observing both important similarities and certain differences between the two traditions. Comparative study at its best provides a basis on which to build from the similarities while placing the differences into meaningful perspective. Further, I have briefly examined how each of these cultural traditions on military ethics has fared in the present day, noting a sharp contrast between the active presence and efforts at application of just war tradition in military contexts in the United States and some other western countries, but a more limited and filtered use of the Chinese historical traditions on just war by the Chinese People's Liberation Army. The question now is what this all implies for future work on comparative military ethics.

First, I would say again that while there has been a great deal of writing on the western idea of just war, much recent work on this idea has frequently proceeded without taking account of the historical tradition or with only a limited and distorted knowledge and understanding of it. Prominent examples of work on just war have effectively engaged in reinventing the just war idea rather than working with the tradition so as to understand its contemporary implications. More attention should be paid to the historical just war tradition so as to understand it better and to better follow out its possible contemporary implications.

Second, even with this said, the western historical tradition on just war has been more deeply explored and is better understood in its context and development than the comparable Chinese historical traditions. Serious cross-cultural comparative work requires a thick understanding of the traditions being compared on both sides. As I note above, Lo and Twiss's 2015 book provides an excellent basic first step in examining the Chinese traditions, and the present volume extends that examination. Yet the need for deeper investigation and explication remains. For doing this I have underscored the importance of direct engagement with the original sources of the relevant historical traditions, which implies that western scholars interested in such a comparative study should develop facility in Chinese so as to work with the tradition on *zizhan* in its original form, just as Chinese scholars interested in examination of the western tradition of just war should have facility in Latin, the language of just war tradition in its classic form, and the various European languages in which sources related to just war tradition are found.

Third, a comparative conversation on military ethics across the lines dividing these two cultures in the present day is very problematic to carry through, given the differences between the understandings of military ethics on each side. The specific ideological cast of the Chinese publications on war surveyed by Lo presents a real problem for such a conversation, and I suggest that a possible way forward would be for greater emphasis on scholarly study of the Chinese traditions on just war coupled with knowledgeable and focused exploration of their implications for contemporary uses of military force. While this might simply broaden the gap between the military and the non-military scholars, it might alternatively draw the former into closer engagement with the work of the latter. That this is what has happened in the United States and other western countries provides a positive indication for what may be possible in China.

The fourth and, for present purposes, last remark I would make on the topic of this chapter is that to "do" military ethics requires engagement and dialogue between and among representatives of all the disciplines involved. In particular, for those of us who are academics in various fields, this means not only that we need to establish regular and substantive contacts not only across the academic disciplinary lines that divide us but that also there is a special need to establish such contacts with professional military persons and with people in government involved in military affairs. I have been particularly fortunate in such dialogue with military and policy people in my own career, and my thinking on just war has benefited greatly from it, but my experience has been different from that of the majority of other academics I know who have worked on the topic of just war or issues related to it. Making such connections is not easy, but it needs to be done. The openness of western societies makes such interactions possible and tends to facilitate them. Conditions and institutional structures are different in the PRC; so possibly the best venue for such interactions is at the level of international institutions, including academic societies involved in the study of ethical traditions on war and military ethics. There are also limited interactions between professional military personnel from the two cultures, and seeking to build on these represents a second possibility. To forge such bonds remains, though, a task to be pursued, and there is no certain road map toward success.

Rather than to end on such a pessimistic note, let me return to the more positive possibilities before us as academics: that we can work separately and together so as better to understand our respective ethical traditions on just war, their agreements and disagreements, and draw out what we see to be their implications for present-day military affairs. This too is a challenge, but it is one over which we implicitly have greater control.

Note

1 For the case of Islam the work of John Kelsay continues to add to knowledge of both the historical tradition on jihad of the sword and prominent examples of its contemporary interpretation and use. As with the case of the just war idea, though,

there is a considerable literature on jihad that is concerned only with contemporary conceptions and applications of the jihad idea (notably those of radical Salafist military movements).

Bibliography

Aho, James A. 1981. *Religious Mythology and the Art of War.* Westport, Connecticut: Greenwood Press.

Brekke, Torkel, ed. 2006. *The Ethics of War in Asian Civilizations.* London and New York: Routledge.

Brunstetter, Daniel R., and Cian O'Driscoll, eds. 2018. *Just War Thinkers.* Abingdon, Oxon, and New York: Routledge.

Ferguson, John. 1977. *War and Peace in the World's Religions.* London: Sheldon Press.

Huntington, Samuel P. 1993. "The Clash of Civilizations?" *Foreign Affairs* 72, no. 3 (Summer): 22–49.

Johnson, James Turner. 1975. *Ideology, Reason, and the Limitation of War.* Princeton and Guildford, Surrey: Princeton University Press.

_____. 1981. *Just War Tradition and the Restraint of War.* Princeton and Guildford, Surrey: Princeton University Press.

_____. 2008. "Thinking Comparatively about Religion and War." *Journal of Religious Ethics* 36, no. 1 (March): 157–79.

_____. 2014. *Sovereignty: Moral and Historical Perspectives.* Washington, DC: Georgetown University Press.

Johnson, James Turner, and Eric D. Patterson, eds. 2015. *The Ashgate Research Companion to Military Ethics.* Farnham, Surrey: Ashgate Publishing Ltd., and Burlington, Vermont: Ashgate Publishing Company.

Journal of Military Ethics. 2002–present. Oslo, Norway, and Basingstoke, UK: Taylor & Francis.

Lo, Ping-cheung, and Sumner B. Twiss, eds. 2015. *Chinese Just War Ethics: Origin, Development, and Dissent.* London and New York: Routledge.

Popovski, Vesselin, Gregory M. Reichberg, and Nicholas Turner. 2009. *World Religions and the Norms of War.* Tokyo, New York, and Paris: United Nations University Press.

Reichberg, Gregory, Henrik Syse, and Endre Begby, eds. 2006. *The Ethics of War: Classic and Contemporary Readings.* Malden, Massachusetts; Oxford, UK; and Carlton, Victoria, Australia: Blackwell Publishing.

Sorabji, Richard, and David Rodin, eds. 2006. *The Ethics of War: Shared Problems in Different Traditions.* Aldershot, Hants., and Burlington, Vermont: Ashgate Publishing.

Wright, Quincy. 1942. *A Study of War.* 2 vols. Chicago: University of Chicago Press.

3

CLAUSEWITZ VS. SUNZI

Comparing Western and Chinese Ways of War and their Ethics[1]

C. Anthony Pfaff

Introduction[2]

In their now oft-cited book *Unrestricted Warfare,* two Chinese colonels, Qiao Liang and Wang Xiangsui, described the future of warfare in ways that would be at once very familiar and very strange to a Western, particularly American audience.[3] As in the West, their discussion invoked both Clausewitz and Sunzi; however, the way of war they advocated was very different from the Western understanding, and explicitly so. They argued after the U.S. victory over Iraq in 1991 the character of warfare had changed, and the U.S. military had effectively made itself obsolete. The military overmatch the United States demonstrated in the Gulf War ensured that no competitor would confront it on conventional terms.

The result, they argued, was that the Western Way of War, inspired as it was by Clausewitz, was a thing of the past and that a new way of war, which followed more closely the teachings of Sunzi, would not only emerge, but also prevail. Despite the fact that Sunzi is also taught in Western military institutions, the way of war they proposed seemed very different, not only in how it conceived of war, but also in the norms that governed it (Qiao and Wang 1999).

What they saw passing into history was a way of war that emphasized military force as the primary, if not exclusive, means to impose one's will on the enemy. In its place would be one that would rely on combinations of military, political, economic, social, and informational means to compel an enemy to accept one's interests (Qiao and Wang 1999, 7). This distinction is not uniquely Chinese. American Cold War scholar Thomas Schelling, in fact, described the difference between "the power to hurt and the power to seize" as a way to differentiate warfighting from international competition (Schelling 2020, 6). Where the power to seize entails taking an adversary's choice away,

DOI: 10.4324/9781003336372-4

the power to hurt is how one bargains with adversaries to incentivize choosing behaviors that work to one's advantage.

What makes *Unrestricted Warfare* more distinctively Chinese, however, is its blurring of the boundaries between war and competition. As Ofer Fridman, author of *Russian Hybrid Warfare*, observed, Qiao and Wang offered an "intellectual conceptualization of future conflicts," intended to transcend the boundaries of warfighting to combine military and non-military means to succeed against an adversary (Fridman 2018, 12–13). Predictably, as Fridman also observed, this view was received in the United States as promoting a deceptive, illegal, and probably immoral way of war that entailed undermining the international peace-time order (Fridman 2018, 12). In fact, in an English-language version published in the United States, the title was modified to *Unrestricted Warfare: China's Master Plan to Destroy America* and its approach described in commentary as deceitful, subversive, and manipulative (Birnes 2017; Qiao and Wang 2017; Fridman 2018, 12–13).

The characterization of *Unrestricted Warfare* as unethical, while arguably unfair, underscores the connection between ways of warfare and their ethics. There is often a tension between what is perceived as necessary and what is perceived as just, and it is the role of professional militaries to find an appropriate balance. In fact, both East and West have long moral and ethical traditions regarding when and how to fight wars, which still inform their practice today. With this point in mind, this chapter does not provide a comprehensive account of either tradition - East or West. Rather, it focuses on figures like Clausewitz and Sunzi because they have a demonstrable, if not entirely measurable, impact on how both approach war and its ethics.

Of course, the fact of a moral tradition does not entail ethical behavior; however, it is also not the point of this chapter is to assess how well China and the United States live up to their own ethical ideals. Nor is it to assess how well each tradition comparatively addresses the moral challenges war poses. Rather it is to point out that the common experience of war forces both the United States and China to grapple with war's destruction and determine both the most effective paths to victory as well as the most appropriate restraints on its practice. In doing so, their differing conceptions of the character of war yield different conceptions of its ethics. The ethic that results from imposing one's will is largely deontological, where one must adjudicate the competing imperatives of military necessity, non-combatant immunity, and force protection. The ethic that results from compelling interests is more one of virtue, which may be more permissive regarding means and ends but which bounds them, in the ideal at least, by the demands of justice and benevolence.

When these two different systems interact, the result is often a great deal of misunderstanding and unnecessary provocation. Thus, understanding these differences and, more importantly, how they are justified by deeper philosophical commitments regarding what war is and what it is for, should be a first step towards not only avoiding such misunderstandings in the future, but also in crafting foreign policies that avoid war while seeking to achieve legitimate national interests.

The Nature and Character of War

One of Clausewitz's many contributions to the Western canon of warfare was to introduce a confusion regarding the nature of war, which he claims is both "complex and changeable" but still the same thing (Gray 2010, 7). To resolve the confusion, the British strategist Colin Gray differentiates between war's nature and character, where the former is "universal and eternal" while the latter is constantly in flux (Gray 2010, 6). For him, the nature of war is organized, violent, and political. The political ends of war distinguish it from other sorts of violence, such as recreational and criminal, and it is violence that distinguishes it from other sorts of international competition (Gray 2010, 6). The character of war, on the other hand, can change depending on the circumstances in which wars are fought, which include political, social, cultural factors as well as technology and tactics (Strachan 2013, 511; Mewett 2014). Those factors drive choices about how actors will organize, equip, and train their militaries to be effective against any putative enemy. What emerges from these decisions is a "way of war" that specifies what kinds of forces and what kinds of strategies actors will pursue to achieve battlefield success.

Given Gray's concept of the nature of war, the U.S. and Chinese concepts, as so far articulated here, may appear to be addressing two different things. While both traditions agree on the political and organized nature of war, the Chinese view, at least as articulated in *The Art of War*, sets bloodless victory as an ideal and, as will be discussed later, prioritizes non-violent over violent means (Sun 1971, 77). This priority is not only reflected in other works like *Unrestricted Warfare,* but also in other Chinese doctrine, such as the "Three Warfares," which emphasizes information, psychological, and legal operations as a way of achieving the ideal of "winning without fighting" (Ota 2014).

To the extent violence is incidental, rather than inherent, in the Chinese view, it is worth asking if it is better described as a view of war or simply a preference for international competition over war. If the latter, then comparisons will not likely be any more useful than a comparison of a way of chess with a way of checkers. If war and international competition are two different things, then naturally we would expect them to be governed by different rules. However, if the former, one must then ask the question whether war can indeed have different natures depending on who is waging it.

At first glance, it is not clear that it can. From a Western perspective, a view of war in which violence played no necessary role would simply be a view of another form of conflict. Schelling, as an example of Western strategic thinking, made this point when he differentiated war from coercion not only as the power to seize and the power to hurt, but also as the difference between "taking what you want and making someone give it to you" (Schelling 2020, 2). The former is accomplished by brute force and the latter by incentivizing compliance. Moreover, where brute force is unconcerned with an enemy's interest, incentivizing compliance is "the very exploitation of an enemy's wants and fears" (Schelling 2020, 3).[4] So rather than the power to simply destroy, coercion is about the power to bargain (Schelling 2020, 8–9). To the extent bargaining entails de-prioritizing violence in favor of a mixed approach of imposing costs and selective cooperation, it may be indistinguishable from the Chinese view, at least those inspired by Sunzi.

But reducing the Chinese view of war to a Western view of coercion does not fully account all that their view encompasses. In prioritizing preservation over destruction, Sunzi, for example, does not de-emphasize violence as much as relocates it. In his hierarchy of strategies, he states that attacking an enemy's plans was best, their alliances second best, their army third best, and then, worst of all, their cities. As Samuel B. Griffith notes in his translation, these strategies represent general policies not specific to the art of generalship (Sun 1971, 77–78). As such, they apply to the adversarial relationship as a whole and not simply one particular activity - war - in which adversaries might engage. Moreover, where they apply to war, violence is less de-prioritized as much as it is devalued. Attacking the cities may be the worst of the three other options; however, it is still better than losing. Thus, where there are no alternatives, logically, violence would be the first resort.

Thus, rather than viewing war differently, it is probably more accurate to say coercion and war form a dichotomy in the Western view and a unity in the Chinese view. While this may seem like a small distinction, it can have interesting implications. To the extent the nature of an activity determines the rules that govern it, then if coercion and war are two different activities, they would reasonably be governed by different sets of rules. If, on the other hand, they are a unity, it would be more reasonable to believe they are governed by a single set. With this difference in mind, it will be easier to discern how the two views diverge in crafting their respective ways of war as well as the norms that govern it.

The Western Way of War

Clausewitz famously envisioned war as a wrestling match where wrestlers ground down each other's resistance, seeking, as stated above, to impose their will on the other (von Clausewitz and Rapoport 1968, 101). This view of war, as Michael Walzer observes, is one of continuous, reciprocal escalation limited only by the physical means employed (Walzer 2015, 23). In this context, imposing one's will depend on eliminating an enemy's capability to resist faster than they can eliminate one's own. Put simply, this elimination requires a strategy of annihilation - or at least attrition - that seeks the destruction of the enemy's forces. It is not incidental that in *The American Way of War*, Russel Weigley describes American strategy for most of its history in precisely these terms (Weigley 1977, 475).

As Schelling also points out, the United States has used military force to hurt, as opposed to seize, in the past, citing General William T. Sherman's "March to the Sea" during the American Civil War and the punitive raids against "unruly" Native American tribes in the middle of the 19th century as examples. It is also worth pointing out the efforts the U.S. military has gone to adapt to the demands of counterinsurgency and nation-building. However, whatever imprint they may have made on how the U.S. wages war, their lack of success and the current emphasis on Great Power Competition suggests they will not be lasting. As Schelling further argued, despite these exceptions, the dominant American approach to war seems to remain seeking out and destroying the enemy's military forces to achieve a crushing victory. In this way, the use of military force was generally an alternative to coercion, not a part of it (Schelling 2020, 14–16).

Thus, in the Western view, war and coercion are not so much on different sides of a threshold or ends of a spectrum, but different ways of going about realizing strategic interests and objectives. This difference does not mean kinetic operations –operations where an actor employs physical violence - cannot play a role in coercion or that non-military measures cannot be effective in war. However, what this difference does entail is that all kinetic operations are not the same. Those intended to take, seize, or otherwise remove the enemy's choice would be examples of brute force. Those intended to encourage an opponent to make certain choices would be examples of coercive force, at least as the term is used here.

Because the destruction of enemy forces entails the inability of the enemy to resist, there is little gap between military and political objectives: the defeat of the enemy's military forces entails the achievement of the political objective. This emphasis on destruction of enemy military capability unsurprisingly yields an ontology of war also comprised of dichotomies (Paret 1968, 395). Actors are either allies or enemies; actions in war entail resistance or surrender; and the end-state of war is victory or defeat. It is true that enemies may fight to a stalemate, but such a state of affairs is not stable, representing only a suspension of hostilities until the sides decide to fight again. The state of war itself continues until one side has imposed its will on the other.

The Western Ethic of War

The Western Just War tradition arguably gets its start with the ancient Greeks, especially Aristotle, who is generally credited with coining the term. Writing in the aftermath of the destructive Peloponnesian War, Aristotle argued that in making decisions about war and peace, statesmen should defer to a common law of nations as their standard of justice. For him that meant either a defensive war or an offensive war that benefitted the conquered (Reichberg, Syse and Begby 2006, 37). The Roman philosopher and statesman Cicero, carried on this tradition. In *On Duties,* he writes that justice in warfare is an important part of public affairs and recommends extending mercy to enemies who fought without cruelty or savagery (Cicero 2013, 14–15). Perhaps more to the point, Rome itself, for all its brutality, also followed rules for declaring wars that included norms such as just cause and last resort (Orend 2006, 11).

Augustine, arguably the best known of the early Just War thinkers, sought to reconcile the demands of war with Christian values. Writing as a newly Christianized Rome came under siege, Augustine argued that wars may only be declared by a rightful authority, waged only for a just cause, and fought with the right intention, which precluded brutality (Temes 2003, 62–63; Reichberg, Syse and Begby 2006, 77–84). Later, Thomas Aquinas would synthesize Augustine's work with others to provide a more systematic account of the justice of war (*jus ad* bellum) and justice in war (*jus in* bello). *Jus ad bellum* provisions included just cause, right intention, and public declaration, whereas *jus in bello* provisions included non-combatant immunity, right intention, prohibited weapons, and some protections for soldiers who surrender. Aquinas also added proportionality to both *jus ad bellum* and *jus in bello,* as it seemed to follow from Aristotle's original attempt to moderate war that one should avoid force in excess of the harm it is responding to as well as force in excess of the battlefield objective (Orend 2006, 15).

Just War norms also evolved in response to crises where the destruction of war became self-defeating or simply excessive. For example, before Aquinas, the "Peace of God" and the "Truce of God" were established by Catholic clergy in the Middle Ages to control widespread feudal wars that not only destroyed Church property, but also devasted local communities to the point there was little left over which to fight. These edicts established protections for noncombatants and their property and outlawed certain kinds of weapons, in some cases - like poison tipped arrows and blades - because they caused suffering unnecessary to achieve the military objective (Orend 2006, 14; Reichberg, Syse and Begby 2006, 93–96).

Another significant milestone occurred when the Swiss businessman Henri Dunant - moved by the suffering he witnessed in the aftermath of the Battle of Solferino - helped established the first Geneva Convention, that specified belligerent responsibilities toward wounded soldiers. This first convention set precedent for futures ones, contributing to many Just War norms being codified into International Humanitarian Law (Corn, Watkin and Williamson 2016, 77). The destruction of World Wars I and II inspired further refinement, including the 1949 Geneva Convention IV, which strengthened legal protections for noncombatants (Roberts and Guelff 1989, 271–272).

This account is certainly not comprehensive nor complete. However, it does illustrate how the ethics of war evolves in response to shared values as well as self-defeating and excessive practices. By accepting these norms, belligerents can alleviate suffering while not losing advantage. Of course, belligerents may cheat, but that does not diminish the importance of these norms as long as they identify behaviors and practices for which belligerents can be held accountable, whether formally in court or informally by means of reprisals or other measures.

Because these norms evolve in tension with the demands of war, it would be a mistake to view them as simply a set of constraints to be applied as a checklist prior to taking military action.

Rather, defense of the state is itself an ethical imperative particularly in service to a just cause. Thus, the moral experience of war, at least in the Western view, is better conceived as balancing the imperatives of military necessity, force protection, and noncombatant immunity that accepts risk to each. In balancing these imperatives, combatants are generally obligated to accept risk and engage in operations to defeat enemy forces. In doing so, they naturally undertake measures to make them less vulnerable; however, when those measures make collateral harm more likely, they are obligated to take additional risks to reduce the chance of that harm. However, because military necessity is itself an imperative, they are not obligated to take so much risk that they the mission will fail, or they will not be able to continue the fight (Walzer 2015, 157).

Thus, this ethics not only informs the way the West wages war, but also harmonizes with it. This balance of imperatives does not interfere with the successful waging of war; moreover, by limiting the damage to civilian lives and property, this ethic of war limits post-conflict grievances and facilitates the transition to peace. Of course, there have been times when Western militaries have attacked purely civilian targets. But here is where the exception proves the rule. Even Air Force General Curtis LeMay, who was responsible for the indiscriminate strategic bombing campaign against Japan, believed that he would be prosecuted as a war criminal had the U.S. lost the war (Rhodes 1995).[5]

So, while there is certainly pressure from within the Western ethic to increasingly limit war's destruction, the imperatives to defeat the enemy and protect the force limit the effect that pressure can have. As mentioned earlier, the escalatory nature of war ensures the incentive is to use more force than less. Thus, the Western normative ideal of war can be expressed as using *the most force permissible given the restrictions of noncombatant immunity, proportionality, and causing unnecessary suffering.*

The Chinese Way of War

As Ping-Cheung Lo remarks, contemporary Chinese discourse on *The Art of War,* is "dominated by PLA authors" (Lo 2012, 124). So, it should be no surprise that *The Art of War* has had an analogous impact on the thinking of the PLA as *On War* has had on the American military. Of course, Sunzi is not the only influence. Works of other masters also became relatively well-known and were later assembled by Sung period scholars in the 11th century CE into the *Seven Military Classics,* which included *The Art of War* (Sawyer 1993). However according to Andrew Seth Myer, Sunzi's materialist and instrumentalist approach eventually dominated ancient Chinese military thought because it displaced religion and aristocratic honor as factors governing war. By the first century BCE new military texts either incorporated it or responded to it in some way (Meyer 2012, 8–22).

As noted above, Sunzi's view of war appears to be more expansive than Clausewitz's. If the purpose of military force is to shape an enemy's interest, it falls to the enemy to determine how much pain they will bear. Thus, unlike in the Western view, there is sometimes a gap between what military force can do and the desired political objective. While it is possible to reasonably determine how much force is required to annihilate or attrit an enemy force, there is no predictable relationship between how much force one uses and whether an enemy changes their minds regarding a particular policy. History, in fact, is replete with stories of weaker opponents resisting stronger powers at great costs, despite the stronger power's use of overwhelming force (Sullivan 2007, 498–500).

Because political and military objectives are more distinct, success depends on bringing to bear other elements of national power and integrating them into the military effort. This feature necessarily draws in non-violent means of competition that emphasizes coercion and deception as opposed to brute force and attrition. This point does not suggest that brute force does not have its place; however, its use reflects a failure of the ideal and thus should be the exception rather than the rule (Lo 2012, 122). Moreover, this view connects the nature and quality of civil governance with military capability as it is the civil government that wields the full range of the tools of conflict. In this context, it is worth noting that Confucian writers also prefaced their works on military affairs with discussions of statecraft (Lo 2015, 8).

So, from this shift in ends emerges a view of war that expands on Clausewitz, changing war's scope. Friend and enemy are joined by collaborator and competitor; resistance and surrender are replaced by acceptance and rejection; and victory and defeat are replaced by success and failure. Further, friend and enemy do not refer simply to states, but to sub-state and non-state organizations as well. Additionally, such conflicts are not zero-sum. If one can achieve one's interests by benefiting the enemy, or some subgroup within the enemy's community, so much the better. Such an approach was offered by Jin Canrong, a close advisor to President Xi. In speeches made in 2016, he detailed a strategy comprised of six sequential "moves" that involve intertwining U.S. and Chinese interests while establishing alternatives to the United States, like the Belt and Road Initiative, to eventually displace its influence, first regionally then globally (Li 2021). The idea seems to be that by cooperating in some areas, China will be in a better position to challenge in others.

Moreover, there is also a different sense of *how* each way of war seeks to realize the ends of war. In *A Treatise on Efficacy: Between Western and Chinese Thinking,* Francois Jullien contrasts Western "means-ends" thinking with Chinese "opportunism." He observes that Western military thinking determines the right objective and then tries to organize the resources available to achieve that objective. In this way, a general engaging in battle is much like a ship's captain undertaking a voyage. As Jullien observes, "both operate within constantly shifting fields, full of unpredictabilities, to the very end never certain of triumphing over the enemy or making it to port" (Jullien 2004, 40–41). Success is never guaranteed: friction and chance can always undo the best of plans.

Chinese thinking, on the other hand, analyzes a situation in terms of one's potential to gain from it. The generals' role here is to understand what factors promote their interests and then act when those factors align. Sunzi captures this insight in Book Four, where he states, "The victorious troops thus begin by winning and only then engage in battle, whereas the defeated troops begin by engaging in battle and only then try to win" (Sun 1971, 55; Jullien 2004, 43).

Like the Western view, the Chinese view acknowledges that there are always factors outside one's control; however, rather than try to control or minimize their effects on achieving the intended objective, the Chinese view does not set an objective in the first place. Rather, it seeks to identify and act on opportunities those factors provide to improve one's situation relative to the enemy. As Jullien further describes the Chinese view:

> "We know that circumstances may often be unforeseen, even unforeseeable, and unprecedented, which is why it is not possible to draw up a plan in advance. Rather, they contain a certain potential from which, if we are agile and adaptable, we can profit."
>
> *(Jullien 2004, 39)*

One can see the approach described here reflected in Chinese doctrine. In response to superior U.S. conventional capabilities, the People's Liberation Army (PLA) has developed the concept of *shashoujian* or 'Assassin's Mace,' which is an umbrella term for doctrinal development and acquisition of weapons systems aimed at enabling the 'inferior' to defeat the 'superior' (Krepinevich 2009; Bruzdzinski 2004). It takes its name from small and light weapon from Chinese antiquity that required considerable skill and deception to employ effectively. It was designed to be concealed in its bearer's cloak and was effective against larger swords as well as armor. So, like the use of this weapon, the Assassin's Mace doctrine relies on surprise as well as deceptive and unorthodox methods "unknown to the adversary." The means employed under this doctrine, such as those described above, are intended to achieve the effects of deterring, decapitating, blinding, paralyzing, or disintegrating enemy forces (Bruzdzinski 2004, 314, 344).

The concern from the Western perspective, of course, is that this blurring of international competition and war is too expansive and brings elements of peaceful international competition into the scope of war. The danger here, of course, is that by doing so this view expands what could count as a just reason to go to war, since in this view violence is always a practical, if not normative, possibility in international competition. This expansion can be especially concerning if one accepts Zheng Wang's analysis in *Never Forget National Humiliation*, where he argues that China's perceived past humiliation by the West serves a barrier to negotiation and trust and places the West and China in conflict from which there is no apparent way out (Wang 2012, 132). As I will discuss in the next section, however, this concern is likely overstated. Just as Clausewitz's view gives rise to norms that limit the scope of violence, so does Confucius and Sunzi's.

The Chinese Ethic of War

On the surface, it will seem odd to base an ethic of war on Sunzi's work. As Meyer points out, *The Art of War* was not only an apparent affront to the aristocratic notions of warfare at the time, its emphasis on deception and profit gave rise to protest by Confucian and Daoist masters who objected to its instrumentalist and unvirtuous approach (Meyer 2012, 19–21). However, as David A. Graff notes, a number of Warring States period (475–221 BCE) writers also often did not address ethical concerns and when they did, it was often to denigrate them. For example, in *Zhanguoce*, which represents hundreds of stories regarding military and diplomatic schemes of the warring kingdoms, decisions to go to war needed no other justification than interest or advantage. Where moral concepts like righteousness were mentioned, it was either to denounce it as ineffective or employed cynically to obscure one's true agenda (Graff 2010, 196–97). Sunzi's work seems to fall into this category of amoral analysis on warfare. In fact, of the seven military classics his is best known for the absence of ethical thought and was criticized for that absence by contemporary Confucian thinkers (Lo 2015, 115).

It is in that discussion that Confucian thought makes an important contribution to Chinese military ethical traditions. For example, Mencius, an early Confucian thinker, condemned aggressive wars intended to expand territory, and similar to Western concepts of *jus ad bellum*, "encouraged wars of self-defense against aggression" as well as "punitive expeditions," undertaken by rightfully authorized state actors to address some wrong. Mencius also advocated for *jus in bello* notions of immunity for non-combatants and limiting destruction. Other thinkers, like Xunzi, also contributed concepts such as "legitimate authority," which includes justice and humanity as a pre-requisite for that legitimacy that would be compatible with Western views (Stalnaker 2012, 98). Moreover, this justice and humanity extends from the character of the ruler to the conduct of the war itself (Graff 2010, 200). In this way, the just (*yi*) and benevolent (*ren*) character of the ruler constrains not only when he can fight, but how (Lo 2015, 8–9).

Chinese thought also tends to see ethics as an enabler rather than as a restraint. Confucian thinker Xunzi, for example, argued that good government that embodies the Confucian values of ritual, humaneness, and justice will motivate the kind of loyalty and deep commitment to follow their leadership, even to war (Stalnaker 2012, 100). In more recent formulations, PLA scholar Zhao Feng argued for "moral warfare," where one seeks to be morally superior to one's enemy in order to influence the attitudes and behaviors of allies and adversaries. It is carried out by making others perceive that one is more moral than one's enemy (Zhao 2003, 86–88).[6] Such a view may appear on the surface as a particularly cynical form of information warfare. However, for moral warfare to be effective, practitioners must make an honest attempt to *be* moral, or failing that, present a stronger moral case than one's enemy.

While Confucian contributions arose out of concerns that views like Sunzi's did not adequately address ethical concerns, the interpretation of *The Art of War* as amoral may not be the best one as prudential as well as moral norms are found throughout the entirety of the text. Emphasizing restraint, Sunzi argues at the beginning, that war should be avoided, but when that fails one should only rely "on smart strategies and tactics rather than a display of maximum force" (Lo 2012, 118). While this restraint could be dismissed as merely prudential, Sunzi's further writing on the conduct of war suggests he sees restraint as a norm whose justification lies outside the demands of waging war successfully. For example, in Chapter 3 "Planning Offensives," Sunzi argues preservation is to be preferred over destruction, even when considering enemy formations (Lo 2012, 118). As Lo states:

> Contrary to an amoral realist, Sunzi advocates self-imposed restraint and limits in the employment of lethal violence. For Sunzi, even if we are living in a gladiatorial world, we should try to disable the opponent and force the enemy to submit ingeniously.
>
> *(Lo 2012, 118)*

It is difficult, however, to assess the motivation of this preference for preservation. As Meyer observes:

> The Sunzi insists that *every* casualty of battle represents a loss for the victorious commander and his ruler: every friendly soldier killed was of course an asset lost, but every enemy soldier killed, every enemy provision destroyed and every enemy fortification razed were also potential assets forfeited.
>
> *(Meyer 2012, 13)*

In this regard, it is not surprising that Sunzi's work is viewed as "coldly pragmatic" (Graff 2010, 201). A major concern for him was that armies consumed resources so it was imperative that their use generated more resources than they consumed. Otherwise not only the war, but the state, would be unsustainable (Meyer 2012, 14). However, here it is important to understand the continuity between governance and military affairs described above is not generally present in Western thought. For both Sunzi and Confucian thinkers of the time, "virtue, benevolence, and righteousness" served a "force multiplier" that increased one's chance for, if not guaranteed, victory. In fact, it is on this last point that Sunzi may have diverged from Confucian thinkers. Sunzi saw moral superiority as an advantage, but one that was neither a necessary nor sufficient requirement for victory (Graff 2010, 201). Mencius and Xunzi, like others at the time, did (Stalnaker 2012, 101; Graff 2010, 200). Still, Sunzi's preference for restraint seems to arise from previous moral commitments to a just and benevolent order and not simply as a means of effective resource management.

Thus, Lo sees Sunzi's ethics as similar to those argued for by Henry Sidgwick whose utilitarian "two-fold rule" insisted that force should only be used in pursuit of objectives that will lead to victory, which serves a just cause, and then only as much as is proportional to the value of that objective (Lo 2012, 122; Walzer 2015, 129). While such an ethic does not fully account for all intuitions regarding the morality of war, it does prohibit attacking purely civilian targets as well as disproportionate uses of force, regardless of what argument one might make regarding the overall effect on ending the war successfully. Moreover, as Walzer notes, Sidgwick's rule, and by extension Sunzi's preferences, at least establish there are rules of war, which further distinguishes it from murder or robbery, for which there are no rules (Walzer 2015, 128).

But Sunzi's account is not entirely utilitarian in the same way Sidgwick's is. Maximizing victory in service to a just cause says nothing about the justice of an act. In fact, a major criticism of Sidgwick's view is that it permits too much: any act, no matter how vicious or indiscriminate, would be permissible as long as it contributed to victory and the harm committed proportional to that contribution. However, if we accept that Sunzi's view is, in fact, compatible with Confucian concerns, then it will not be enough that acts of violence and coercion maximize a just and benevolent order as that would not be too different from Sidgwick's view. Rather, to the extent acts of violence and coercion serve justice and benevolence, they must retain the character of those virtues.

So, while Sunzi's ethics may not be as fully developed as other thinkers of the time, the privileging of just and benevolence governance as the authority for force suggests this commitment to restraint is not merely pragmatic but rests on a deeper notion of human flourishing. Moreover, over time, others did find ways to harmonize both. In the 14th century, for example, Wang Yang-ming, a scholar, statesman, and general, applied an approach to war that blended Sunzi's strategies that emphasized careful calculation, espionage, deception, and surprise with Confucian values that emphasized justice and benevolence. The result was an ethic that emphasized humanitarian concerns for the people, benevolent and just treatment of defeated enemies, and accountability for the aggressor (Twiss and Chan 2015, 161).

Confucians were not the only ones to blend Sunzi with humanitarian and ethical concerns. The Daoist Liu An, the king of Huainan wrote the *Huainanzi* to distill all the knowledge a monarch would need to know, which included military matters. In doing so, the work tried to reconcile *The Art of War* with Daoist ethical concerns by placing military operations on ethical grounds. Where Sunzi emphasized economic and strategic motives for war, the *Huainanzi* emphasized establishing a just order. In fact, the conqueror was not allowed to keep the conquered community's resources but instead was required to form a new political structure that would employ those resources to the benefit pf those conquered. Moreover, rather than clever planning and deception, the *Huainanzi* operated under the principles of "sustaining the perishing, reviving the extinct" (Meyer 2012, 45–46. This ethic of war is similar to Aristotle's, mentioned earlier, who believed wars to expand empire were permissible if they improved the conditions of the conquered.

So, blending Sunzi and Confucian views on the conduct of war, one ends up with an approach that more closely resembles neo-Aristotelian virtue ethics, which emphasizes the agent as opposed to the act. Like consequentialist accounts, virtue ethics does not rule out any act; however, unlike consequentialism it rules in character traits associated with the human good (Hursthouse 1999, 25). These traits thus would rule out acts that maximize a good if those acts did not also reflect the virtue. Since agents must discern the best way to achieve an end, practical reason is inseparable from virtue (Sherman 1989, 3–4). Thus, what matters is what the virtuous person would do given the alternatives at hand. If all the alternatives are terrible, then the virtuous person may do terrible things, though would generally choose the least terrible alternative.

Similarly, a Confucian ethic requires that the authority that wages war must itself have the "mandate of heaven," which requires rulers themselves to possess the virtues of justice and benevolence and rule in (Ivanhoe 2004, 272) Such virtues, when applied to war, may appear utilitarian in application, however, they are constrained by non-utilitarian humanitarian concerns associated with the relevant virtue. To the extent such an ethic is followed, it will place constraint on action that Sidgwick does not. When it is not followed, it can still serve as a cause for criticism, if not remorse.

Such an ethic of virtue harmonizes with a way of war that emphasizes shaping interests and choices. Shaping interests requires there remains with the adversary sufficient capability and capacity to act in a desired way. Widespread destruction would seem to undermine that end. By shifting the emphasis of the end of war from imposing one's will to changing others minds, emphasis of effort shifts from the destruction of enemy combat power to engaging decision-makers and even whole societies. That engagement is best facilitated where one is perceived as virtuous, but equally capable of hurting, as Schelling might put it. Thus, where the Western ethic seeks to employ the most force possible, given the limits of discrimination and proportionality, the Chinese view may be characterized as *using the least force possible, given the demands of a just and benevolent order.*

Conclusion

As Christopher P. Twomey points out, "when nations have different doctrines and hold different beliefs about what kinds of military strategies and capabilities may be effective, diplomacy and signaling will be more difficult, and this can cause escalation or conflict" (Twomey 2010, 18). For example, different - and dismissive - views of each other's way of war encouraged escalation on the Korean peninsula in 1950. For its part, the United States' experience in World War II led it to view war as general, rather than limited, and thus worried little about how to win while containing the conflict.

Moreover, while U.S. leadership was dismissive of Chinese military capabilities, their reliance on airpower, mechanized forces, and combined arms tactics limited what it could do against a Chinese force that relied heavily on manpower, which could be more easily concealed than large equipment-heavy forces but concentrate in large numbers to overwhelm U.S. and its allies' forces. The Chinese made similar errors. They viewed U.S. forces as militarily weak because of the large logistical effort it took to sustain heavy forces. Moreover, they overestimated their own combat capability because they dismissed U.S. tactical capabilities out of a belief that the U.S. military was not good at close combat, night battles, and bayonet charges (Twomey 2010, 76–86).

What is true for doctrine can also be true for ethics. Mao, for example, believed U.S. soldiers would not be motivated to fight because they were the invaders interfering in an otherwise just "People's War" (Zhang and Yao 1996, 211; Twomey 2010). For its part, the United States' intervention would have counted as coming to the defense of a victim of aggression (Dwight D. Eisenhower Presidential Library n.d.). Thus it seems a comparative understanding of these ways of war and how they inform the resulting ethic will not only bring a greater understanding of how to address a range of national security issues, it will also provide greater awareness regarding the moral motivations behind the security-related decisions each community makes. Such awareness should reduce the risk of misunderstanding as well as create space for better communication.

Of course, this has been a discussion of ideals within two rich and complex traditions. So the utility of this analysis would be limited to the extent views like Clausewitz's, Sunzi's, and others' have impacted Western and Chinese views. The literature suggests their impact has been influential; however, human difficulty with ideals also ensures there will be numerous exceptions. Moreover, war is evolutionary and no way of war or its ethic is static or enduring. Thus, while a way of war, like Sunzi's, that relies on coercion and deception may prevail over one that emphasizes firepower and attrition, a new way will evolve in response. It will be up to those who develop that way to ensure that the resulting ethic is one that harmonizes not just with the requirement to win, but also with demands of human flourishing.

Notes

1 Disclaimer: The views expressed in this chapter are the author's and not necessarily those of the United States Government. Draft only, not for attribution.
2 This chapter has been adapted from (Pfaff 2022, 73–88).
3 There is an asymmetry in comparing the "West" with China as the former represents a diversity of cultures, societies, and states. To simplify the discussion, I will focus on U.S. interpretations.
4 Schelling's emphasis on the power to hurt does not diminish the role cooperation can play in international competition. In fact, the power to hurt is also the power not to hurt, that is to cooperate with an adversary's interest as long as they comply with the relevant demand.
5 In the full context of his remarks, General LeMay argued that war is itself immoral, implying that restrictions on the use of force that prolonged it were also immoral. Thus, while recognizing the illegality of the bombings, it also fairly clear he did believe these actions were immoral.
6 Quoted in (Metcalf 2019, 12).

Bibliography

Birnes, William J. 2017. *Forward to the Shadow*. Lawn Press.

Bruzdzinski, Jason E. 2004. "Demystifying Sha Shou Jian: China's 'Assassin's Mace' Concept." In *Civil Military Change in China: Elites, Institutes, and Ideas After the 16th Party Congress*, edited by Andrew Scobell and Larry Wortzel. Carlisle, PA: Strategic Studies Institute.

Cicero. 2013. *On Duties*, edited by M. T. Griffin and E. M. Atkins, pp. 14–15. Cambridge: Cambridge University Press.

von Clausewitz, Carl and Rapoport, Anatol, eds. 1968. *On War*, p. 101. Middlesex: Penguin Books.

Corn, Geoffrey, Ken Watkin, and Jamie Williamson. 2016. *The Law in War: A Concise Overview*, p. 77. New York: Routledge.

Dwight D. Eisenhower Presidential Library. n.d. "Korean War." Accessed October 7, 2022. www.eisenhowerlibrary.gov/research/online-documents/korean-war.

Fridman, Ofer. 2018. *Russian Hybrid Warfare: Resurgence and Politicization*, pp. 12–13. Oxford: Oxford University Press.

Graff, David A. 2010. "The Chinese Concept of Righteous War." In *The Prism of Just War: Asian and Western Perspectives on the Legitimate Use of Military Force*, pp. 196–197. Routledge.

Gray, Colin S. 2010. "War—Continuity in Change, and Change in Continuity." *Parameters* XXXX (Summer): 7. Accessed November 17, 2022. https://press.armywarcollege.edu/parameters/vol40/iss2/5.

Hursthouse, Rosalind. 1999. *On Virtue Ethics*, p. 25. Oxford: Oxford University Press.

Ivanhoe, Philip J. 2004. "'Heaven's Mandate' and the Concept of War in Early Confucianism." In *Ethics and Weapons of Mass Destruction: Religious and Secular Perspectives*, edited by Sohail H. Hashimi and Steven P. Lee, 272. Cambridge: Cambridge University Press.

Jullien, François. 2004. *A Treatise on Efficacy: Between Western and Chinese Thinking*, translated by Janet Lloyd, 40–41. Honolulu, HI: University of Hawai'I Press.

Krepinevich, Andrew F. 2009. "The Pentagon's Wasting Assets." *Foreign Affairs* 88, no. 4 (July 1).

Li, Manyin. 2021. "*What China Really Wants: A New World Order.*" *National Review*, March 7. www.nationalreview.com/2021/03/what-china-really-wants-a -new-world-order.

Lo, Ping-cheung. 2012. "Warfare Ethics in Sunzi's Art of War? Historical Controversies and Contemporary Perspectives." *Journal of Military Ethics* 11, no. 2 (August): 124.

Lo, Ping-cheung. 2015. "Varieties of statecraft and warfare ethics in early China." In *Chinese Just War Ethics: Origin, development, and dissent*, edited by Ping-cheung Lo and Sumner B. Twiss, 8. New York: Routledge.

Metcalf, Mark. 2019. *Moral Warfare: Weaponizing Ethics to "weaken, divide, and smash the enemy*. Unpublished Manuscript, p. 12.

Mewett, Christopher. 2014. "Understanding War's Enduring Nature alongside its Changing Character." *War on the Rocks*, January 21, 2014. Accessed October 7, 2022. https://warontherocks.com/2014/01/understanding-wars-enduring-nature-a longside-its-changing-character.

Meyer, Andrew Seth, trans. 2012. *The Dao of the Military: Liu An's Art of War*, pp. 8–22. New York: Columbia University Press.

Orend, Brian. 2006. *The Morality of War*, p. 11. Petersborough, ON: Broadview Press.

Ota, Fumio. 2014 "Sun Tzu in Contemporary Chinese Strategy." *Joint Forces Quarterly* 14 (April 1). Accessed October 3, 2022. https://ndupress.ndu.edu/Media/ News/Article/577507/sun-tzu-in-contemporary-chinese-strategy.

Paret, Peter. 1968. "Education, Politics, and War in the Life of Clausewitz." *Journal of the History of Ideas* 29, no. 3 (July–September): 395.

Pfaff, C.A. 2022. "Chinese and Western Ways of War and Their Ethics." *Parameters* 52, no. 1 (2022): 73–88, doi:10.55540/0031-1723.3130.

Qiao, Liang and Wang, Xiangsui. 1999. *Unrestricted Warfare*. Beijing: PLA Literature and Arts Publishing House. Accessed October 7, 2022. English translation available at www.c4i.org/unrestricted.pdf.

Qiao, Liang and Wang, Xiangsui. 2017. *Unrestricted Warfare: China's Master Plan to Destroy America*. Shadow Lawn Press.

Reichberg, Gregory M., Syse, Henrik and Begby, Endre, eds. 2006. *The Ethics of War: Classic and Contemporary Readings*, 37. Malden, MA: Blackwell Publishing.

Rhodes, Richard. 1995. "The General and World War III." *The New Yorker*, June 19, 1995. Accessed October 7, 2022. www.newyorker.com/magazine/1995/06/19/ the-general-and-world-war-iii.

Roberts, Adam, and Guelff, Richard, eds. 1989. *Documents on the Laws of War*, pp. 271–272. Oxford: Clarendon Press.

Sawyer, Ralph D., trans. 1993. *The Seven Military Classics*. Boulder, CO: Westview Press.

Schelling, Thomas. 2020. *Arms and Influence*, p. 6. New Haven, CT: Yale University Press.

Sherman, Nancy. 1989. *The Fabric of Character: Aristotle's Theory of Virtue*, pp. 3–4. Oxford: Clarendon Press.

Stalnaker, Aaron. 2012. "Xunzi's Moral Analysis of War and Some of its Contemporary Implications." *Journal of Military Ethics* 11, no. 2 (August): 98.

Strachan, Hew and Scheipers, Sibylle, eds. 2013. "Strategy in the 21st Century." In *The Changing Character of War*, p. 511. Oxford: Oxford University Press.

Sullivan, Patricia. 2007. "War Aims and War Outcomes: Why Powerful States Lose Limited Wars." *Journal of Conflict Resolution* 51, no. 3 (2007): 496–525. Accessed November 17. 2022. doi:10.1177/0022002707300187.

Sun, Tzu. 1971. *The Art of War*, edited by Samuel B. Griffith, 77. Oxford: Oxford University Press.

Temes, Peter S. 2003. *The Just War: An American Reflection on the Morality of War in our Times*, p. 62–63. Chicago, IL: Ivan R. Dee.

Twiss, Sumner B. and Chan, Jonathan K. L. 2015. "Wang Yang-ming's ethics of war." In *Chinese Just War Ethics: Origin, development, and dissent*, edited by Ping-cheung Lo and Sumner B. Twiss, p. 161. New York: Routledge.

Twomey, Christopher P. 2010. *The Military Lens: Doctrinal Difference and Deterrence Failure in Sino-American Relations*, p. 18. Ithaca, NY: Cornell University Press.

Walzer, Michael. 2015. *Just and Unjust Wars*, 5th Ed., p. 23. New York: Basic Books.

Wang, Zheng. 2012. *Never Forget National Humiliation: Historical Memory in Chinese Politics and Foreign Relations*, p. 132. New York: Columbia University Press.

Weigley, Russel. 1977. *The American Way of War*, p. 475. Bloomington, IN: Indiana University Press.

Zhang, Junbo and Yunzhu Yao. 1996. "Difference Between Traditional Chinese and Western Military Thinking and Their Philosophical Roots." *Journal of Contemporary China* 5, no. 2 (July): 211.

Zhao, Feng. 2003. "Military Ethics Should Investigate the Scope of Moral War." *Journal of Nanjing Institute of Politics* 19, no. 4: 86–88. [Published in Chinese: 趙楓〈軍事倫理學應當研究道德戰範疇〉《南京政治學院學報》。]

4

EAST/WEST JUST WAR DIALOGUES

Reflections on the Larger Implications

Martin L. Cook

In this chapter, I want to reflect on an ever-broadening set of realizations about the nature of the international system that has emerged (for me at least) as a result of participation in East-West dialogue. When this group[1] first began to meet in 2011, the focus was a fairly narrow scholarly examination of historical Chinese thinkers about war and comparison of those thinkers to historical Western thinkers' ways of framing the same issues. This forum brought together experts in historical Chinese thought with Western scholars who were knowledgeable in the tradition of Just War as it evolved in the European traditions of the Christian church and, after the Reformation, in secular international law. Those early efforts produced some excellent work on historical traditions of Chinese and Western thought about war and use of military force more generally. For me personally, participation in this dialogue has caused me to not only to reflect on uniquely Chinese ways of thinking about war, but more broadly to appreciate the cultural and historical context and evolution of particular ways of framing issues of war, the nature of states, and assumptions about the shape and history of world order. Specifically, it provided an excellent opportunity to reflect more deeply on the historical and culturally specific origins of ostensibly "universal" standards of international law and the ethical principles of the just war tradition. It prompted a deeper reflection on the specifically Western, European and largely Christian roots of those traditions and how, owing to the cultural dominance of those traditions over the planet for the past few centuries, those traditions came to have the universal appearance they do. Lastly, and most importantly, it prompted deeper reflection on the increasing emergence of competing cultural traditions and narratives. That reemergence, coupled with the declining dominance of the West among the powers of the world, precipitated reflection on the implications for a possible reopening of cross-cultural dialogues about the

DOI: 10.4324/9781003336372-5

principles of international relations and war that once seemed, at least in theory, to be resolved into a single globally agreed international system of laws, principles and institutions created after World War II.

From my position at the Naval War College and seven years of deep immersion in the thoughts and fears of the United States Navy, a parallel train of thought emerged and increasingly prompted an expansion of the range of considerations bearing on these issues. During the decade our group has been meeting, the relationship of modern China to the rest of the world has been changing rapidly – in particular, in ways that have deeply engaged and concerned the U.S. military and especially the U.S. Navy. As is commonly known, China has dredged up material to build up shoals throughout the South China Sea and has proceeded rapidly (despite President Xi's promising not to do so in 2015) to build military installations, runways, and missile installations on those shoals.[2] Although China is a signatory to the Law of the Sea Convention (UNCLOS)[3], it has acted in clear and obvious violation of the terms of that Convention – in particular the provision that no object which is naturally underwater at high tide can be used as a basis for asserting territorial claims (UNCLOS, Art. 121, Part VIII). Indeed, on July 12, 2016, the Arbitration Tribunal established by Annex VII to the 1982 United Nations Law of the Sea Convention ruled that China's claims had no legal basis and that the "Nine-Dash Line" which China was relying on to make its claims had no historical validity.[4] China refused to participate in the arbitration process as required by the Law of the Sea Convention and rejected the ruling after it was issued.[5] Further, China is creating a historical narrative of dubious validity to support its territorial claims in the South China Sea based on the historical voyages by Zheng He in the 15th century.[6]

Of course, one way of accounting for China's behavior is straightforward cynicism: no one is in a position or willing to advocate use of force to stop China from doing what it is doing and therefore this is a simple example of power politics overruling ostensibly binding international law. But there is another and more intriguing possibility and one that participation in this East-West dialogue over a number of years has brought into focus. Might it be the case that, as China regains its major place in the international system, treaties and legal conventions they went along with while enfeebled are now being supplanted as deeper and distinctively Chinese ways of understanding China's own proper role and rules for behavior are reemerging? Perhaps those culturally deep-rooted narratives are supplanting what had appeared to be universal international standards

As one begins to entertain that thought, one quickly realizes that China is not the only example of reemerging cultural traditions challenging existing international order. The most glaring example besides China is, of course, Russia. Russia's actions in Ukraine, Crimea, and Syria and globally in the cyber domain with election interference and the sowing of dissension and discord in the domestic politics of democracies throughout the world.[7] These actions are pushing against international norms and suggest that Russia, too, may be relying on its own counternarrative to allow it to violate previously well-established rules of international behavior, despite the fact that Russia's predecessor state, the Soviet Union, directly participated in the creation of the post-World War II international order and especially the United Nations.

Further, the Westphalian organization of the globe into sovereign states has never fit comfortably into Islamic ways of thought. This was clear as the world witnessed the rise of the Islamic State and its claims to be reestablishing a single unitary Caliphate in principle for all Muslims.[8] Clearly, the appeal of that classical Islamic ideal should not be underestimated, as was evident in the ability of the Islamic State to draw in idealistic Muslims from around the world. The normative division of the world into the *dar al harb* and the *dar al Islam* has a much deeper cultural resonance in the minds of Muslims than somewhat arbitrary borders established by departing Colonial powers in the Middle East.[9] Further, imposition of nominally Westphalian states on areas of former colonial dominance, especially in Africa, fails to track culturally with indigenous power structures.[10]

This chapter builds on the East-West dialogue to explore these larger and deeper questions. Might the Liberal international order created by Europe and the United States, largely over the course of the 20th century, in fact be losing its apparent universality? Clearly, the actions of both China and Russia are violations of what had seemed to be settled international law[11]. But more importantly, it appears that both China and Russia have a narrative about themselves that genuinely permits or requires them to consider their contest-able actions as legitimate and appropriate. This raises an important and intriguing possibility: As China and Russia and others find themselves in positions of sufficient military and political power to reassert alternative understandings and to guide their own behavior, might the world be facing once again a historical inflection point, perhaps as significant as the one that created the Westphalian system in the 17th century

Europe created the Westphalian order after the wars precipitated by the Reformation (Encyclopedia Britannica 2022). In the 20th century, the world (led by the U.S. and European powers) attempted to build legal and institutional structures above the sovereign state with the creation of the League of Nations after World War I and the United Nations after World War II. In other words, history shows that there is no "natural" shape to world order and that particular modes of organization are created by human actions and agreements in particular places and times and in response to specific historical circumstances. Such historical observations lead inevitably to the conclusion that there are inflection points in global order from time to time in which new powers with differing cultural traditions and changing historical circumstances reshape the normative understanding of how in their view global order is to be understood and maintained. The relatively long period of the dominance of the Westphalian system of sovereign states can blind us to the realization that it, too, is a human creation at a particular place and time, in response to specific historical circumstances of Europe in the 17th century with additional strata created largely by Western powers up through the 20th century. There is widespread evidence that the established international order is fraying and that even the states that created it are losing faith in those international insti-tutions.[12] That loss of faith, combined with the rise of different cultural tra-ditions attached to rising world powers may well portend the need for novel international and cross-cultural dialogue if a new system of agreed upon international order is to emerge. It is important to remember that shared understandings of global order are always human creations and responses to particular historical circumstances such as the collapse of Christendom in Europe in the 16th and 17th centuries or the colossal destruction of the two world wars of the Twentieth Century. The fact that Liberal world order was stamped across the planet is in fact a consequence of several centuries of colonialism and European and American dominance. That world order was imposed on parts of the world where it is not a natural "fit," most notably in Africa, the Middle East, and Asia.

One useful result of our years of collaboration in this East-West comparative study was to gain an understanding of how the Chinese tradition has under-stood China itself and the legitimate uses of military force. Concepts such as punitive expeditions, China as the Middle Kingdom expecting deference and tribute from its neighbors, and the "century of bad treaties" which made China nominally subject to a system of international order imposed on it from outside set the grounds for beginning to understand contemporary Chinese behavior where it appears to be in violation of that order (Twiss and Chan 2012).

In addition to China, it is increasingly clear that Vladimir Putin's vision of Russia's place in the world and of legitimate uses of military force fall well outside norms that until fairly recently seemed well-established and secure. As he famously remarked in an April 2005 state of the nation address to Russians: "Above all, we should acknowledge that the collapse of the Soviet Union was a major geopolitical disaster of the century. As for the Russian nation, [this collapse] became a genuine drama. Tens of millions of our co-citizens and co-patriots found themselves outside Russian territory. Moreover, the epidemic of disintegration infected Russia itself."[13] Certainly ten years ago it would have seemed absurd to suggest that a state would take territory by force in Europe and even more unlikely that such a state would assert the ethical and legal legitimacy of such actions.

Participation in this comparative just war study group has provided a long historical perspective from which the Liberal international order that has appeared to provide a universal and unquestioned shape to the international system reveals itself as it really is: a human creation. Further, that perspective increasingly makes clear that the historical dominance of the Liberal model might well be fading, and a robust and more truly multicultural dialogue may be beginning from which some new and not clearly foreseen agreed international system may be beginning to emerge. If that is indeed increasingly the situation in international affairs, it also raises a troubling but inevitable question as to whether a new and commonly shared set of rules of international behavior can emerge peacefully for sustained cross-cultural discussion. The reality that previous global inflection points culminating in new international norms have often emerged from horrific conflict (e.g., Westphalia after the Thirty Years War, and the formation of the United Nations and other post-World War II institutions) also raises the sobering question of whether whatever the new shape of the international system turns out to be can emerge before a major international conflict.

The "rising new powers and declining great powers" has often been called "the Thucydides Trap," reflecting Thucydides' comment in Book I of his history that "the truest cause" of the Peloponnesian War was the rise of Athens' power and the fear it caused in the Spartans.[14] Examples of peaceful waning/rising national power do exist, notably, the relatively smooth transition from British to American dominance of the international scene. But the prospects of major power conflict certainly increase in such times of transition. The weakening of the institutions of the Liberal world order that have contributed to the lack of major power conflict since World War II threatens to deprive the world community of resources to prevent conflict in the present. If the decline of those institutions continues, this threat becomes even more the case extended into the future. In other words, this particular "Thucydides trap" moment arrives at a historical point when the mechanisms to help avoid major conflict are weakened and appear to be in decline.

The central thesis of this chapter is that the dominance of specific ideas of the shape and "rules" of the international order that developed in Europe over the course of five centuries and were then imposed across the planet can no longer be taken for granted. Despite their appearance as universally accepted international legal frameworks and standards for several centuries, there is no reason to assume they will retain that status. There are many indications that they are being challenged even among the countries and cultures that created them and even more so from rising cultures that never fully embraced their central values in the first place.

Other cultural norms and narratives are now reasserting themselves in the international system. Further, these are not merely competing cultural traditions and narratives but come with backing of sufficient military and economic power to vie openly to supplant the established international system with powerful counternarratives. As Samuel Huntington put it well years ago, "The very notion that there could be a 'universal civilization' is a Western idea, directly at odds with the particularism of most Asian societies and their emphasis on what distinguishes one people from another." (Huntington 1993, 41) If that is right and the West's ability and willingness to strenuously support and defend the universality of that very civilization is uncertain, it is quite possible that we are entering another of those historical inflection points and we should expect at a minimum a period of heated multicultural dialogue about what takes its place. Of course, a worse possibility cannot be ruled out either: that significant military conflict lies ahead as the globe undergoes the violent shifting of the cultural tectonic plates before they settle into a new agreed upon order or, failing that, a return to more or less continual great power conflict for some time to come.

I turn therefore to an attempt to articulate some of the major competing narratives which at this historical moment come with sufficient political, economic and military power to be able to compete and put forward alternative visions of legitimate use of force and of world order, if indeed we are in the midst of another significant cultural inflection point.

China's Narrative

Professor Zheng Wang's excellent book, *Never Forget National Humiliation: Historical Memory in Chinese Politics and Foreign Relations*, provides a deep analysis of China's "master narrative" of its modern history and its traditional sense of its proper place in the international system (Wang 2012; Ward 2019; Pilsbury 2019). The title of the book concisely summarizes the master narrative that China was subject for several centuries to "national humiliation" at the hands of Western powers, as well as of course the Japanese. But underlying the humiliation, Wang argues, there is a deeper narrative of what China's rightful place ought to be: the Middle Kingdom, respected and deferred to by at least other Asian states and the major power among Asian nations.

An example implicitly evoking this narrative can be found in a speech of President Xi in 2015 at the Fifth National Border and Coast Defence Work Meeting. In that speech, Xi invoked specifically the past as a time in which China was "poor and weak." But he further went on to argue that China now is strong and must place the "highest priority" on safeguarding "sovereign territorial integrity and maritime rights and interests." It will do so, he insisted, by building an "impregnable wall [literally, "a wall of copper and iron"] for border and ocean defense."[15] It is in this light that one must understand Chinese concepts such as "active defense," which explains and justifies its actions in the South China Sea. What looks to most of the world, and certainly to international law (especially the Law of the Sea Convention) like naked aggression genuinely seems to the Chinese mind to be unquestionably justified as merely restoring China to its proper boundaries and status. This s particularly clear when one examines Chinese claims to all land and maritime space within its "nine-dash line," which it views as legitimate claims.

The implications of this way of thinking are enormous for the question this chapter is addressing: the future and stability of the international system as it has been developed over the past four or five centuries. The clear implication is that China does not really see itself as one Westphalian state among others, but to be truly an exception to that system. Chinese thinkers will grant that this way of thinking deviates from the international system and even from the plain language of treaties to which the country was signatory. But they dismiss such observations by invoking a "century of bad treaties" made when China was weak, but which, now that it is strong, they feel free to ignore or abrogate totally.

Wang tells a fascinating anecdote that perfectly illustrates the disconnect between how China sees itself in contrast to the European model of sovereign states. Wang writes:

> "In March 1839, Charles Elliot, the British superintendent of trade, wrote a letter to Lin Zexu, the Qing commissioner on the opium trade, regarding the trade issues between the two countries. In this letter, he referenced Britain and China as 'two countries.' This angered the commissioner so much that he returned the letter, saying, 'no place under heaven can be referred to as equal to our celestial empire.'"
>
> *(Wang 2012, 71)*

The central point here is that, if one is to understand Chinese behavior in the international order, one probably must fully take on board the idea that China simply does not accept a status as one equal sovereign state among others. Behaviors that look manifestly illegal and at variance with international norms become in Chinese thinking perfectly reasonable. Whether one focuses on Chinese "island building" and militarization of shoals in the South China sea[16], its "belt and road" initiatives expanding its involvement and investment of resources throughout the world (China Power Team 2017), or its current detention and forceable "reeducation" of Uighurs internally[17], one must explain Chinese behavior as, in its mind, returning to its authentic self-understanding. China sees itself as shaking off the "bad treaties" and episodes of national humiliation that forced it to conform to behaviors imposed on it when it was weak.

The fact that it is succeeding in building "facts on the ground" that most of its neighbors and the rest of the international system view as illegal and unwarranted suggests that its assertion of an alternative framework is likely to continue to expand. No nation or coalition has shown any appetite for using force to enforce the ostensible norms against such behavior. That would suggest that the robust enforcement of international norms envisioned by the UN Charter is rapidly becoming a dead letter in Asia. This behavior constitutes one strong example of a possibly successful assertion of a competing vision of legitimate behavior in the global system and supports the conclusion that the Liberal international order is rapidly losing its claims to universality and global norms. It is clear evidence that, to use Kagan's phrase, "the jungle is growing back."

Of course, this clash is not merely one of ideology. As the United States and other Asian powers continue to conduct freedom of navigation operations and other challenges to Chinese territorial claims, the possibility of inadvertent military clashes is ever-present. Needless to say, even a small clash might well be the trigger for full-scale warfare.

The Putin/Russia Narrative

"The breakup of the Soviet Union was the greatest geopolitical tragedy of the 20th century."

Vladimir Putin[18]

Unlike China, which was barely a bit player in the creation of the Liberal international order and the establishment of the United Nations and the post-World War II norms (recall that the "China" at the time of the creation of the UN was Taiwan!), the Soviet Union was a major player in those events. Obviously, it was enshrined as one of the "Big Five" powers as a permanent member of the Security Council of the United Nations. It was centrally involved in the conflict termination process following World War II. Throughout the Cold War, despite (or because of!) its enormous military machine and nuclear arsenal, it refrained from initiating a major power war [19]

In recent years, however, there is steadily mounting evidence that Putin's Russia has no commitment to even the partially hypocritical respect for the norms in international behavior and international law. Its interventions in Ukraine with "little green men" and hybrid forms of unacknowledged war, its annexation of Crimea, its direct participation in the brutal bombing of Syrians, its full-scale invasion of Ukraine in 2022, and its nefarious cyber activities of not yet fully understood scope all point to Russia's departure from international norms.

Despite a quite weak economy (Miller 2019), Russia has spent vast sums of money modernizing its armed forces, building new weapons systems, and conducting large scale and enormously expensive military training exercises (The World Bank 2020). By claiming a kind of concern and jurisdiction over Russian language speakers wherever they may be, Russia has established a framework under which it could claim justification for interventions almost anywhere in the old Warsaw Pact nations, since Russian speakers spread far and wide within the old Soviet sphere.

Just to take the most obvious case, the idea that national territory within Europe would be taken by force as in Crimea and Ukraine would have seemed beyond the realm of the possible only a few years ago. Further, the idea that such forcible acquisition of territory would succeed without significant and robust military or other response until the most recent invasion of Ukraine (and even then stopping well short of direct intervention by other states) would have seemed to violate the fundamental principles that were established in Europe after World War II.

In 2015 Russia released a new National Security Strategy document. As one report summarized the central claims of the NSS, it "consistently refers to creating and supporting a 'polycentric' international order, one which would make the Russian state a partner equal to the United States, the European Union, and China" (Akin 2017).

In the long run (assuming we get to a long run), Russia's efforts to destabilize the international system will probably have less permanent effect than what China is doing. The Russian economy is certainly not strong enough to sustain permanent great power status. And the demographics of Russia's aging population (Adamson 1997) does not portend a long-term bright future for Russia. Nevertheless, it is obvious that Russia under Putin has serious intentions of doing everything in its power to destabilize democratic states where possible, to foment extremist elements in European and American societies, to destabilize the European Union and to spread distrust of Western institutions among Western populations (Ioffe 2018).

Russia's rhetoric and behavior suggest a fundamental rejection of the rules-based order the United States and Europe believed they had established after World War II. Of course, it is impossible to predict how many more incursions and takeovers Russia is likely to attempt in future years, but it seems likely that there will be more. On the other hand, given the poor performance of the Russian military in Ukraine and the enormous losses of life and materiel during this invasion, Russia's appetite for further adventures of this sort may be tamed. Still, short of a publicly observable and massive use of military force against a NATO member, there seems to be little willingness to use force collectively to repulse any such Russian efforts. Indeed, the reliability of the bedrock of NATO found in Chapter V's pledge of mutual military aid has been publicly questioned by the then American president (Sullivan 2018). That possible combination of Russian assertiveness in its "near abroad," combined with uncertainty of NATO's will (and perhaps ability) to resist it have the possibility of bringing Europe back to the kind of great power politics practiced in the 19th century and thus unravelling the supposedly stable international order created over the last century.[20]

In short, in Russia as in China, we see a major power deeply dissatisfied with the rules-based international order and acting in violation of it in an attempt to regain a deeply culturally grounded sense of a lost proper place in the world. Rather than viewing individual violations in isolation from the others, this suggests that we may indeed be moving into a major world-historical inflection point in the international system. Perhaps the most apt metaphor is the increasing strength and frequency of tremors in advance of an earthquake. As with that situation, it is clear that pressures and tensions are building toward some sort of upheaval. But as with the earthquake, it is impossible to predict the realignment of the terrain that will emerge after the tectonic plates have shifted.

The Classical Islamic Narrative

The third major civilization that was brought into the Westphalian/Liberal order by forces beyond its control is the entire Islamic world. The apparent structure of the nation-state system one sees on the map of the Middle East is the familiar common international structure. But that appearance is somewhat misleading since it was imposed on the region by departing colonial powers. Indeed, one very astute analysis of the region by Charles Glass is aptly titled, "Tribes with Flags: Adventure and Kidnap in Greater Syria." An instructive alternative is the map of the Middle East as conceived by Lawrence of Arabia, who was able to draw on detailed knowledge of the cultures and divisions of the region to imagine what a more culturally appropriate division of the area might look like.[21]

Even more fundamentally, however, the European-style nation state really has no basis whatsoever in normative Islamic thought. In that tradition, the world ideally should be divided into the *dar al Islam*, the sphere of Islam, and the *dar al Harb*, the world of struggle beyond the borders of the Islamic sphere. Ideally, the *dar al Islam* should be ruled by a single political head, the *Caliph*, and ideally the *dar al Islam* should be expanded regularly until it fulfills its destiny of spanning the globe (Encyclopedia Britannica 2011). In the eyes of someone steeped in this vision of the world, the existing countries and governments in the Islamic world lack legitimacy, not just because of failures of governance, but in their very essence.

The revival of this political model in the modern age is often attributed to the writing of Sayyid Qutb. Qutb saw the model of true freedom was to be found only in returning to the founding vision of Islam and rejecting all Western influences on Muslim cultures. From his prolific writing in the early 20th century, his ideas have sprouted in the Muslim brotherhood organization, in *al Qaeda*, and most recently in the Islamic State (Von Drehle 2006; Hamid 2018). If one were to doubt the powerful attraction to some Muslims of that vision, one need only to look at the success of Islamic State's ability to recruit adherents from all over the world, most notably of individuals who on the surface appear to be completely successful and integrated into Western societies (National Bureau of Economic Research 2016; Cohen 2015). The vision of a "pure" Islamic society, ruled by a Caliph and organized in accordance with *shari'a* law, while perhaps realistically only a utopian vision, nevertheless has demonstrated its powerful appeal for more than a century throughout Muslim communities all over the world.

Conclusion

Participation in this East-West just war study group over now many years initially had the effect of bringing me to better understand deep-rooted cultural differences between East and West regarding not merely warfare, but legitimate governances, alternative understandings of the world order and, in particular, China's expectation of its proper place in the world.

Those insights combined with trends developing throughout the world in these years have prompted some deep reflection on the ways in which particular conceptions of international order are formed, how they achieve broad recognition, and how they sometimes decay or collapse. World order has no "natural" shape. It is a human creation. More specifically, it is a creation at specific places and times, in response to particular historical circumstances and often resulting from an attempt to solve very particular issues and challenges that seemed most salient at the time. Once created, these rules of international order may persist for a matter of centuries and therefore seem as if they are permanent and fixed. Further, a model of order created in a specific place and time, depending on the political, economic and military dominance of the cultures that framed a particular order, may be imposed or accepted by many other places and cultures that had no role in framing that order.

That has been the case, it is now apparent, with the European Westphalian state system, augmented, of course, over the past nearly four centuries by international law, treaties, customary international law, and international organizations and structures such as the United Nations, NATO, and other regional security organizations.

There are many reasons to doubt the stability and "givenness" of that order in the future. The cultures and peoples that created that order are no longer as dominant as they became shortly after Westphalia and through the 20th century. The imposition of its structures on the rest of the planet never "fit" especially well in much of the world and show signs of fracturing and realigning into more culturally appropriate contours. Further, even in the heart of those founding cultures in the United States and Europe, there is increasing resistance to sustaining and building up institutions such as the United Nations and NATO and indeed the entire Liberal world order. Cultures that were forced into nominal compliance with that system are emerging as important global powers with military and economic clout, most notably China.

The combination of rising powers with different cultural ways of framing global questions and the relative decline of powers that initially built, sustained, and advocated for the system globally suggest a shifting normative landscape in which competing and differing cultural traditions will need to reopen questions previously viewed as settled. This would suggest we are entering a period characterized upheavals similar to those in Europe that led to the Westphalian system.

It took a long and bloody struggle before the states of Europe were forced to create the Westphalian system as a hoped-for means of ending perpetual religious war in Europe. As was suggested above, a rising China, a more assertive Russia, and unrest throughout the Islamic world may well portend a dramatic reworking of the structure and agreed principles of world order. Some suggest that this "Thucydides trap" scenario will probably result in major military conflict before the major parties are forced to negotiate new and probably different rules and structures of global order.

What is already clear is that China, Russia, and parts of the Islamic world are already acting on principles and expectations that seem appropriate and powerful to them, even if much of what they do deviates from previously agreed-upon global principles. It is also clear that they are not deterred by having such deviations pointed out to them or by the disapproval of much of the international system. It is also clear that the powers that created the order are manifesting a lack of will, and increasingly, the capability and political leadership, to use force to prop up that order in effective ways in the face of newly emboldened rising military, economic and political powers.

There are powerful reasons to believe that we are in the midst of a world-historical cultural shift in which previously taken-for-granted rules of international behavior cannot be assumed to hold into the future.

It clearly is no longer adequate to cajole conformity with those standards, when China, Russia and much of the Islamic world clearly see those rules as imposed on them when they were weak in a "century of bad treaties" (to use the Chinese phrase). One can hope that wise leaders will respond to these challenges, foster a true and effective international dialogue and hammer out near agreements acceptable to diverse cultural traditions. On the other hand, it would be historically and politically naive to rule major power conflicts out of the question as the earthquake reshuffles the global tectonic plates until a new and stable landscape emerges and the shifting terrain settles down. It may be that, if the existing international institutions and understandings are truly in retreat, the dark vision of Kagan's book may prove to be the case, and the jungle really will grow back. It increasingly appears that both ideas and leaders competent to rebuild an agreed-upon framework for international order into the future are lacking. We may well be entering into a very dark passage in global affairs.

Notes

1 The "War and Peace in Chinese Thought" research group was convened by scholars at the Hong Kong Baptist University in 2011, which invited a number of American and British scholars to join in the subsequent years. See Lo and Twiss 2015, xvii.
2 For current discussion of China in the South China Sea, see The Center for Preventive Action 2022.
3 For the official overview of UNCLOS, see UN Division for Ocean Affairs and the Law of the Sea 1982.
4 Arbitration Between the Republic of the Philippines and the People's Republic of China. PCA Case No. 2013–19, Award (July 12, 2016). www.pca-cpa.org.
5 Foreign Ministry Spokesperson Hong Lei Hosted a Regular Press Conference on February 19, 2013, *People's Republic of China Ministry of Foreign Affairs Press Release*, February 19, 2013.
6 For a brief overview, see Kahn 2005.
7 NATO's official website's position on Russia's recent actions can be found at NATO 2022). For an overview from the Brookings Institute, see Aleksashenko 2018.
8 For an overview of ISIS's objective, see RAND; Wood 2015.
9 Further discussion, see Danforth 2015; Mason 2016; Wright 2016.
10 Further discussion, see Fisher 2012; T. 2016.
11 For further discussion, see Coyle 2018; The Editorial Board 2018.
12 Several recently published provocative books explore this theme of the construction and possible unraveling of the Liberal International order. In Hathawayu, Oona A.

and Shapiro, Scott J. 2018. *The Internationalists: How a Radical Plan to Outlaw War Remade the World.* New York: Simon & Schuster. the authors argue that the commonly dismissed Kellogg-Briand Pact/Treaty of Paris which declared aggressive war illegal had a far greater effect on reducing conflict in the world than is generally agreed. But the book concludes with an exploration of whether that near century of international consensus is now breaking down and speculating about the effects were that to occur. Robert Kagan, in *The Jungle Grows Back: America and our Imperiled World* (2018) New York: Alfred A. Knopf, writes even more ominously about the indicators that the United States and Europe are failing to defend the Liberal Order they created after World War II and paints a bleak picture of the state of the world if that order were to be abandoned. See also Edward Luce. 2018. *The Retreat of Western Liberalism.* London: Abacus and Michael Mandelbaum. 2019. *The Rise and Fall of Peace on Earth.* New York: Oxford University Press. Whether these and other similar diagnoses are correct in every detail should not distract us from recognizing the central point: that it appears the global community is at a world-historical inflexion point in which the apparent verities of the international system are no longer secure and in which what comes next is unclear, uncertain, and remains to be created.

13 See Putin 2005 and Sanders 2014.
14 This is a theme explored in depth by Graham Allison's many books in which he explores the question whether war with China is virtually inevitable as a global reaction to China's assertive and militarized expansion into maritime areas patrolled regularly by the U.S. Navy since World War II. See especially Allison 2017.
15 Cited in Martinson 2016, 27.
16 Congressional Research Service, U.S.-China Strategic Competition in South and East China Seas: Background and Issues for Congress, R42784 (2022). https://fas.org/sgp/crs/row/R42784.pdf.
17 For a series of articles by various authors in *The New York Times* on the Uighurs in China, see www.nytimes.com/topic/subject/uighurs.
18 See note 19 (Putin 2005).
19 Of course, it is obvious that throughout the Cold War, both the USSR and the U.S. engaged in many, many proxy wars and meddled in internal affairs of countries all over the globe. Nevertheless, the absence of a direct confrontation between great powers and their allies is a considerable achievement.
20 See Erlanger 2018 for a discussion of this point.
21 A brief discussion of the map and a digital reproduction can be found in Amos 2005.

Bibliography

Adamson, David M. and DaVanzo, Julie. 1997. "Russia's Demographic 'Crisis': How Real Is It?" Santa Monica, CA: RAND Corporation. www.rand.org/pubs/issue_papers/IP162.html.

Akin, Andy. 2017. "What do we know about Russia's 'Grand Strategy'?" *The Washington Post*, May 2, 2017.

Alexsashenko, Sergey et al. 2018. "Restoring equilibrium: U.S. policy options for countering and engaging Russia." *Brookings*, February 2018. www.brookings.edu/research/restoring-equilibrium-u-s-policy-options-for-countering-and-engaging-russia.

Allison, Graham. 2017. *Destined For War: Can America and China escape Thucydides's Trap.* Boston, MA: Houghton Mifflin Harcourt.

Amos, Deborah. 2005. "T.E. Lawrence's Middle East Vision." *NPR*, October 2005. www.npr.org/templates/story/story.php?storyId=4967572.

Arbitration Between the Republic of the Philippines and the People's Republic of China. 2016. PCA Case No. 2013–19, Award (July 12, 2016). www.pca-cpa.org.

China Power Team. 2017. "How Will the Belt and Road Initiative Advance China's Interests?" *China Power.* May 8, 2017. Updated August 26, 2020. https://chinapower.csis.org/china-belt-and-road-initiative.

Cohen, Jared. 2015. "Digital Counterinsurgency: How to Marginalize the Islamic State Online." *Foreign Affairs,* November/December 2015. www.foreignaffairs.com/articles/middle-east/digital-counterinsurgency.

Congressional Research Service, U.S.-China Strategic Competition in South and East China Seas: Background and Issues for Congress, R42784. 2022. https://fas.org/sgp/crs/row/R42784.pdf.

Coyle, James J. 2018. "Will Russia Reinterpret International Law and Get Away With It?" *Atlantic Council,* February 6, 2018. www.atlanticcouncil.org/blogs/ukrainealert/will-russia-reinterpret-international-law-and-get-away-with-it.

Danforth, Nick. 2015. "The Middle East that might have been." *The Atlantic,* February 13, 2015. www.theatlantic.com/international/archive/2015/02/the-middle-east-that-might-have-been/385410.

Encyclopedia Britannica. 2011. Editors of Encyclopaedia. "Dār al-Islam." August 29, 2011. www.britannica.com/topic/Dar-al-Islam.

Encyclopedia Britannica. 2022. Editors of Encyclopaedia. "Peace of Westphalia." January 23, 2022. www.britannica.com/event/Peace-of-Westphalia.

Erlanger, Steven. 2018. "Is the World Becoming a Jungle Again? Should Americans Care?" *The New York Times,* September 22, 2018. www.nytimes.com/2018/09/22/world/europe/trump-american-foreign-policy-europe.html.

Fisher, Max. 2012. "The Dividing of a Continent: Africa's Separatist Problem." *The Atlantic,* September 10, 2012. www.theatlantic.com/international/archive/2012/09/the-dividing-of-a-continent-africas-separatist-problem/262171.

Foreign Ministry Spokesperson Hong Lei Hosted a Regular Press Conference. February 19, 2013, *People's Republic of China Ministry of Foreign Affairs Press Release,* February 19, 2013. www.mfa.gov.cn/web/fyrbt_673021/jzhsl_673025/t1014798.shtml.

Hamid, Shadi. 2018. "Muslim Brothers: The Rivalry That Shaped Modern Egypt." *Foreign Affairs,* September/October 2018. www.foreignaffairs.com/reviews/review-essay/2018-08-14/muslim-brothers.

Hathawayu, Oona A. and Shapiro, Scott J. 2018. *The Internationalists: How a Radical Plan to Outlaw War Remade the World.* New York: Simon & Schuster.

Huntington, Samuel. 1993. "The Clash of Civilizations." *Foreign Affairs* 72, no. 3 (Summer): 41.

Ioffe, Julia. 2018. "What Putin Really Wants." *The Atlantic,* January/February 2018. www.theatlantic.com/magazine/archive/2018/01/putins-game/546548.

Kagan, Robert. 2018. *The Jungle Grows Back: America and our Imperiled World.* New York: Alfred A. Knopf.

Kahn, Joseph. 2005. "China has an ancient mariner to tell you about." *The New York Times,* July 20, 2005.

Lo, Ping-cheung and Twiss, Sumner B. 2015. *Chinese Just War Ethics: Origin, Development, and Dissent.* London: Routledge.

Luce, Edward. 2018. *The Retreat of Western Liberalism.* London: Abacus.

Mandelbaum, Michael. 2019. *The Rise and Fall of Peace on Earth.* New York: Oxford University Press.

Martinson, Rayan D. 2016. "Panning for Gold: Assessing Chinese Maritime Strategy from Primary Sources." *Naval War College Review* 69, no. 3(Summer): 27.

Mason, Paul. 2016. "Paul Mason on Sykes-Picot: how an arbitrary set of borders created the modern Middle East." *The New Statesman*, May 9, 2016. www.newstatesman.com/world/middle-east/2016/05/paul-mason-sykes-picot-how-arbitrary-set-borders-created-modern-middle.

Miller, Chris. 2019. "Can Putin Fix Russia's Sputtering Economy?" *Foreign Affairs*, March 13, 2019.

National Bureau of Economic Research. 2016. "Where Are ISIS's Foreign Fighters Coming From?" *The Digest*, June 2016. www.nber.org/digest/jun16/w22190.html.

NATO. 2022. "Relations with Russia." www.nato.int/cps/en/natolive/topics_50090.htm.

Pillsbury, Michael. 2019. *The Hundred-Year Marathon: China's Secret Strategy to Replace America as the Global Superpower*. New York: Henry Holt and Company.

Putin, Vladimir. 2005. "Annual Address to the Federal Assembly of the Russian Federation." Address by the President of Russia, April 25, 2005. http://en.kremlin.ru/events/president/transcripts/22931.

RAND Corporation. "*The Islamic State (Terrorist Organization)*." www.rand.org/topics/the-islamic-state-terrorist-organization.html.

Sanders, Katie. 2014. "Did Vladimir Putin call the breakup of the USSR 'the greatest geopolitical tragedy of the 20th century?" *Politifact*, March 6, 2014. www.politifact.com/punditfact/statements/2014/mar/06/john-bolton/did-vladimir-putin-call-breakup-ussr-greatest-geop.

Sullivan, Eileen. 2018. "Trump Questions the Core of NATO: Mutual Defense, Including Montenegro." *The New York Times*, July 18, 2018. www.nytimes.com/2018/07/18/world/europe/trump-nato-self-defense-montenegro.html.

The Center for Preventive Action. 2022. "Territorial Disputes in the South China Sea." *Global Conflict Tracker*. www.cfr.org/interactive/global-conflict-tracker/conflict/territorial-disputes-south-china-sea.

The Economist. 2016. "Why Africa's borders are a mess." November 17, 2016. www.economist.com/the-economist-explains/2016/11/17/why-africas-borders-are-a-mess.

The Editorial Board. "Russia Attacks Ukrainian Ships and International Law." *The New York Times*, November 26, 2018. www.nytimes.com/2018/11/26/opinion/russia-ukraine-attack-ships-crimea.html.

The New York Times. "The Uighurs in China." www.nytimes.com/topic/subject/uighurs.

The World Bank. 2020. "Military expenditure (% of GDP) – Russian Federation." https://data.worldbank.org/indicator/MS.MIL.XPND.GD.ZS?locations=RU.

Twiss, Sumner B. and Chan, Jonathan. 2012. "Classical Confucianism, Punitive Expeditions, and Humanitarian Intervention." *Journal of Military Ethics* 11, no. 2, 81–96. doi:10.1080/15027570.2012.708177.

UNCLOS. Part VIII, "Regime of Islands." Article 121. www.un.org/depts/los/convention_agreements/texts/unclos/closindx.htm.

UN Division for Ocean Affairs and the Law of the Sea. United Nations Convention on the Law of the Sea, December 10, 1982. www.un.org/depts/los/convention_agreements/convention_overview_convention.htm.

Von Drehle, D. 2006. "A Lesson in Hate: How an Egyptian student came to study 1950s America and left determined to wage holy war." *Smithsonian Magazine*, February 2006. www.smithsonianmag.com/history/a-lesson-in-hate-109822568.

Wang, Zheng. 2012. *Never Forget National Humiliation: Historical Memory in Chinese Politics and Foreign Relations*. Columbia University Press.

Ward, Jonathan D. T. 2019. *China's Vision of Victory: and Why America Must Win*. The Atlas Publishing and Media Company.

Wood, Graeme. 2015. "What ISIS Really Wants." *The Atlantic*, March 2015. www. theatlantic.com/magazine/archive/2015/03/what-isis-really-wants/384980.

Wright, Robin. 2016. "How the curse of Sykes-Picot still haunts the Middle East." *The New Yorker*, April 30, 2016. www.newyorker.com/news/news-desk/how-the-curse-of-sykes-picot-still-haunts-the-middle-east.

PART II

Chinese Thinking about War and Peace

5

JUST CAUSE IN MENGZI AND GRATIAN

Similar Ideas, Different Receptions and Legacies

Ping-Cheung Lo

Introduction

In this chapter, I elaborate on the different intellectual traditions on just war in China and the West, focusing on the just causes articulated by Mengzi (Mencius) and Gratian. I take three distinct approaches: (1) I examine the different ways that Mengzi textually expresses the term "war waging"; (2) I consider the literary/oratorical contexts of Mengzi's discourses on war; and (3) I compare the subsequent Confucian and Christian (Catholic and Protestant) traditions of warfare ethics. In terms of Confucianism, I examine classical commentaries on the *Mengzi* (e.g., Zhao Qi, Zhu Xi, Jiao Xun), and for the subsequent Christian traditions I focus on high medieval and early modern thinkers in the European "just war" tradition (Francisco Suarez, Hugo Grotius, and Samuel Pufendorf). In the final section I specifically explain why the rudimentary just cause ideas in the *Mengzi* have not been developed, refined, and revised, unlike those in Europe.

Just Cause in the *Mengzi*

Mengzi (Mencius, c. 372–289 BCE), the best-known follower of Kongzi (Confucius) during the Warring States period, provides substantial and nuanced discussions of war through his extended dialogues with various rulers. Many of the English-speaking scholars that examine Mengzi's warfare ethics are over-reliant on English translations of the text and are thus misled by the wording of the translations. Thus, I first note the different words that are translated as the phrase "war waging" in the book of *Mengzi*:

DOI: 10.4324/9781003336372-7

a   *zheng* (征, 16 occurrences);
b   *fa* (伐, 26 occurrences); and
c   *tao* (討, 3 occurrences).

These three words all refer to wars that are proactively waged rather than in self-defense. *Fa* (伐) differs markedly from the other two, as it is used by Mengzi as a morally neutral term; such a war is neither morally right nor wrong. However, *zheng* (征) and *tao* (討) always have a positive moral connotation in the *Mengzi*. *Zhu* (誅; 14 occurrences) is also used in the *Mengzi* and is related; *zhu* literally means "to kill as punishment for crime." This can be used in the context of war and always implies moral approval. The act of execution is the culmination of a justified war—to put the culpable party to death as a penalty.[1]

Most English-language discussions (e.g., Bell 2006; Glanville 2010; Kim 2010, 2017; Twiss and Chan 2015) follow the translation of D. C. Lau (*Mencius* 2003) and understand *zheng* (征) as "punitive expedition." Lau's translation, however, is somewhat free and some occurrences of *zheng* are translated as "to march." Irene Bloom's improved translation (*Mencius* 2009) consistently renders *zheng* as "punish," "punishment," or "pursue the work of punishment." However, this also reflects a one-sided and inadequate understanding of *zheng*. The literary contexts indicate that punishment is only one aim of such an expedition. Sticking to the phrase "punitive expedition" does not acknowledge either Mengzi's general ethics of warfare or his views of the just cause. Bryan W. van Norden (*Mengzi* 2008) retains "punitive expedition" in only one passage; in all other passages, he renders *zheng* as "attack" or "invasion," which is a faithful literal translation. However, van Norden does not consider the positive moral connotation given to *zheng* by Mengzi. In this section I examine the ethical discourse around all 16 occurrences of *zheng* and discuss how the just causes denote more than punishment.

Expeditions of Zheng and Benevolent Governance

Three passages in the *Mengzi* (1B11, 3B5, 7B4) refer to the historical precedent of the military campaigns of *zheng* waged by Tang (湯) against Jie (桀) in the late Xia dynasty. This military campaign is an important morally paradigmatic precedent for Mengzi, and thus he references it in three of his audiences. Out of the 16 occurrences of *zheng*, 13 appear in these three passages. As I have analyzed these three passages in great detail in another article (Lo 2020), I present only a summary and a further analysis here.

In 1B10 and 1B11, Mengzi discusses with King Xuan of the state of Qi (齊) his waging of war with the state of Yan (燕). Mengzi agrees that for the people of Yan, life under their tyrannical ruler was like living in "deep water and scorching fire" (i.e., in great misery: 1B10). Mengzi then raises the historical precedent of the military campaigns (*zheng*) of Tang (湯) against the

Earl of Ge (葛), who had neglected his proper political duties. These expeditions were greatly welcomed by the local people.

> With this he gained the trust of the Empire, and when he marched [*zheng*] on the east, the western barbarians complained, and when he marched [*zheng*] on the south, the northern barbarians complained. They all said, "Why does he not come to us first?" The people longed for his coming as they longed for a rainbow in time of severe drought … He punished [*zhu*, executed] the rulers and comforted the people, like a fall of timely rain, and the people greatly rejoiced. (1B11)[2]

Thus, the military campaigns of Tang were more than simply punitive expeditions of retributive justice. They were also successful humanitarian rescue missions of rectifying justice, to the extent that oppressed peoples in various regions of China trusted him and longed for him to "liberate" them. In appealing to the past just war of Tang, Mengzi's intention is not to suggest that King Xuan should initiate a similar just war. Instead, he uses it to reprimand the king for the unjust war he had waged, a war that Mengzi disapproved of from the very beginning.

In Mengzi's view, oppressed people everywhere long for liberation, and a political leader who rescues them would manifest "benevolent governance," which is the key tenet of Mengzi's political philosophy. However, instead of rescuing oppressed people "from water and fire," King Xuan was only interested in annexing Yan into his territory. Thus, Mengzi reprimands him as follows: "Even before this, the whole Empire was afraid of the power of Qi. Now you double your territory without practicing benevolent government. This is to provoke the armies of the whole Empire" (1B11). Thus, King Xuan's *post bellum* deeds illustrated his true intention; he did not act out of benevolence at all.

In the long commentary tradition of the *Mengzi*, the scholarly consensus is that three commentaries stand out as significant (Editorial Committee 2002, 18). Zhao Qi's (108?–201 CE) was the first commentary on the entire text of the *Mengzi* and is still highly regarded. Succinct summaries are provided after each unit of discourse, and are still consulted today (cf. Li 2010, 139–158). The collected glosses edited by Zhu Xi (1130–1200 CE) were published at the peak of the neo-Confucian movement in the Song Dynasty. The *Mengzi* thus became a distinct canon within the Confucian canon. The commentary by Jiao Xun (1763–1820 CE) is also noteworthy for its meticulous scholarship and is informed by numerous studies of the *Mengzi* (Lin 1995, Li 2010, 284–307). The corrupted, classic commentary by Zhao Qi was restored and included to ensure a continuity of exegesis and interpretation. It is noteworthy that none of these major commentators read this passage (1B11) as an ethical discourse on war, but as an exhortation for moral cultivation and the statecraft of benevolent governance. Zhao Qi observes that the oppressed people throughout the empire trusted Tang in his expeditions of *zheng* because of his renowned virtues. An act of humanitarian rescue is both "combating evil (of the tyrant) and cultivating goodness (of the rescuer)."[3] Zhu Xi concurs that Mengzi refers to Tang's humanitarian rescue not to encourage its re-enactment, but to indict King Xuan for not practicing the virtue of benevolent governance (Zhu 1983).

In another passage (3B5), Mengzi's student Wanzhang asks him a question. The small state of Song (宋) was willing to practice benevolent governance, but two strong neighbor states (Qi and Chu) were eyeing Song as a possible easy target and could attack it at any time. What could be done about this situation? Mengzi replies by again referring to the historical precedent of Tang's military expeditions, using similar language. Tang launched 11 expeditions (*zheng*), but barbarian tribes throughout the empire complained that Tang did not come to them first. "The people longed for his coming as they longed for rain in time of severe drought ... He punished [*zhu*, executed] the rulers and comforted the people, like a fall of timely rain, and the people rejoiced greatly" (3B5). Later, when the state of You (攸) resisted Tang's rule, an expedition (*zheng*) was again sent, and "[t]he King of Zhou [周, i.e., Tang] rescued the people from water and fire and took captive only their cruel masters." Again, in this extended discourse there are two related just causes: executing culpable tyrants and rescuing the people from the miserable plight of tyranny,[4] and the former is a means to the latter. However, this entire discourse is not to encourage the ruler of Song to launch an expedition of *zheng*. Mengzi's conclusion is straightforward:

> It is all a matter of failing to practice Kingly governance. If you should practice Kingly governance, all within the Four Seas would raise their heads to watch for your coming, desiring you as their ruler. Qi and Chu may be big in size, but what is there to be afraid of?
>
> *(3B5)*

Again, it was Tang's virtuous approach that mattered, rather than encouraging ventures similar to his expeditions of *zheng*. If the ruler of Song continued to cultivate virtues and practice benevolent governance ("Kingly governance" in this passage), he would have nothing to fear in terms of national security. Thus, this discourse on war begins and concludes with an emphasis on benevolent governance. Zhao Qi's general comment on this section is explicit: "Cultivated virtues would not stay weak and wickedness would not stay strong. Toward the end of the Xia and Shang Dynasties, the general population was longing for Tang and Wu. It was not up to these two men to decline kingship."[5]

In the third passage (7B4), in which Mengzi appears to be teaching his students, he once again invokes Tang's humanitarian rescue missions (*zheng*) and notes that they were welcomed and longed for by many. Mengzi also refers to a similar military campaign by Wu against the tyrant Zhou (紂), in which he draws on the etymological meaning of the word *zheng* and declares:

"To wage a war of *zheng* (征) is to rectify (*zheng* 正)" (7B4).[6] Here, he is indicating that the morally justified military expedition of *zheng*, which is modelled upon those of Tang and Wu, will be a war of righting wrongs, of rectifying a wrongful situation, or of correcting something gravely unjust. Contrary to Lau's and Bloom's translations of "punitive expedition," the just cause invoked in *zheng* is not primarily punitive; rather, it is primarily to rectify injustice perpetrated by the wicked ruler.[7] The punitive execution of the tyrant is only a means to the "liberation" of the oppressed people. Mengzi then concludes that those who are oppressed welcome such campaigns of rectification and submit voluntarily: "There is no one who does not wish himself rectified. What need is there for war?" Thus, Zhu Xi observes in his commentary that "the people were tormented by the tyrant; they all wanted a benevolent person to come rectify the wrongful situation of their states" (Zhu 1983, 365). This conclusion supports Mengzi's earlier remark in this section: "If the ruler of a state is drawn to benevolence, he will have no match in the Empire." Thus, for the third time, Mengzi does not invoke the moral exemplar of the *zheng* military campaigns by Tang and Wu to encourage their enactment. Rather, he offers the moral lesson of cultivating virtues and practicing benevolent governance.

In addition, we should note that Mengzi's speech in 7B4 is a response to the appalling suggestion that "'I am expert at military formations; I am expert at waging war.' This is a grave crime." Mengzi never suggests that lethal violence should be used as an instrument of morality, and his references to Tang and Wu are not to encourage the waging of just wars, but the cultivation of virtues. Thus, his emphasis here and other similar passages is that a ruler will have "no match" if he is willing to cultivate virtues such as benevolence, and thus "what need is there for war?" Zhao Qi's succinct summary of this passage clearly shows how subsequent Confucians read it: "People long for an enlightened ruler in the same way as people in drought long for rain. Who would not be delighted to see [him/her] combatting tyranny with benevolence? ... There is no need to have an expertise in warfare; hence Mengzi called it a grave crime" (Jiao 1987, 965).[8]

Proper Authority to Declare and Conduct a War of Zheng

In addition to these three passages that invoke the historical examples of Tang's and Wu's military campaigns of *zheng*, two other passages in the *Mengzi* refer to wars of *zheng*; both are concerned with the proper authority. In 7B2, Mengzi discusses who has the proper authority to initiate campaigns of *zheng*, and he concludes that peer rulers do not. In 2B8, Mengzi makes it very clear that any tyrant, such as the ruler of Yan, deserves to be attacked, but not every rival ruler is morally qualified to carry out this mission. If the ruler of Qi waged war on Yan, it would simply be "one Yan attacking another

Yan," or one tyrant warring with another, which he disapproves of. Thus, the word for "war" used throughout this passage is *fa* (伐), not *zheng* (征).[9]

In 1A5, Mengzi seems to encourage King Hui of Liang to launch a war of *zheng* against a peer state ruler. However, Mengzi first urges King Hui to practice "benevolent governance" in his own state. Mengzi then advises him to launch military expeditions (*zheng*) against the tyrant. "If you should go and punish [*zheng*] such princes, who is there to oppose you? Hence it is said, 'The benevolent man has no match.' I beg of you not to have any doubts" (1A5). Once again, Mengzi's emphasis is on cultivating virtues and practicing benevolent governance. A sufficiently virtuous and benevolent ruler will face no opposition. Such a ruler is thus the proper authority to launch humanitarian rescue missions against neighboring states and will have the aims of relieving the suffering and improving the well-being of the people. Such a virtuous leader will have "no match" (i.e., no enemy) because their benevolent domestic policies will be renowned and welcomed by oppressed people everywhere. Any military rescue missions such a leader embarks on will thus be prompted not by dubious aims but by the moral intention of rectifying injustice, which the people who are suffering will recognize. Thus, through "people's power," this virtuous leader will succeed without serious opposition. (As I clarify in the last section of this chapter, commentators suggest that such a benevolent ruler who is without enemies is not simply better than average, but actually "an agent sent by Heaven.")

To summarize this discussion, I present a list of the 16 occurrences of *zheng* and D. C. Lau's and my translations. Mengzi's ethics of war is more understandable if we regard *zheng* as meaning "a military campaign of justice."[10]

Mengzi offers two views of the nature of the just cause in his discussions of *zheng*. First, resorting to war to punish a tyrant represents retributive justice. Second, using force to rescue the people from the "deep water and scorching fire" they suffer under the tyrant represents rectifying justice. I intentionally refer to "humanitarian rescue" rather than "humanitarian intervention," as state boundaries were fluid in the Warring States period; territorial sovereignty and sovereign states were not relevant issues. Thus, the just cause of the military campaigns to which *zheng* refers is not simply that such a war is morally justified but that it represents justice.

We should also note that Mengzi's discourses on expeditions of *zheng* are always embedded in his broader discourses concerning the practice of benevolent governance, which is a manifestation of virtuous leadership. A ruler who aspires to win all wars, according to Mengzi, should begin by cultivating moral virtues in general and the virtue of benevolence in particular. The Confucian tradition bequeathed by the *Mengzi* has a very clear understanding of this bigger picture: an expedition of *zheng* is only a tree, while moral self-cultivation and benevolent governance are the forest.

TABLE 5.1 The usage of the word 征 (zheng) in the Mengzi and its English translations

	D. C. Lau's translation	My translation in terms of ethics
1A5	If you should go and punish such princes	If Your Majesty goes and starts a just military campaign against such princes
1B11	In his punitive expeditions Tang began with Ge	Tang's just military campaigns began with Ge
1B11	When he marched on the east	When his just military campaign headed east
1B11	When he marched on the south	When his just military campaign headed south
1B11	When you went to punish Yan which practiced tyranny over its people	When Your Majesty went and started a just military campaign against Yan's ruler who practiced tyranny over his people
3B5	When an army was sent to punish Ge for killing the boy, the whole Empire said, "This is not coveting the Empire but avenging common men and common women"	When Tang started a just military campaign against the Earl of Ge for his killing of the boy, the whole Empire said...
3B5	Tang began his punitive expeditions with Ge	Tang began his just military campaigns with Ge
3B5	In eleven expeditions he became matchless in the Empire	In eleven just military campaigns he became matchless in the Empire
3B5	When he marched on the east	When his just military campaign headed east
3B5	When he marched on the south	When his just military campaign headed south
3B5	The King went east to punish it	The king led a just military campaign to the east
7B2	A punitive expedition is a war waged by one in authority against his subordinates	A just military campaign is a war waged by one in authority against his subordinates
7B2	It is not for peers to punish one another by war	Rivaling states are not in the position of waging just military campaigns against one another
7B4	When he marched on the south	When his just military campaign headed south
7B4	When he marched on the east	When his just military campaign headed east
7B4	To wage a punitive war is to rectify	To wage a just military campaign is to rectify

Just Cause in Gratian'S *Deretum* and Its Subsequent Development

As explained in an earlier chapter, Gratian can be regarded as the European pioneer of discussions of the notion of a just war and predates Thomas Aquinas. He provides a focused, comprehensive, and dialectical assessment of warfare ethics in

Causa 23 of Part Two of the *Concordantia Discordantium Canonum* (*The Harmony of Discordant Canons*), or *Decretum Gratiani* for short. In this section, I summarize Gratian's discussions of just cause examined in another article (Lo 2020) and then explain how his rudimentary ideas evolved in the High Middle Ages and early modern period to establish the Christian just war traditions, both Catholic and Protestant. The notions of just cause expounded in the *Decretum* and the *Mengzi* diverge in these periods, and thus require some explanation.

In Question 2 of Causa 23, Gratian provides definitions of the notion of a just war from Isidore of Seville and Augustine. From Isidore he inherits the idea that the justice in a just war is similar to that of a judicial edict and sentencing to punishment. From Augustine he gets a definition of just war, popularized by Thomas Aquinas later, "Just wars are usually defined as those which have for their end the avenging of injuries, when it is necessary by war to constrain a nation or a city which has either neglected to punish an evil action committed by its citizens, or to restore what has been taken unjustly" (q. 2, c. 2). In both definitions, the just cause is a rectifying response to a wrongdoing; punishment is therefore a salient motif. A morally justified war can thus be regarded as a war of justice, as it administers both retributive and rectifying justice. The idea of punitive justice is further emphasized in the discussion of "vengeance" in Question 4. Gratian elaborates on Augustine's moral justification of "vengeance" (*vindicta*) against the evildoer: "From all this we gather that vengeance is to be inflicted not out of passion for vengeance itself, but out of zeal of justice; not in order that hatred be vented, but that evil deeds be corrected" (q. 4, c. 54, d.p.c.). Thus, proper vengeance is grounded in both retributive and rectifying justice. Gratian also makes it very clear in Question 5 that killing in a just war is morally equivalent to an executioner carrying out a judge's sentence of capital punishment on a criminal. This again emphasizes the punitive nature of a just war.

As noted earlier, Mengzi also regards the just cause for military campaigns of *zheng* as consisting of both retributive and rectifying justice. The degree of agreement between the *Mengzi* and the *Decretum* concerning just cause is substantial. However, Mengzi's discussions of just cause in warfare did not lead to a subsequent and continuing "just war tradition" in China. In contrast, Gratian's discussions have had a major influence in Europe. The discussions regarding just causes for war found in High Medieval and early modern Christianity can be regarded as elaborations, refinements, and revisions of Gratian's just war ethics. In this section I summarize three key themes of these discussions, *viz*, punishment or retributive justice, humanitarian rescue or rectifying justice, and revolutionary war against tyranny or regime change, as representing just cause.

Punishment or Retributive Justice As a Just Cause

War has long been regarded as an instrument for punishment in the premodern Western tradition. As Rory Cox explains: "The punishment of injustice was paramount for Gratian and for most medieval writers on the just war. Since injustice could be found at both home and abroad, one's obligation to combat injustice did not stop at one's geographical borders, which were often in flux during the Middle Ages" (Cox 2018, 44). As in the time of Mengzi, in a war of punitive justice "no limitation was imposed by concepts of the inviolability of sovereignty or territory that are the hallmarks of post-Westphalian international law" in the High Middle Ages (Cox 2018, 44).

Cajetan (or Thomas de Vio, 1468–1534), the great Italian Dominican commentator on Aquinas' *Summa Theologiae*, argues that punishment is central to the just cause, insisting that "a just combat is an act of vindicative justice" (*actus vindicativae iustitiae*) (from Reichberg, Syse and Begby 2006, 247). The Protestant thinker Hugo Grotius (1583–1645) agrees. His *On the Law of War and Peace* (1625) is the first comprehensive theory of just war in early modern Europe. The entire second book of this lengthy treatise is devoted to discussions of just causes, which can be either defense- or offense-oriented, and include "recovery of property, punishment … obtaining of what is owed to us" (Reichberg, Syse, and Begby 2006, 402). He devotes chapters 20 and 21 of Book II to how punishment can be a just cause. Grotius agrees with earlier Catholic thinkers that "wars are usually begun for the purpose of exacting punishment. Most of the time, however, this cause is joined with a second, the desire to make good a loss, when the same act was both wicked and involved loss" (Reichberg, Syse, and Begby 2006, 406). However, Grotius also cautions against resorting to war too quickly, noting that even trivial and common wrongdoings are not punished in criminal law. If the wrong warrants a punishment that only war can mete out, "the punishment is joined either with a precaution against future harm … or protects injured dignity, or checks a dangerous example" (Reichberg, Syse, and Begby 2006, 406). Thus, the punishment is not only a backward-looking form of retribution but also a forward-looking deterrence (cf. Lang 2018, 137–138).

Early modern European thinkers also caution against implementing retributive justice through inter-state wars. For example, Francisco Suárez (1548–1617), a Spanish Jesuit theologian, argues in his *De Bello* that to conceive of a just war as punishment is actually a violation of justice, as "the same party in one and the same case is both plaintiff and judge, a situation which is contrary to natural law" (from Reichberg, Syse, and Begby 2006, 350). The post-Westphalian Protestant German jurist and political philosopher Samuel Pufendorf (1632–1694) also objects to the notion of a universal right to wage a punitive war. In *The Law of Nature and Nations* (1672) he explains that "the evils inflicted by right of war have properly no relation to punishments, since they neither proceed from a superior as such, nor have as their direct object the reform of the guilty party or others, but the defence and assertion of my safety, my property, and my rights" (from Reichberg, Syse, and Begby 2006, 461).

Thus, High Medieval and early modern European thinking regarding just warfare initially inherited Gratian's emphasis on punishment or retributive justice. However, this view was gradually subjected to internal critique and revision. Western just war thought continues to wrestle with the notion of punishment. One suggestion is that the proper role of punishment in a just war does not lie in *ius ad bellum*; rather, it should lie in *ius post bellum*.[11]

A Comparison with the Classical Confucian View

Extensive overlap can be found between the Mengzian–Confucian, High Medieval, and early modern Christian just war views on punishment or retributive justice as a just cause. Both the *Mengzi* and Gratian's *Decretum* are pioneering works. However, China after Mengzi witnessed no just war tradition comparable to that emerging in Europe after Gratian. Mengzian-Confucian just war ethics remained largely undeveloped (see chapter 4) and were even misunderstood by otherwise reasonable scholars. In his introduction to the English translation, D. C. Lau offers a very particular view of Mengzi's ethics of war. He suggests that "war is to a state – and so to the ruler – what punishment is to the criminal … No war other than punitive wars are justified" (*Mencius* 2003, xliv). Along with his misunderstanding of *zheng*, Lau reveals his ignorance of relevant Western discussions regarding the complex relationship between war and punishment. Thus, contemporary Confucian discussions about punishment as just cause must address various issues already raised by western thinkers. First, the twin ideas of retributive and rectifying justice as just causes for war must be disentangled. Fighting a punitive or vindictive war is currently too dangerous, particularly in the shadow of nuclear weapons. Confucians may therefore need to accept the prudent relocation of punishment from *ius ad bellum* to *ius post bellum*. As explained in another article (Lo 2020), the great neo-Confucian philosopher Zhu Xi insisted on the need to launch a counter-offensive against the Jurchen (Jin) during the national security crisis of the South Song dynasty. Both the recovery of lost land and the necessity for punishment were regarded as just causes. Zhu Xi insisted that the enmity (*chou* 仇) of Song China against the Jurchen was so deep that "we will not share the same sky" (不共戴天), and thus there was a score to settle. Contemporary Confucians should be cautious about precisely this kind of punitive mentality. They should also renounce the punitive execution (*zhu*, 誅) component of the military campaign of *zheng*. To insist on attempting to kill "the bad guy" will only strengthen his resistance and increase the casualties on both sides. A regime change will serve the purpose of relieving people's suffering under tyrannical rule. However, many people in mainland China today still hold the dangerous view that countries that offend or humiliate China should be severely punished and people who do so may even warrant execution.[12] The misunderstanding of *zheng* as solely a "punitive expedition" can have disastrous consequences. We should also be aware that the last war in which the PRC government was involved, the Sino–Vietnamese War of 1979, was officially justified as a punitive war waged to "to teach Vietnam a lesson" (Scobell 2003, 125, 127, 132, 135, 136, 142).

Humanitarian Rescue or Rectifying Justice As a Just Cause

Although rectifying justice was accepted as a just cause after Gratian's discussion, the rectification in subsequent discourses is more about stolen property and land than about oppressed human beings. It was only in the early modern period that rectifying tyrannical oppression through a just war received proper attention, particularly among Protestant thinkers.

Grotius argues that war *may* rightfully be undertaken on behalf of (a) subjects, (b) allies, (c) friends, or (d) anyone, because of "the mutual tie of kinship among men, which of itself affords sufficient ground for rendering assistance" (II.XXV. VI: from Reichberg, Syse, and Begby 2006, 416). Grotius poses the question in the style of a scholastic disputation, and asks "whether there may be a just cause for undertaking war on behalf of the subjects of another ruler, in order to protect them from wrong at his hands" (Reichberg, Syse, and Begby 2006, 417). The answer may be negative because of the political sovereignty of other rulers, but it may be positive. "If, however, the wrong is obvious ... the exercise of the right vested in human society is not precluded." Grotius argues for the latter with a scholastic distinction: "though it be granted that even in extreme need subjects cannot justifiably take up arms ... it would still not follow that others may not take up arms on their behalf." What is forbidden for a subject is not forbidden for those who are not subjects. In spite of the Westphalian Peace treaty, Grotius' strong argument for humanitarian intervention is surprising, and it resembles Mengzi's argument in many ways. However, unlike Mengzi, Grotius is not a moral–political idealist. He concludes his case for humanitarian intervention with a caution about human ulterior motives. "We know, it is true, from both ancient and modern history, that the desire for what is another's seeks such pretexts as this for its own ends; but a right does not at once cease to exist in case it is to some extent abused by evil men ... " (Reichberg, Syse, and Begby 2006, 417).

Pufendorf similarly declares that in addition to defending their own subjects, states should also defend their allies, their friends, and any "oppressed party who makes a plea for assistance" (Reichberg, Syse, and Begby 2006, 459). The justification for defending oppressed subjects of other states, as Grotius also notes, is our common humanity. He also echoes Grotius when cautioning against such endeavors because they can be easily abused. He therefore articulates a proviso:

> Indeed, it is also contrary to natural equality to thrust oneself forward unasked as the arbiter of human affairs, as it were; not to mention the fact that this would be open to great abuse, since there is hardly anyone against whom war could not be undertaken on such a pretext. Therefore, an injury

inflicted on another can provide us with a sufficient reason for war only when the one affected by it calls us to his aid, so that we do whatever we undertake here in the name of the one injured and not our own.

(Pufendorf 1994, 260)

Thus, the previous and continuing request from those who are oppressed is a necessary condition.

This cause of humanitarian intervention was eventually taken up by the United Nations. The International Commission on Intervention and State Sovereignty put forward the principle of "responsibility to protect" (or R2P) in a report entitled *The Duty to Protect* in 2000. This principle was subsequently endorsed in the 2005 World Summit hosted by the UN General Assembly. The UN Security Council also authorized an armed humanitarian intervention in Libya in 2011 after endorsing the R2P in principle.

Comparison with the Classical Confucian View

Mengzi is a strong advocate of armed humanitarian rescue, but he is also mindful of the easy abuse of this just cause. He builds two important provisos of such military action into his discourse.

First, a rescue mission of *zheng* must be welcome by the oppressed people. On three occasions (1B11, 3B5, 7B4) Mengzi explains that when Tang, a benevolent ruler, initiated his armed expedition phase by phase, those who were oppressed but not yet "liberated" complained: "Why does he not come to us first?" Mengzi says that these oppressed people "longed for his coming as they longed for a rainbow in time of severe drought" (1B11). This implies their continuing request. He told the King of Qi that when Qi invaded Yan, "the people thought you were going to rescue them from water and fire, and they came to meet your army, bringing baskets of rice and bottles of drink" (1B11). Mengzi's proviso here is similar to Pufendorf's, in terms of the continuing request from those who are oppressed.[13] However, given that most societies are today relatively pluralistic and politically divided, assessing the existence of a "continuing request" from oppressed people is not an easy task.

Second, armed humanitarian rescue should be the initiative of a virtuous ruler. A ruler who practices benevolent governance is expected to extend his benevolence toward the people suffering under wicked rulers, because he cannot bear to see the suffering of the oppressed. If a ruler is known for his or her virtues, he or she will be welcomed and trusted by those who are oppressed.[14] Mengzi emphatically does not endorse such a mission if it is launched by rulers who have ulterior motives (see the earlier discussion on *Mencius* 2B8). However, the noble moral ideal of a virtuous leader trusted by oppressed people who can provide action guidance in the post-Westphalian world is a Confucian project that is still in progress, as virtuous and trustworthy politicians remain rare. The lofty idealism of Mengzi's warfare ethics is clear (see also the relevant discussions in Lo 2020).

In Mengzi's historical context, any feudal lord who was sufficiently virtuous and benevolent could be an agent sent by Heaven, and thus could lead a punitive rescue expedition against any wicked or recalcitrant feudal lord. Peer-to-peer political entities simply disappeared in the re-unification of the Chinese nation, and neighboring "barbarian" nations were not treated as peers. Thus, Mengzi's theory of just warfare could not be embodied in his time and was not historically embodied in the subsequent millennia.

Revolutionary War Against a Tyrant or Regime Change As Just Cause

Again, this cause for just war was not discussed in the Middle Ages, but early modern just war theorists in Europe did give it proper attention. Suárez advocates waging offensive war to overthrow a tyrant. "The commonwealth as a whole, however, may rise in war against such a tyrant; and this uprising would not be a case of sedition in the strict sense [with a connotation of evil] … the commonwealth, as a whole, is superior to the prince, for the commonwealth, when it granted him his power, is held to have granted it upon these conditions: that he should govern politically, and not tyrannically; and that, if he did not govern thus, he might be deposed" (from Reichberg, Syse, and Begby 2006, 369). This view is consistent with his acceptance of humanitarian intervention as a just cause for war and is also similar to Mengzi's view.

Similarly, Pufendorf suggests that the right of humanitarian intervention follows from the oppressed subjects' right of armed resistance: "The safest principle to go on is, that we cannot lawfully undertake the defence of another's subjects, for any other reason than they themselves can rightfully advance, for taking up arms to protect themselves against the barbarous savagery of their superiors" (from Glanville 2018, 151).

In contemporary Western just war discourses, a revolutionary war against a tyrant is considered a civil war and so receives much less attention than interstate warfare. However, Brian Orend does discuss this issue in his contemporary restatement of just war ethics, and introduces the idea of a "minimally just political community." He suggests that a minimally just state "1) is recognized as legitimate by its own people and most of the international community; 2) avoids violating the rights of other legitimate states; and 3) makes every reasonable effort at satisfying the human rights of its own citizens" (Orend 2013, 87). He then sets forth his case for revolt as a just cause.

> There simply must be a baseline of physical security, and freedom from severe and systemic violence, if there is to be a political community at all … . When, if ever, is it just to fight a civil war? One case would be when

the central government fails any of the criteria of minimal justice. Any such government has no moral reason to exist, and so to resist its tyranny with force of arms is permissible."

(Orend 2013, 87)

Orend regards a government that turns tyrannical as violating the social contract and thus it perpetrates "internal aggression," and thus as for external aggression, armed resistance is appropriate. Orend's discourse illustrates how Mengzi's moral endorsement of revolutionary wars against tyrants is not unknown in Western attitudes to just war.

Comparison with the Classical Confucian View

As explained in the beginning of this chapter, the armed rebellions of Tang and Wu represented moral paradigms in classical Confucian political thought. Mengzi strongly suggests that the common people have a moral priority over the ruler: "The people are of supreme importance; the altars to the gods of earth and grain come next; last comes the ruler" (7B14). This is similar to Suárez's view that "the commonwealth, as a whole, is superior to the prince." Thus, it follows for Mengzi that tyranny everywhere should be overthrown and punitive action should be taken, domestically or otherwise. Mengzi explicitly asserts that tyrannicide is not regicide (1B8), which is similar to Suárez's view that overthrowing a tyrant through a revolutionary war is not sedition.

Mengzi uses the same word (*zhu*, or "punitive execution") to refer to killing domestic tyrants and those elsewhere (1B8, 1B11, 3B5). Like Pufendorf, Mengzi suggests that overthrowing a tyrant is legitimate as it is an act of self-defense by the people, in both a revolutionary war and a humanitarian rescue mission requested by those oppressed in other states. This is also similar to Orend's idea of "internal aggression," but Mengzi again has his proviso: the revolutionary war must be led by a virtuous leader as not everyone has the right to start a just revolt.[15] The guidance this lofty idealism can provide in today's complex world is again a project in progress.

Mengzi's endorsement of revolutionary war against tyrants was a cause for alarm for subsequent emperors. The version of Confucianism adopted after China became unified was an ideology of empire, in which the supremacy of the emperor was emphasized. The *Mengzi* was not taken seriously by the government until the emergence of the neo-Confucian movement in the 12th century. The views in the *Mengzi* were politically sensitive and hotly debated in the imperial court. At one time the *Mengzi* was even censored, and common people were only allowed to read an abridged edition. Recent research (Huang 2017) indicates that the same intense debate and controversy about the notion of tyrannicide in the

Mengzi occurred in premodern Japan and Korea. Thus, this just cause for war espoused by Mengzi was not given proper attention in subsequent just war discussions.

Mengzi's Just War Ideas Bequeathed a Very Different Legacy in China

Military Campaigns of Zheng *As Exhortation for Benevolent Governance*

Mengzi's general ideas about just cause in war did not lead to a just war tradition or legacy. They were not extensively considered in subsequent Confucian traditions, and so were not debated, refined, and reformulated as Gratian's were in premodern and early modern Europe. I have provided some general explanations in another article (Lo 2020),[16] and in this section I focus on the military campaigns of *zheng* extolled by Mengzi.

I explain in the first section of this chapter how Mengzi's discussions about the expeditions of *zheng* are always embedded in wider discourses about the practice of benevolent governance, which is a manifestation of virtuous leadership. Mengzi suggests that a ruler who aspires to win all wars should begin by cultivating moral virtues in general and the virtue of benevolence in particular. This interpretation is faithful to the text and is supported by the subsequent Confucian tradition. In addition, as the statecraft of benevolent governance is the central tenet in Mengzi's political thought, it seems that Mengzi brings up the topic of humanitarian rescue not primarily to offer a just cause for future wars, but to illustrate his theory of benevolent governance. He partly attempts to persuade rulers of the benefits of this theory by arguing that it is straightforward to implement, as it is rooted in human nature. Mengzi is well known for arguing that human nature is predisposed toward moral goodness. He suggests that human beings are by nature endowed with the seeds of the virtues of benevolence, righteousness, propriety, and wisdom, and that of benevolence is "the mind (moral sentiment) which cannot bear to see the suffering of others" (i.e., the sentiment of commiseration, 惻隱之心). Mengzi then provides his famous example.

> When I say that all people have the mind which cannot bear to see the suffering of others, my meaning may be illustrated thus: Now, when people suddenly see a child about to fall into a well, they all have a feeling of alarm and distress, not to gain friendship with the child's parents, nor to seek the praise of their neighbors and friend, nor because they dislike the reputation [of lack of humanity if they did not rescue the child]. From such a case, we see that a person without the feeling of commiseration is not a person …
>
> *(2A6; Chan 1963, 65; trans. modified)*

Mengzi is of the opinion that for individuals, a good life necessarily involves nurturing these innate seeds of moral goodness in us and cultivating the four cardinal virtues. For society, good governance is also a matter of manifesting the virtue of benevolence that we are naturally inclined toward, i.e., benevolent governance (*ren zheng*). Mengzi argues as follows:

> All men have the mind which cannot bear [to see the suffering of] others. The ancient kings had this mind and therefore they had a government that could not bear to see the suffering of the people. When a government that cannot bear to see the suffering of the people is conducted from a mind that cannot bear to see the suffering of others, the government of the empire will be as easy as making something go round in the palm."
>
> *(2A6; Chan 1963, 65)*

I propose that it is this very ideal of "a governance that could not bear to see the suffering of the people" that leads Mengzi to affirm that military campaigns to rescue people who are suffering under tyrants are morally justified. Just as every person would rescue a child about to fall into a well, every virtuous ruler would, if necessary, rescue those suffering under tyrants, whether near or far, through military campaigns of *zheng*.

Highly Idealized Campaigns of Zheng

As noted above, Mengzi argues that benevolent governance is "as easy as making something go round in the palm" for a sufficiently virtuous ruler (2A6). Similarly, winning a war of humanitarian rescue will be very easy for such a ruler. If we read the *Mengzi* carefully, we cannot fail to notice that Mengzi idealizes benevolent governance and just war in his discourses of *zheng*, as he is aiming to persuade the rulers they are directed at. Even the campaigns of *zheng* by Tang and Wu are mythologized, and have the following dramatic features.

1. *Matchlessness and invincibility.* Mengzi uses the phrase "matchless in the world" (*wudi yu tianxia* 無敵於天下) to refer to the campaigns five times. As explained in the last chapter, Mengzi uses the phrase ambiguously. It suggests "invincible" when he uses it to describe the past expeditions of *zheng* by Tang and Wu,[17] and "without enemies" when explaining his alleged superior political solution to the inter-state bloody wars of his time.[18] Thus, it broadly implies that a virtuous ruler who practices benevolent governance is guaranteed success both politically and militarily. Perhaps Mengzi is suggesting to the ruler he is talking to that both senses of *wudi* should be considered. Mengzi's message to ambitious political leaders was simple and consistent—cultivating virtues and implementing benevolent policies offers a sufficient condition for victory in all senses. The popular support would be so overwhelming that this virtuous leader's ascendancy would be unstoppable, and thus warfare would not be

necessary. This is the "way of the True King" (*wangdao* 王道), in contrast to the "way of the hegemon" (*badao* 霸道), who habitually resorts to lethal force. In another passage, Mengzi is even more explicit about invincibility in just wars.

> Hence it is said, it is not by boundaries that the people are confined, it is not by difficult terrain that a state is rendered secure, and it is not by superiority of arms that the Empire is kept in awe. One who has the Way [*Dao*] will have many to support him; one who has not the Way will have few to support him ... Hence either a gentleman [*junzi*] does not go to war or else he is sure of victory, for he will have the whole Empire at this behest, while his opponent will have even his own flesh and blood turning against him."
>
> *(2B1)*

The eminent commentator Zhao Qi summarizes this passage well: "The ruler with the Dao of course longs for peace as the Dao of the gentleman is renouncing warfare. If he has to wage war, he is sure of victory" (Jiao 1987, 254).

2. *Surgical precision in operation*. In two separate passages, Tang's campaigns of *zheng* are said to be so efficient that he did not disturb the life of the common people when taking out the tyrant. "Those who were going to market did not stop; those who were ploughing went on ploughing. He punished [*zhu*, executed] the rulers and comforted the people, like a fall of timely rain, and the people greatly rejoiced" (1B11), and "Those who were going to market did not stop; those who were weeding went on weeding. He punished [*zhu*, "executed"] the rulers and comforted the people, like a fall of timely rain, and the people rejoiced greatly" (3B5). Tang was supposed to be engaged in a revolutionary war, but these two passages make it sound like a special forces stealth operation. Fu Peirong remarks that "This is hard to imagine" (Fu 2007, 107), and although the American marines took down Osama bin Laden with surgical precision as he was living as a civilian, they could not do the same to Saddam Hussein as the ruler of a country.

3. *Prompt success with no bloodshed*. Mengzi firmly believes that a campaign of *zheng* led by a benevolent ruler that has overwhelming popular support should be on the whole bloodless and should be victorious with little effort. He insists that Wu's campaign of *zheng* was devoid of bloodshed, although historical documents suggest otherwise. Mengzi said:

> If one believed everything in the *Book of History*, it would have been for the *Book* not to have existed at all. In the *Wucheng* chapter I accept only two or three strips [i.e., a few pages]. A benevolent man has no match in the Empire. How could it be that 'the blood spilled was enough to carry staves along with it,' when the most benevolent waged war against the most cruel?
>
> *(7B3)*

The *Wucheng* chapter of the *Shang Shu* (*Book of History/Documents*), an ancient history book, suggests that the revolutionary war waged by Wu against Emperor Zhou led to much bloodshed. Mengzi dismisses this historical record as fake news, but he does not cite new facts. Rather, he again refers to his conviction that "a benevolent person must have no match in the Empire," that is, no opposition. Wu was a "supremely benevolent person" and Emperor Zhou the exact opposite, so Mengzi infers that Wu's revolutionary war could never have met stiff resistance and thus bloodshed could not have been the result. This campaign of *zheng* should have resulted in rapid success. However, the Battle of Muye (牧野之戰 c.1045 BCE) was renowned in ancient Chinese history and was subsequently studied in great detail. This was a watershed battle in the campaign, and chances are much blood was spilled. However, many Confucians later came to Mengzi's defense, citing a technical explanation. One influential interpretation is that there was indeed widespread bloodshed as recorded in the *Shang Shu*, but this was not caused by Wu's army. Rather, as Wu's revolutionary army approached the palace of Emperor Zhou, the Emperor's officials mutinied, and the bloodshed was caused by their infighting. Hence Mengzi was technically correct (Zhu 1983, 365). This revisionist view prompted much debate (Jiao 1987, 959–961). Fu Peirong, a contemporary Confucian scholar in Taiwan, is willing to accept another scenario. He appeals to the common-sense view that a tyrant seldom rules alone; he has a small inner circle of die-hard supporters who also benefit from his rule. Thus, the argument is that out of self-interest, this inner circle fiercely defended Emperor Zhou and resisted the revolt led by Wu, which led to the bloodshed (Fu 2007). Modern historians concur that the Battle of Muye represented a bloody rout by the outnumbered coalition army of Wu (about 45,000 soldiers) of the much bigger army of Zhou (about 170,000 soldiers). It was a decisive battle that altered the course of war.[19] In Mengzi's moralism, Wu was "the most benevolent" whereas Emperor Zhou was "the most malevolent," and thus a benevolent ruler transforms enemies into supporters and so on, which led him to disregard the historical facts. Other instances of Mengzi's moral outlook leading him to manipulate historical facts can be identified (Chen 1980; cf. He and Shi 2015). Thus, to rhetorically strengthen his arguments for benevolent governance, Mengzi highly idealizes and even mythologizes such a just war.

4. *Oppressed people unanimously invited foreign liberation.* In the three narratives of campaigns of *zheng* by Tang and Wu [1B11, 3B5, 7B4], oppressed people everywhere are portrayed as unanimously inviting the virtuous rulers to rescue them. The following narrative in one passage is similarly repeated in two others. "[A]nd when he marched [*zheng*] on the east, the western barbarians complained, and when he marched [*zheng*] on the south, the northern barbarians complained. They all said, 'Why does he not come to us first?' The people longed for his coming as they longed for a rainbow in time of severe drought" (1B11). Mengzi speaks of the oppressed awaiting liberation as one

people; they suffered greatly under the wicked ruler and in one voice demanded liberation by foreign soldiers. This scenario again appears to be highly idealized. It suggests that only the tyrant and his family benefitted from the status quo and that "the oppressed" were everyone else. This has little resemblance to the complex world we live in today. The modern-day predicament of one ethnic minority suffering intensely under a racially insensitive regime with majority support would be too complicated for the world view of Mengzi's time. The request by the minority for foreign intervention would be fiercely opposed by the society's majority population. Most societies today are pluralistic, and even if many people suffer the same plight, a foreign and virtuous leader's contemplated humanitarian rescue will be simultaneously sought after by the have-nots and fiercely resisted by the haves.

5. *Timely deliverance by "an agent sent by Heaven."* Another indication of Mengzi's idealized account is his view that it is only "an agent sent by Heaven" (*tianli* 天吏) who would qualify to launch a military campaign of *zheng* and deliver the oppressed people from tyranny (2B8). In only one instance does Mengzi himself apparently encourage a ruler of his time to launch a just war against another ruler of state (1A5), and here an ancient commentator observes that it is "an agent sent by Heaven" who has the authority to do so (quoted in Zhu 1983, 206). In Mengzi's view, such an entity epitomizes the practice of benevolent governance and the soft power of winning people's heart and mind, which can lead to unstoppable political success. As Mengzi explains, "If you can truly execute these five measures [of benevolent governance], the people of your neighboring states will look up to you as to their father and mother … In this way, you will have no match in the Empire. He who has no match in the Empire is an agent sent by Heaven, and it has never happened that such a man failed to become a true King" (2A5). As explained earlier in this chapter, when Mengzi invokes the moral exemplar of the successful military campaigns of *zheng* by Tang and Wu (3B5, 1B10, 7B4), he is not recommending that the rulers re-enact such campaigns. Instead, he urges the rulers first and foremost to focus on cultivating virtues and practicing benevolent governance. A ruler can then become "an agent sent by Heaven" and win it all.[20]

6. *Binary mode of thought.* As a motivational speaker, Mengzi often takes a simple presence/absence approach. This passage is typical:

> The Three Dynasties won the Empire through benevolence and lost it through non-benevolence [*bu ren*, i.e., malevolence]. This is true of the rise and fall, survival and collapse, of states as well. An Emperor cannot keep the Empire within the Four Seas unless he is benevolent; a feudal lord cannot preserve the altars to the gods of earth and grain unless he is benevolent; a Minister or a Counsellor cannot preserve his ancestral temple unless he is benevolent; a Gentleman or a Commoner cannot preserve his four limbs unless he is benevolent.
>
> *(4A3)*

Thus, a person of benevolence does everything with great success and a person without benevolence will be an utter failure at everything.[21] This view informs Mengzi's portrayal of the campaigns of *zheng* by Tang and Wu as complete successes. In this binary approach, everything hinges on the presence or absence of virtues. Hence, Mengzi can claim that "The great person alone can rectify the evils in the ruler's heart. When the ruler is benevolent, everyone else is benevolent; when the ruler is righteous, everyone else is righteous; when the ruler is correct [rectified], everyone else is correct. Simply by rectifying the ruler one can put the state or a firm basis" (4A20; translation modified). In Mengzi's ethics of just war, the most important factor is not the just cause but the proper authority, which is the moral rather than political authority to declare and wage war. When the ruler is morally rectified and virtuous, everything he does will be right and successful.[22] As discussed earlier in this chapter, Mengzi views a military campaign of *zheng* as a campaign of rectification (7B4), but this passage (4A20) points out that rectification begins with the ruler himself. "Simply by rectifying the ruler one can put the state on a firm basis." Thus, 7B4 concludes, "What need is there for war?" Such a benevolent ruler is "matchless in the Empire" in the sense of having no enemy and will therefore always win without a fight.

Guozhu Jiang is a People's Liberation Army officer and a respected scholar of Chinese philosophy. His chapter on Mengzi in his five-volume *A History of Chinese Military Thought* concludes that "Mengzi's key teachings about warfare are as follows: opposing warfare, governance with benevolence, loving and protecting the people, victory without war" (Jiang 2006, 368). This conclusion of a lifelong military scholar matches my observation above.

The Mengzi *Does Not Provide Normative Warfare Ethics*

Mengzi's lofty idealism was noted by politicians of his time, and his political philosophy was not generally well-received. Sima Qian (c. 145–90 BCE), in his *The Grand Historian's Records* (*Shi Ji*), narrates the life of Master Meng as follows:

> Meng Ke was a native of Zou. He received instruction from a disciple of Zisi. After he had mastered the Way, he went abroad and served King Xuan of Qi. King Xuan was unable to use him and he went to Liang. King Hui of Liang did not think his speech fruitful. He was thought of as impractical and removed from the reality of events ... Thus wherever he went he did not fit in. He retired and together with disciples such as Wan Zhang discussed the *Odes* and *History* and laid out the ideas of Confucius, composing *Mengzi* in seven books.
>
> *(Nienhauser 1994, 179)*

Mengzi's moral–political thought was regarded by rulers at the time as high-minded and "removed from the reality of events." Does it fare better in today's world? Various English language scholars (Bell 2006, Kim 2010, 2017, Barney and Chan 2015) find that Mengzi's ethics of *zheng* warfare provide numerous implications for just or ethical warfare today, particularly in terms of the ongoing global discussion about humanitarian intervention, or the "responsibility to protect," as articulated by the United Nations. At first glance there are parallels, but Mengzi's ethics of just war are not readily adaptable to our time, because as discussed they are highly idealized and cannot be enacted. Mengzi's most important criterion for the legitimate use of force is not just cause, but the proper authority to declare and conduct the war—his "agent sent by Heaven."[23]

Mengzi did not write political philosophy in the Western sense; nor did he conduct legal analyses as Gratian did. He was primarily an inspirational speaker seeking to convert the rulers of his day. They aimed to wage and win wars, so Mengzi construed the highly idealized warfare of *zheng* to engage his audience. Mengzi is a master of oratorical hyperbole and exaggeration (Chen 1980, 252–256, 329–337), and in my view his talk of invincibility, victory with little effort, surgical precision, and little bloodshed, are oratorical devices designed to persuade a ruler. Mengzi's speeches sound more like inspiring sermons in a religious meeting than a clear guideline that the military of today could follow. What Mengzi does is tantamount to preaching a "wealth and health" gospel, in which the cultivation of virtues always leads to political success. "Hence it is said, 'The benevolent man has no match.' I beg of you not to have any doubts" (1A5). However, a war aimed at regime change and the liberation of oppressed people, as Mengzi describes it, would be almost impossible to implement in today's world. In spite of our advanced technology, warfare usually leads to the widespread killing of combatants and non-combatants and the destruction of property, particularly if regime change is the goal. A virtuous leader with a just cause and morally right intentions cannot guarantee limited bloodshed, surgical precision, victory with little effort, no disturbance to the regular life of common people, or invincibility.

The evidence suggests that many subsequent Confucians agonized over these highly idealized warfare ethics. According to Mengzi, any virtuous leader attempting to re-unite China will have a strong aversion to bloodshed (1A6). How many of the subsequent dynasties set out with an aversion to bloodshed and so had the moral legitimacy of the Way of the True King? Is Mengzi's insistence on little or no bloodshed then doctrinaire, dogmatic, and impractical? Neo-Confucians who wrestled with this issue came up with contrived answers, suggesting that a few emperors did indeed start a new dynasty with little bloodshed, including the "inaugural emperor of our time" (Zhu 1983, 206).

In short, we must bear in mind the different purposes of writing for Mengzi and Gratian. Gratian offers a meticulous analysis of warfare ethics to provide action guidance, seeking to help Christians participate in just wars without moral scruple. However, Mengzi's purpose of writing is not to provide advice on the normative ethics of warfare. As there has been too many alleged just wars already, Mengzi's main concern is to provide a statecraft that can render war either unnecessary or swift and without bloodshed, rather than guidance on waging just wars. Early readers of these two pioneering works understood their respective purposes and so the intellectual traditions developed differently. Thus, it is no surprise that the Confucian political tradition, as mediated by the *Mengzi*, has been preoccupied with discussing benevolent governance. Unlike the just war traditions of Europe, the subsequent Mengzian–Confucian tradition by and large has not significantly extended, refined, or revised the ethics of punitive warfare, a war of humanitarian rescue, or a revolutionary war against a tyrant, beyond the *Mengzi*.

Notes

1 Other words are also war-related (cf. Legge 1970, 478); of note is *zhan* (戰), which is the most common modern Chinese generic term for war.
2 Unless otherwise stated, all English quotations of this book are from *Mencius* 2003. To ensure the style is consistent, all Wade-Giles transliterations of Chinese words are converted to Pinyin.
3 「天下信湯之德 … 伐惡養善」(Jiao 1987, 152, 156).
4 In both passages Mengzi, in hyperbole, articulates that there can be surgical precision in taking out the tyrant without disturbing the life of the common people at all: "Those who were going to market did not stop; those who were ploughing went on ploughing."
5 「言修德無小，暴慢無強。事故夏商之末，民思湯武；雖欲不王，末由也已」(Jiao 1987, 437).
6 As D. C. Lau explains in his translator's note, "the two verbs in Chinese are cognate." (*Mencius* 2003, 158).
7 Thus, James Legge takes this root meaning as the norm and translates *zheng* (征) as "to correct." "'Correction' is when the supreme authority punishes its subjects by force of arms. Hostile States do not correct one another." (Legge 1970, 478). However, in other occurrences Mengzi does not use this term in accordance with its root meaning, but with its secondary meaning; that is, to wage a war.
8 「民思明君，若旱望雨，以仁伐暴，誰不欣喜 … 焉用善戰，故云罪也。」
9 D. C. Lau increases the confusion by translating the word *fa* in this passage as "march on," which he sometimes also uses to translate the word *zheng*.
10 Lin Yutang's *Chinese-English Dictionary of Modern Usage* (available online) is highly regarded. His explanation of the meaning of *zheng* as a verb is as follows: 1. to go on a journey, to travel; 2. to invade, to start a campaign; 3. to collect tax. Hence, D. C. Lau's frequent rendering of *zheng* as "to punish" or "to initiate a punitive expedition" in the English translation of *Mencius* (2003) is an inaccurate moral translation.
11 For example, Brian Orend explicitly re-locates retributive justice from *ius ad bellum* to *ius post bellum*. After the war, in addition to public apology, demilitarization, and war crime trial, there should also be: "1) a backward-looking system of compensation payments, from Aggressor to Victim(s); and 2) a forward-looking system of sanctions, placed by Victim and the international community on Aggressor" (Orend 2013, 202; for details see 202–205). However, Orend ultimately finds punishment as retribution less compelling than punishment as rehabilitation (hence

no compensation payments, no sanctions, forcible regime change, and aid with post-war reconstruction for 10–15 years; Orend 2013, 205–206, 215–240).

12 As the popular saying goes, "犯我華者，雖遠必誅" ("For those who offend or humiliate China, we will punitively execute you no matter how far away you are").

13 McMahan also invokes this proviso: "One reason why the American invasion of Iraq in 2003 was not a justifiable instance of humanitarian intervention is that there was no evidence that ordinary Iraqis wanted to be freed from the Ba'athist dictatorship *by the United States*" (McMahan 2005, 13; emphasis his).

14 Twiss and Chan suggest that the contemporary equivalent of a virtuous leader is "a good competent political leader" (Twiss and Chan 2015, 126). This would not be sufficient, as the proviso of the oppressed people's trust is absent.

15 Thus, I agree with Tiwald's thesis that in the *Mengzi* common people are not regarded as having the right of rebellion (Tiwald 2008).

16 To supplement my general explanation, I draw on the findings of Lo (2022). Major intellectual movements and new education institutions emerged both in Europe and in China around the 12th century CE, with scholasticism in universities and neo-Confucianism in private Confucian colleges, respectively. European universities were developed as alternative institutions of higher learning to monastic schools. Learning could thus be legitimately pursued for learning's sake and in terms of professional development. Private Confucian colleges, however, were akin to monastic schools because the intended learning outcome was spiritual–moral formation. European just war ethics continued to develop because it was sustained by the division of knowledge in universities. Confucian just war ethics, although it predated those of Augustine by at least 700 years, remained rudimentary and undeveloped because such discussions were deemed irrelevant to spiritual–moral cultivation, which were aimed at the goals of sagehood or sainthood. Only military officers would engage in sustained discussions concerning just warfare; it was not considered a sufficiently serious subject to be discussed and debated in the prestigious Confucian colleges.

17 Tang was "matchless in the Empire" in all 11 of his military expeditions (*zheng*) against Ge, because he was a virtuous leader who practiced benevolent statecraft (3B5). Wu's revolutionary war against Emperor Zhou succeeded because "a benevolent man has no match in the Empire" (7B3). In responding to the awful rumor that Mengzi is an expert in the art of war, he declares, "If the ruler of a state is drawn to benevolence he will have not match in the Empire. 'When he marches (*zheng*) on the south, the northern barbarians complained … '" (7B4). In a fourth passage the phrase is used slightly differently: "If you should go and punish [*zheng*] such princes, who is there to oppose you? Hence it is said, 'The benevolent man has no match.' I beg of you not to have any doubts" (1A5).

18 After implementing the five benevolent domestic policies recommended by Mengzi, a ruler's benevolence will be well known abroad. "In this way, you will have no match in the Empire … and it has never happened that such a man failed to become a true King" (2A5). Mengzi further explains with certainty that "He who models himself on King Wen will prevail over the whole Empire, in five years if he starts with a big state, and in seven if he starts with a small state … Confucius said, 'Against benevolence there can be no superiority in numbers. If the ruler of a state is drawn to benevolence, he will be matchless in the Empire'" (4A7).

19 "Although later Confucian moralists refused to concede that a confrontation between the 'extreme virtue' of King Wu of Zhou [dynasty] and the 'extreme vice' of King Di Xin of Shang [dynasty] could possibly have resulted in any bloodshed at all, there is little doubt that the Zhou [dynasty] did win a decisive military victory" (Loewe and Shaughnessy 1999, 310; cf. Editorial Team 2002, 24–34).

20 Bell suggests that "potentially virtuous" contemporary rulers can qualify as those with the proper authority to launch campaigns of human intervention (Bell 2006, 39). However, according to Mengzi all human beings are potentially virtuous, and

as such a ruler will not have established virtuous policies this moral bar is too low. To remedy this significant defect Bell adds that this potentially virtuous ruler "must have some moral claim to have the world's support." With such a "trust of the world," he asserts, "even rulers with bad track records may be regarded as potentially virtuous leaders who can bring peace to the world" (Bell 2006, 39–40). This represents a poor contemporary rendition of a Confucian virtuous person. The virtuous character in Mengzi is reduced to a mere persuasive talker.

21 As Daqi Chen points out, in other discussions Mengzi does allow for different degrees of benevolence and non-benevolence (Chen 1980), and not simply its presence or absence.

22 "Like Confucius, Mencius regarded the transformative power of a cultivated person as the ideal basis for government" (Shun 1997, 163).

23 Sungmon Kim (2017), unlike Bell, notes this salient feature of Mengzi's theory. He proposes to incorporate a democratic theory to revise his virtue ethics, and thus to indicate that humanitarian interventions are less subject to abuse. However, Kim does not give due consideration to the highly idealized and mythologized accounts of campaigns of *zheng* that make it almost impossible for them to be re-enacted. Kim also does not sufficiently consider that Mengzi is a contingent pacifist who, except for self-defense, does not recommend any warfare to rulers; his talks of just and invincible wars are only bait to persuade them to accept his political philosophy. Mengzi's central concern in these war talks is thus not normative warfare ethics at all.

Bibliography

Bell, Daniel A. 2006. "Just War and Confucianism: Implications for the Contemporary World." In *Beyond Liberal Democracy: Political Thinking for an East Asian Context*, 23–51. Princeton: Princeton University Press.

Chan, Wing-tsit. 1963. *A Sourcebook in Chinese Philosophy*. Translated and compiled. Princeton: Princeton University Press.

Chen, Daqi 陳大齊. 1980. *Perplexities in Reading the Mengzi*《孟子待解錄》. Taipei: Commercial Press.

Cox, Rory. 2018. "Gratian," in *Just War Thinkers: From Cicero to the 21ˢᵗ Century*, edited by Daniel R. Brunstetter and Cian O'Driscoll, 34–49. London: Routledge.

Editorial Committee of the Dictionary of the Thirteen Classics 十三經辭典編纂委員會. 2002. *Dictionary of the Thirteen Classics: the Mengzi*《十三經辭典：孟子卷》. Xi'an: Shaanxi People's Press.

Editorial Team of Chinese Military History 中國軍事史編寫組. 2002. *A History of Chinese Military Strategy*《中國歷代軍事戰略》. Two volumes. Beijing: People Liberation Army Press.

Fu, Peirong 傅佩榮. 2007. *Inclination Toward Moral Goodness: Fu Peirong on the Mengzi*《人性向善：傅佩榮談孟子》. Taipei: Bookzone.

Glanville, Luke. 2010. "Retaining the Mandate of Heaven: Sovereign Accountability in Ancient China." *Millennium: Journal of International Studies* 39 (2): 323–343.

Glanville, Luke. 2018. "Samuel Pufendorf," in *Just War Thinkers: From Cicero to the 21ˢᵗ Century,* edited by Daniel R. Brunstetter and Cian O'Driscoll, 144–155. London: Routledge.

He, Tieshan, and Xinggu Shi 何鐵山、施杏姑. 2015. *Reading the Mengzi with Chinese Linguistic Approaches*《漢字學視域下的〈孟子〉》. Hangzhou: Zhejiang University Press.

Huang, Chun-chieh 黃俊傑. 2017. *East Asian Confucian Discussions on Ren: Historical Discussions*《東亞儒家仁學史論》. Taipei: National Taiwan University Press.

Jiang, Guozhu 姜國柱. 2006. *A History of Chinese Military Thought*, volume 1, *Pre-Qin*《中國軍事思想通史》卷一:先秦卷. Beijing: Chinese Academy of Social Sciences Press.

Jiao, Xun焦循. 1987. *Corrected Interpretations of the Mengzi*《孟子正義》. Two volumes. Beijing: Zhonghua Press.

Johnson, James Turner. 1975. *Ideology, Reason, and the Limitation of War: Religious and Secular Concepts 1200–1740*. Princeton: Princeton University Press.

Johnson, James Turner. 1981. *Just War Tradition and the Restraint of War: A Moral and Historical Inquiry*. Princeton: Princeton University Press.

Kim, Sungmoon. 2010. "Mencius on International Relations and the Morality of War: From the Perspective of Confucian *Moralpolitik*." *History of Political Thought* XXXI (1, Spring): 33–56.

Kim, Sungmoon. 2017. "Confucian Humanitarian Intervention? Toward Democratic Theory." *The Review of Politics* 79: 187–213.

Lang, Jr., Anthony F. 2018. "Hugo Grotius," in *Just War Thinkers: From Cicero to the 21st Century*, edited by Daniel R. Brunstetter and Cian O'Driscoll, 128–143. London: Routledge.

Legge, James. 1970. *The Works of Mencius: Translated, with Critical and Exegetical Notes, Prolegomena, and Copious Indexes*. New York: Dover (reprint of 1895 edition).

Li, Junxiu 李鈞岫. 2010. *Mengzian Studies in the Han and Tang Dynasties*.《漢唐孟子學述論》Jinan: Qilu Press.

Lo, Ping-cheung. 2015a. "The Art of War Corpus and Chinese Just War Ethics Past and Present," in *Chinese Just War Ethics: Origin, Development, and Dissent*, edited by Ping-cheung Lo and Sumner B. Twiss, 29–65. London: Routledge.

Lo, Ping-cheung. 2015b. "Legalism and Offensive Realism in the Chinese Court Debate On Defending National Security 81 BCE," in *Chinese Just War Ethics: Origin, Development, and Dissent*, edited by Ping-cheung Lo and Sumner B. Twiss, 249–280. London: Routledge.

Lo, Ping-cheung. 2020. "Gratian and Mengzi: Pioneer Works in Christian and Confucian Just War Traditions," *Journal of Religious Ethics*, 48:4 (Dec 2020): 689–729.

Lo, Ping-cheung. 2022. "Scholastic Universities, Monastic Schools and Confucian Colleges: Historical Tensions in Whole Person Education and Prospective Solutions," in *Whole Person Education in East Asian Universities: Perspectives from Philosophy and Beyond*, edited by Benedict S. B. Chan and Victor C. M. Chan, 31–60. London: Routledge.

Loewe, Michael, and Shaughnessy, Edward L., edited. 1999. *The Cambridge History of Ancient China: From the Origins of Civilization to 221 B.C.* Cambridge: Cambridge University Press.

McMahan, Jeff. 2005. "Just Cause for War." *Ethics & International Affairs* 19 (3): 1–21.

Mencius. 2003. Revised edition. Translated from the Chinese with an Introduction and Notes by D. C. Lau. London: Penguin Books.

Mencius. 2009. Translated by Irene Bloom. Edited and with an Introduction by Philip J. Ivanhoe. New York: Columbia University Press.

Mengzi: with Selections from Traditional Commentaries. 2008. Translated with Introduction and Notes by Bryan W. van Norden. Indianapolis: Hackett.

Orend, Brian. 2013. *The Morality of War*, second edition. Peterborough, Ontario: Broadview Press.

Pufendorf, Samuel. 1994. *The Political Writings of Samuel Pufendorf.* Edited by Craig L. Carr and translated by Michael J. Seider. New York: Oxford University Press.

Reichberg, Gregory M., Syse, Henrik and Begby, Endre, eds. 2006. *The Ethics of War: Classic and Contemporary Readings.* Oxford: Blackwell.

Scobell, Andrew. 2003. *China's Use of Military Force: Beyond the Great Wall and the Long March.* Cambridge: Cambridge University Press.

Shun, Kwong-loi. 1997. *Mencius and Early Chinese Thought.* Stanford: Stanford University Press.

Tiwald, Justin. 2008. "A Right of Rebellion in the Mengzi?" *Dao: A Journal of Comparative Philosophy,* 7 (3, September): 269–282.

Twiss, Sumner B. and Jonathan K. L. Chan. 2015. "Classical Confucianism, Punitive Expeditions, and Humanitarian Intervention," in *Chinese Just War Ethics: Origin, Development, and Dissent,* edited by Ping-cheung Lo and Sumner B. Twiss, 117–134. London: Routledge.

Zhu, Xi. 1983. *The Collected Glosses on Passages and Chapters in the Four Books* 《四書章句集注》. Beijing: Zhonghua.

6

SEVEN MILITARY CLASSICS

Martial Victory through Good Governance

Yvonne Chiu[1]

In an increasingly tumultuous global political landscape, many states are asking how they can retain or regain their international credibility, influence, and competitiveness. Liberal democracies have not been immune from the widespread democratic backsliding during the past decade or so, and some are now deciding that an important part of the answer is to "get your own house in order," presumably because having more moral purchase abroad will make the state more effective geopolitically.

Why would the quality of domestic society matter so much for international competition? Classical Chinese philosophy as found in the *Seven Military Classics* (武經七書) offers a compelling argument for how good governance contributes to military and geopolitical victory.

Good governance also speaks to the contemporary normative debate about humanitarian intervention, both militarized and non-militarized. Even as wars since the mid-19th century have become more ideological and concerned with rendering justice, international legal justifications for war have narrowed, and the United Nations' Charter permits wars only in "self-defence" by individual or collective states (Art. 51).[2] As a result, the international laws of war have effectively ceded questions of *jus ad bellum* and instead focus primarily on *jus in bello*—how one may fight *during* a war—independent of the justice of one's cause (Chiu 2019, 193–233).

This is problematic for states, some of which will not have others to come to their "self-defence," but even more so for oppressed people within states, as the UN charter is premised on the idea of state sovereignty. State sovereignty—including a state's right to oppress its own people without interference—and the limited international legal framework for legitimate warfare are continually challenged, however, by human rights and other claims, and a

DOI: 10.4324/9781003336372-8

classical Chinese "good governance" argument offers a productive angle for revisiting justified causes for intervention.

Good governance as legitimate cause for war

While there is enormous variation in Chinese thought on political philosophy and military ethics,[3] the *Seven Military Classics* weave military strategy and virtue ethics together with detail and nuance and, taken collectively, their arguments can be reconstructed into a conception of "good governance" which maintains that good governance at home actually wins wars abroad. A contemporary reconstruction from these canonical texts of Chinese military philosophy offers potential lessons on the international ramifications of domestic virtues. Especially with the empirical correlation between democratic governance and military effectiveness, there may be much to learn from this comprehensive approach.

The *Seven Military Classics* are representative of creative and formative periods in Chinese history, but they and other early Chinese military works do not constitute a unified school or theory. Each of the seven was written at different times in different contexts,[4] and later "canonized" as a set for military education in imperial China (during the 11th century, Song Dynasty); they have different foci and exhibit varying mixes of Confucian, Mohist, Daoist, and Legalist influences,[5] so it is necessarily somewhat artificial to analyze them as a group.

Still, a common thread of non-military sources of security runs through these works, specifically: righteous, just governance both is legitimate cause for war and strengthens and secures the state, by increasing the morale of its soldiers so they will fight harder and by deterring the enemy with the ruler's "awesomeness" (威 *wei*). In each of these works, philosophy of governance is integral to military strategy in a way that is largely missing from modern Western military strategic thought (with the notable exception of its counter-insurgency doctrines).[6]

Legitimate cause for war (jus ad bellum)

Chinese classical tradition generally holds that wars waged for the sake of "righteousness" are morally acceptable.[7,8] Although wars are abhorrent and always tragic, some are necessary evils in order to stop aggressive wars, restore stability and order, and depose tyrants and end despotic rule (Lo 2012, 414–415). For example, Confucians sanctioned "punitive expeditions" against tyrants, and the *Seven Military Classics* generally concur.[9]

More expansively, wars to instill the necessary virtues in others are permissible. According to *Si Ma Fa* (司馬法), "the Tao for imposing order on chaos" starts with benevolence, then uses credibility, straightforwardness, unity, righteousness, "change [wrought by authority]," and finally "centralized authority".[10] Not everyone will voluntarily submit to civilized rule, however, so force is then justified to bring people to righteousness:

> As for warfare, when upright methods do not prove effective, then centralized control of affairs [must be undertaken]. [If the people] do not submit [to Virtue], then laws must be imposed. If they do not trust each other, they must be unified. If they are dilatory, move them; if they are doubtful, change [their doubts].[11]
>
> Authority comes from warfare, not from harmony among men. For this reason if one must kill men to give peace to the people, then killing is permissible. If one must attack a state out of love for their people, then attacking it is permissible. If one must stop war with war, although it is war it is permissible. Thus benevolence is loved; righteousness is willingly submitted to..."[12]

Domestic disorderliness warrants correction by outside forces, *Si Ma Fa* continues: those who let "fields turn wild and [their] people scatter," do harm to their relatives or the people, "murder the Worthy" or overthrow their ruler, or are otherwise "chaotic and rebellious both within and without their borders" will be "purged," "extinguished," or otherwise "rectified."[13]

As "terrible" as physical coercion is, "rectification and punishment" through military campaigns are sometimes appropriate, says *Questions and Replies between Tang Taizong and Li Weigong* (唐太宗李衛公問對).[14] *Taigong's Six Secret Teachings* (六韜) holds that to "respect the people" requires not only treating those who "submit and accord with you...generously with Virtue," but also "break[ing] with force" those who oppose righteous rule.[15]

As such, both the political ruler and the military leader have crucial roles in "bring[ing] peace to those who are in danger," says *Three Strategies of Huang Shigong* (黃石公三略), for "the essence of the army and the state lies in investigating the mind of the people and putting into effect the hundred duties of government."[16]

Violence as a last resort

Even when justified and necessary, war is problematic and should be a last resort.[17] Multiple works posit that the Sage Kings took no pleasure in military expeditions and thought weapons were "evil," to be used only when all other tools of righteous governance had failed.[18]

Daoism influences some of the seven *Classics*: even when the use of violence is unavoidable and "accords with" the Dao of Heaven, "weapons are inauspicious instruments and the Dao of Heaven abhors them."[19] Not only is success never certain no matter how expert or prepared one is,[20] but even victorious military ventures suffer lost resources, including precious men (human capital);

so one must always seek quick victory, for wars harm everyone including the winners, says Sun Tzu.[21]

To add to the danger, warfare can invigorate and can take on a logic and motivation of its own: while "those who forget warfare will certainly be endangered," perhaps more common, in the words of *Si Ma Fa*, is that "those who love warfare will inevitably perish."[22]

Winning without fighting and the geopolitical endgame

Thus, the greatest military victories are achieved, paradoxically, without fighting at all, as several of *Classics* say. The most famous of these statements comes from Sun Tzu, who declares:

> It is best to keep one's own state intact; to crush the enemy's state is only a second best. ... the expert in using the military subdues the enemy's forces without going to battle, takes the enemy's walled cities without launching an attack, and crushes the enemy's state without a protracted war. He must use the principle of keeping himself intact to compete in the world. Thus, his weapons will not be blunted and he can keep his edge intact.[23]

Threats are only credible if the weapon would be effective, but there is a limit to how battle-ready troops can be without having been battle-tested, for nothing genuinely prepares one for war except war. To extend Sun Tzu's metaphor, the most effective weapons are those that have been sharpened, blunted, then whetted again. Sun Tzu surely knows this, so what should we make of this synecdoche?

It is in the first instance a warning to rulers to use warfare only sparingly, so as to suffer its losses (even in victory) minimally and thus leave the state in a stronger position for future competition. It could also mean that a fearsome military *posture* should do much of the work of persuading the enemy to capitulate.

A third possibility is that keeping one's own state intact carries much more meaning than it appears. If an "intact" state means it is secure, stable, and just, then this is incredibly difficult and complicated to first achieve and then to sustain, as we shall see. It would be a rare accomplishment, so an intact state signals not only the breadth and depth of the ruler's virtues but also a capable and effective complex of political, social, and security institutions, and it is through the use of these capacities that one should subdue an enemy state.

This would buttress good governance's claim that righteousness (an internal attribute) will manifest external effects by defeating the enemy. Rather than being a sport[24] or some other end in itself, war is a tool—and simply one among many, including diplomacy, espionage, propaganda, superior virtue, effective domestic institutions, etc.—for securing peace.[25]

How far does the duty to secure peace extend, geopolitically? Some works in the *Seven Military Classics* could be read to imply that peace is not merely a condition of calm and non-violence between states, but rather much more ambitious: a stable and just system of All under Heaven.[26] Derek Yuen argues that the goal in *Sun Tzu's The Art of War* (孫子兵法), for example, was not to win wars but rather to put All under Heaven, "i.e.,…the rule of one [*just*] state, namely, China."[27] If so, he believes that a particular geopolitical order must be restored, which consists not of sovereign states but rather a unified "All under Heaven."

Leadership virtues and good governance

The best way to win without fighting is to prevent war from happening at all— through one's own good governance. In both classical and contemporary Chinese thought, there is little demarcation between "cold" and "hot" war, and one blends seamlessly into the other. This means that the ethics of warfare are actually located in a primarily different place than in Western thought: classical Chinese writings focus on the period before the war and especially on its prevention.

Classical authors repeatedly advocate the use of "humane governance" in order to prevent warfare. As removing a tyrant is one of the few acceptable justifications for war, the best deterrent is to not be a tyrant and therefore not a legitimate target.

Furthermore, demonstration of superior morality persuades others to join the righteous ruler's kingdom. Advises *Huang Shigong*, "Thus it is said, 'Draw in their men of character and valor and the enemy's state will be impoverished.'"[28] In his 13th-century commentary on this work, Shi Zimei 施子美 adds, "An enlightened king concentrates on expressing virtue, thus the four barbarians submit to his rule. Thus by propagating virtue one can then make those distant submit. What need is there to rely on expanding territory?"[29]

Rightness is not only desirable for its own sake[30] but also efficacious, for if one governs people "according to the forms of propriety [*li*] [and] stimulate[s] them with righteousness, …then the officers will die [for the state]."[31] This allegiance is earned primarily through the ruler's own rectitude, and that is where the state's strength lies. "Fortune and misfortune lie with the ruler, not with the seasons of Heaven,"[32] and a wicked ruler earns no loyalty and thus only endangers his own state and people. *Huang Shigong* admonishes, "One who concentrates on broadening his territory will waste his energies; one who concentrates on broadening his Virtue will be strong."[33] Virtues include benevolence, righteousness, loyalty, trust [good faith], courage, and planning,[34] as well as properly judging other people, for good attracts good and evil begets evil. Rulers will find that dismissing one good man from office will yield exponential losses for their kingdoms, and rewarding one evil man will "draw myriad evils."[35]

Thus, a ruler must attend to his own virtue and heed the advice of the wise,[36] for only a "person of civil virtue [can] bring peace to the empire."[37] In order to impose virtuous living on others, one must display virtue oneself, including a degree of spiritual maturity as evidenced by one's attitudes toward war, and demonstrate the superiority of that way of life. The idea is to get willing submission from others, rather than simple domination over them.[38]

Virtue is a "subtle"[39] strength that conquers not with force, but with the overwhelming appeal of its rectitude: "The government of a Worthy causes men to submit with their bodies. The government of a Sage causes men to submit with their minds," says *Huang Shigong*.[40] For even from "beyond the seas,"[41] people will travel to reside under and give their allegiance to righteous, humane, and virtuous rule.[42, 43]

Beyond the value of virtue's public demonstration, we can also detect an argument that *righteous conduct will win wars*. Outcomes of war will reflect the general quality of warring parties' respective domestic administrations: *whoever is the better ruler will win the war*. This is not simply a blind faith that the gods will reward those who are more virtuous. It is a statement about *good governance*.[44] Those who govern well will win wars because: (a) their population is more satisfied and therefore more willing to work and sacrifice for the sake of the kingdom; and (b) their kingdom is less corrupt and better organized, and can therefore more efficiently and effectively marshal materiel and human resources for the war effort. This is why righteous rulers will win the war, these texts argue—not merely by being righteous, but also because of the effects of their good governance.

Implications for jus in bello

At the same time, classical Chinese thinkers and poets were concerned with how war would burden the population. Offensive wars were criminal, of course, but even righteous wars impose unnecessary suffering on both the rulers' own people and their opponents'. Long before the West talked about winning "heart and minds," classical Chinese thinkers emphasized the military importance of earning the support of the people one seeks to conquer. One must attend to relations between superiors and inferiors, between kings and commoners; to that end, victorious kings must not penalize the common people who did not fight against them and instead focus on punishing those responsible for the war—presumably their unjust and/or uncivilized rulers.[45]

Multiple writings advise different ways to woo the opponent's population. *Six Secret Teachings* says a siege should only sever the city's supply routes and surround and guard the city: it does not include engaging in battle, setting fire to or destroying buildings, cutting down trees, or killing captives or those who surrender. A well-conducted siege simply outlasts the opponent, instead of destroying them or their property. Similarly, harming non-combatants or damaging their property is prohibited, and there are guidelines for treating prisoners of war humanely.[46] *Si Ma Fa* reiterates limiting the pursuit of fleeing and retreating enemies, distinguishing between combatants and non-combatants, treating sick and wounded soldiers and civilians, accepting surrender, and respecting the gods, infrastructure, natural resources, and property rights of the invaded. Furthermore, one must properly signal or declare the start of combat as a gesture of good faith, as well as choose an appropriate time of attack (not when a state is in national mourning or suffering from natural disaster, and in neither summer nor winter), in order "to love both your own people and the enemy's people."[47]

Again and again, the classical texts exhort aspiring conquerors to demonstrate their own virtuousness in contrast with the enemy ruler's viciousness. Careful treatment of the enemy population is a concerted effort to split those people from their unjust rulers: it allows the enemy's people to recognize the invader as righteous and gives them no reason to fight in opposition. This advice has been borne out by comparative surrender rates. When possible, soldiers of autocratic regimes are more likely to surrender to democratic opponents because they believe they will receive better treatment; for example, in World War II, German soldiers fighting American or British troops were more willing to surrender than those facing Soviet troops (Reiter and Stam 2002, 69; Reiter and Stam 1997). As *Six Secret Teachings* advises:

> Show them benevolence and righteousness, extend your generous Virtue to them. Cause their people to say 'the guilt lies with one man.' In this way the entire realm will then willingly submit.[48]

When the enemy's people are presented with a righteous path they can follow, it minimizes the fighting and effort required to subjugate them by turning the invader into the ruler.[49]

While it is certainly pragmatic to treat the enemy population with justice, it is not merely so. Exhibiting one's virtue makes a public statement about one's right to rule. Proper behavior during war is necessary in order to reveal the aspiring conqueror's right intentions as well as his broader righteousness. In this way, the content of *jus ad bellum* and *jus in bello* are tightly tied, with the former guiding the latter.

The banality of governance

If the business of a "righteous army" is to "suppress the violently perverse and rescue the people from chaos,"[50] then it must behave accordingly: not only must the war be won virtuously, but it must lead to future peace and humane and virtuous rule. Therefore, unsurprisingly, righteous behavior must extend to the post-war period as well.[51]

The victor must display correct attitudes before, during, and after battle because the only thing that gives him legitimacy to conquer and rule is his superior virtue. His moral excellence is established through his just actions and the success of his everyday rule, so once conquered peoples become part of his own society, he must demonstrate his righteousness by governing them in that same just manner that gave him the legitimacy to subjugate them.[52]

It is one thing to display flourishes of magnanimity in the tumult and fervor of war, but quite another to be virtuous under tedious circumstances. Daily governance of a complex society requires painstaking attention to dense yet monotonous details, and it can be both incredibly demanding and unimaginably dreary.

A benevolent ruler focuses primarily on administering his own state well, which includes facilitating wealth generation and accumulation, minimizing taxes and other impositions, equitably distributing land, not harming the people, rooting out corruption, instilling a sense of justice and shame in the population, enforcing justice and establishing rule of law, and properly respecting the ancestors and according men with appropriate ranks.[53]

It is much more difficult to act with virtue, rule with righteousness, and establish and maintain a just and stable society over an extended period of time, yet it is no less critical to security than military efficacy. "Being victorious in battle is easy, but preserving the results of victory is difficult," says *Wuzi*, [54] and non-military components such as domestic moral excellence are essential for long-term security. To protect his state, a ruler must govern well.[55] To govern well, a ruler must: attract courageous and "worthy" men to fill his state's offices,[56] prevent the "chaos" that results from officials forming factions that pursue their own interests ahead of the state's;[57] retain his authority over the military;[58] refrain from over-taxing and otherwise economically over-burdening his people;[59] lead the people to be content and "peaceful";[60] and himself resist greed and geopolitical ambition. Instead of coveting more territory and sparking more conflict, he must focus on his existing territory and his own virtue.[61] It is the periods of peace—in between wars—that are most important, but they are also far more difficult to manage.

One can see how the concerns of morality come full circle, in a cycle of virtuousness: virtue determines one's spiritual state, which determines the condition of the political and social entity, which in turn enables one to prevent or win wars and gives one legitimacy to wage war if necessary, thereby improving the spiritual, political, and social circumstances of others.

The relevance of classical Chinese "good governance" for *jus ad bellum*

Classical Chinese military philosophy folds secular virtue ethics of military leadership into their societal context in order to tackle broader questions of governance and rectification. Incorporating some aspects of this "good governance" theory into *jus ad bellum* can: (a) better connect international to domestic principles and (b) offer an alternative to "rights" discourse.

Why classical Chinese just war theory instead of Western?

Other traditions also connect the domestic and the international, so why reach for this school of thought when existing international law of armed conflict (LOAC) is based primarily on Western historical developments and Western military ethics? The legalistic formulation of contemporary LOAC reflects the Westphalian-based international system of state sovereignty by maintaining strict distinctions between domestic and international, but Western just war theory frequently connects the two, for example, medieval Aquinas or contemporary Paul Ramsey who positions *The Just War* (1968) as a part of a comprehensive theory of statecraft. Given LOAC's historically Western roots, these might be less dissonant frameworks from which to draw.

However, this classical Chinese military theory of "good governance": (a) ties moral questions of war to those of everyday politics and governance, while (b) maintaining the distinction between the two, and (c) is more easily secularized. Western just war theory that directly channels domestic principles into its international ones is either religious in nature or, when secular, connects domestic with international by collapsing the distinction between them and effectively erasing the international (e.g., revisionism).[62] Any universalistic theory should seek to bridge international principles and domestic values in some way; at the same time, maintaining a distinction between the two realms, in a secular fashion, is necessary to formulate a more robust ethics for contemporary geopolitical circumstance and its reality of separate and diverse states.

Connecting domestic values and international principles

The contemporary legal framework of LOAC that separates principles of international justice from domestic justice of societies is inconsistent with the governing principles and values of any society that espouses a universal ideology, and of liberal democracies in particular. Liberal democratic countries *should* have an interest in better aligning their foreign policy with their domestic politics and in fighting wars in ways that are informed by their underlying domestic principles—or at least consciously, rather than inadvertently, making exceptions. This is a challenge not only for political and military practice, but also for academic and philosophical study, which usually treats military affairs and just war theory separately from other issues of global justice. The increasingly urgent questions about whether and how to extend domestic systems of governance and justice to the global realm (e.g., in the field of global governance) frequently overlook problems of war, but domestic principles have both normative and empirical relevance for warfare.

Relative to other forms of governance, democracies excel at making their societies richer. Despite contested causation and worrisome recent trends, the majority of the richest forty-percent of countries in the world are still considered "free" according to the Freedom House rubric, and most of the remainder are petro-states (Freedom House 2018; World Bank 2018). This correlation between democracy and wealth may be epistemically revealing: insofar as governments are supposed to improve their citizens' lives, greater wealth and higher standards of living might show that democracies "know" better ways of structuring political society.

Less discussed but no less important is the relationship between democratic governance and military achievement. Contrary to the image of authoritarian societies being better at war because they are more disciplined, democracies have in fact won about eighty percent of the time, during the past two centuries.[63] While some attribute this phenomenon to the greater likelihood of democracies assisting one another (Choi 2004) or their superior ability in marshalling resources from their society to fight,[64] others argue that their system of governance means that democracies usually only engage in wars they are likely to win (selection effect) because their leaders are accountable to the population (in a way, Kant's theory in *Perpetual Peace* in action), that their military leaders are more skilled because they are less likely to have been chosen for political reasons, and that their soldiers are more willing to fight than those of non-democracies as evidenced by surrender rates (Reiter and Stam 2002). If any of these theories are true, then liberal democracies are winning for reasons related to their peculiar constitutional features and institutional arrangements and their distinctive value systems; the implications of good governance for military effectiveness are of enormous theoretical and practical consequence.

This phenomenon at least demonstrates a correlation between good governance and military success—and makes possible the veracity of classical Chinese good governance's argument that the former is the root of the latter.[65] This correlation suggests that virtue ethics[66] needs more integrated attention in contemporary Western just war theory, and classical Chinese treatment of good governance might offer some ideas for how to go about it.

Rights, or the lack thereof

The second advantage of good governance contrasts with the rights-based approaches to just war theory that dominate contemporary Western perspectives. "Rights" are powerful precisely because they are strong statements of individual desert and offer rigorous protections for individuals. A Hohfeldian framework, for example, both separates and relates the myriad, complex components of rights, and one can point to precise duties that people have for a right to be upheld. But things get murkier beyond the formal jural components of rights, for example, in determining the bounds of *collective* rights and duties within societies (including under circumstances of war), and even more so with creating the *conditions* for exercising those rights.

In that respect, rights discourse may be trying to do too much by both philosophically establishing the existence and nature of the relevant rights *and* establishing the circumstances under which those rights can be meaningfully held.

The latter requires a complex of institutions, laws, circumstances, and values, which may include the delineation of other rights, but what a right is and the conditions for the possession of those rights are ultimately different things.

This distinction points to the value of thinking about "good governance" as a legitimate reason for *jus ad bellum*. There is no concept of "rights" in classical Chinese thought and classical Chinese and contemporary Western concepts do not easily map onto each other; but Western conceptions of rights are often overstated, and humanitarian intervention can be justified in other ways, perhaps with "good governance" as developed in the *Seven Military Classics*. [67]

Good governance takes a broad perspective of society: it complements the modern focus on human rights justifications for *jus ad bellum*, but focuses the sprawling concept of human rights in a way that can actually instantiate the implementation of those rights. It also highlights broader institutional questions that must be confronted in human rights violations, as decades of experience have shown us that humanitarian aid alone can not only be insufficient, but may also exacerbate the underlying problems that led to humanitarian crisis. Giving basic humanitarian aid effectively and in a way that does no further harm is difficult enough when just dealing with bad governance, and may be impossible when there is also war to contend with.[68]

"Good governance" is broader yet also more detailed than "human rights" because it encompasses a variety of institutions, practices, and values needed to *sustain* respect for human rights, so it must fold consideration of post-war governance and long-term outcomes into *jus ad bellum*. In doing so, it must account for the just war theory principle of probability of success in a practical way that human rights do not; "good governance" may be a more robust and useful concept to use for *jus ad bellum*, or at least a necessary addendum to it. On the other hand, "good governance" is more difficult to achieve and sustain than the possession of human rights within any given timeframe, which makes success less likely, so the threshold for meeting *jus ad bellum* standards is effectively higher for a good governance approach.

The pitfalls of comparative theorizing

As with all comparative political theorizing, we must tread carefully, as there are dangers in cherry-picking the lessons we find attractive, such as reading into a tradition something that is not there or trying to draw limited lessons that are unsuitable in piecemeal form.[69]

Complex virtues

To begin with, traditions are not always internally consistent. In the midst of the *Seven Military Classics'* repeated admonitions for virtuousness, for example, *Huang Shigong* advises rulers to "use those that have desires":

The *Army's Strategic Power* states: 'Employ the wise, courageous, greedy, and stupid. The wise take pleasure in establishing their achievements. The courageous love to put their will into effect. The greedy fervently pursue profits. The stupid have little regard for death. Employ them through their emotions, for this is the military's subtle exercise of authority.'[70]

Thus, military strategy seems to call for the ruler to intentionally exploit all types of people, including the decidedly dissolute. How should one reconcile the directive to cultivate people's vices for his own ends, however just, with the repeated mandate to guide and educate the people toward their own virtuousness? Perhaps this directive should only apply to the ruler's actions abroad—but *Huang Shigong* is also adamant that the ruler should target his virtues toward the enemy's people as well as his own. Perhaps a utilitarian interpretation or a "dirty hands" approach could go some way toward reconciling this, but these are in high tension with the dominant narrative about the inherent value of the sage's virtuous leadership and that *jus ad bellum* cannot be had without domestic righteousness.

Paternalistic politics

Classical Chinese military thought also resides in a paternalistic and hierarchical system, with all its accompanying dangers. Not only are ranks, honors, and riches to be properly apportioned by occupation, but these occupations reside in an inflexible society where farmers, artisans, and merchants must "dwell solely in [their respective] districts," to prevent "scheming" as well as "confusion" between districts and clans, says *Six Secret Teachings.* [71]

This reflects a paternalistic view of politics that equates the state with parents or elder siblings, and subjects with children or younger siblings, and assigns them those accompanying duties.[72] It is unclear to what extent classical Chinese virtues can be reconstructed for a modern context that largely rejects such political paternalism and hierarchy, and therefore how precisely to translate its features.

Tension with the contemporary international political structure

That righteous warfare in classical Chinese just war theory intends to unite all states under the rule of a single just system[73] is difficult to map onto the contemporary geopolitical reality of many distinct states whose sovereignties are enshrined in international law. The normative continuity between domestic and international that undergirds this good governance argument also makes it harder to accept that there may be many different legitimate centers of power.

While not impossible, an ideal "all under heaven" end goal should at least initially be shelved, as the immediate context of classical Chinese military strategy's good governance argument is the existence of disparate states of varying degrees of righteousness and therefore the constancy of conflict.

This leads us to the next questions about how exactly to apply the argument. For example, given the shortage of Sage Kings, does that argue for a more non-interventionist or pacifist stance? Does the broad definition of "war" tend toward non-military forms of humanitarian aid, rather than coercive intervention? Does the importance of a society's internal orderliness and the ruler's benevolence and righteousness, as both goods in themselves and as persuasive beacons to outsiders, suggest the need for more open borders and easier migration so that people who seek righteous societies may move to settle in them, rather than having a just ruler intervene where the disorder resides?

Implications of an expanded *jus ad bellum*

Finally, I highlight three additional hazards especially relevant for just war theory: (1) abuses of a "good governance" justification; (2) eroding civil/military separation in liberal democracies; and (3) implications for the moral equality of combatants.

The first concern is no less important for its obviousness: expanding justifications for coercive action is ripe for abuse, and history has shown that any opportunity of that sort will be taken. Even the most well-intentioned individuals and governments are corruptible, so it may be better on net to legally restrict intervention more than normative philosophy would permit.

At the same time, one cannot ignore serious human rights abuses across the world; but all intervention, however justified, comes at a cost, which leads to the second crucial problem. One strength of classical Chinese military ethics is that it treats leadership virtues comprehensively and integrates virtuous leadership with questions of governance, but such an approach might upend contemporary liberal democracies who expect military subordination to civilian rule.

In classical Chinese philosophy, ruler's virtues are inseparable from military virtues—they are one and the same—and this feature was a product of their time and political system. Contemporary liberal democracies, however, demand strict civil-military separation, and for good reason, historically. There should be some overlapping virtues between military and political leadership— e.g., service, self-sacrifice, patriotism—but not all of their virtues should coincide, and liberal democracies do not want military personnel in their official capacity to become too concerned with questions of politics or governance.[74] Contemporary Western liberal democracies manifest this separation in their constitutional arrangements and legal restrictions, and contemporary Western just war theory reflects this separation (perhaps unwittingly) by often treating moral questions in war (*jus in bello*) as its own realm, one that draws on but is not wholly governed by the ethical principles relevant to everyday life in peaceful society. As a result, one must be judicious about which virtues one pulls from classical Chinese military thought and how they are articulated under contemporary circumstances, as that has the potential to erode military-civilian separation in liberal democracies.

A third difficulty is that broadening *jus ad bellum* with good governance to justify humanitarian intervention—or any substantive questions of justice beyond self-defense—has knock-on effects for the doctrine of moral equality of combatants.

With modern warfare's "trial by combat" structure,[75] the use of soldiers as proxies of the state has led to a moral equality doctrine that gives combatants on all sides equal rights of self-defense and equal privileges to kill. This moral arrangement is reflected in various established wartime institutions, including medical immunity, medical neutrality, and non-penal POW detention (Neff 2010, 63–64). For example, all POWs must be extended Geneva Convention protections regardless of the justice of their cause because a soldier is considered an agent of his state and only kills as such: as a vessel or tool, his act of killing in war is not personal, criminal, or inherently punishable.

Moral equality of combatants is a critical part of international law's and contemporary Western just war theory's attempts to limit war's destructiveness by confining its scope to settling political disagreements (as opposed to establishing cosmic rightness), and it recognizes that warfighters operate under epistemic limitations that constrain their ability to ascertain the justice of their cause. As a result, even warfighters for an unjust cause are considered moral equals bound in "shared servitude" by their military service, rather than criminals (Walzer 2015, 36–37). While they are responsible for adhering to *jus in bello*, they are not considered directly responsible for determining *jus ad bellum*, which is the responsibility of political leaders,[76] and are permitted acceptable wartime killings in the name of an unjust cause.

Making *jus in bello* dependent on *jus ad bellum* in some way—whether with revisionism's individualist, criminal legal theory-inspired approach to evaluating just war[77] or by using a broad framework of good governance that integrates duties of domestic justice into international policy and war—would consider individual soldiers to be responsible for discerning *jus ad bellum* or righteousness and acting accordingly.

In the context of a classical Chinese-based good governance doctrine, would soldiers and subjects of an unrighteous state have a perfect or imperfect duty to submit to a righteous invader? Would they have a perfect or imperfect obligation to overthrow or abandon their unjust ruler if a righteous invader tries to rectify their state? Would a righteous ruler have a perfect or imperfect obligation to punish tyrants or to impose order on a chaotic state? Because the *Classics* largely treat the people as passive—only rulers and officials seem to have effective agency—and because this reconstructed good governance theory maintains meaningful normative distinctions between domestic and international realms instead of collapsing the two, these respective duties are unclear.

***

The unintended consequences of "good governance" may pose significant challenges: broadening contemporary international law's narrow self-defense justifications for war and injecting more substantive claims of justicial right into the hard-won geopolitical *modus vivendi* may open the door to greater pursuit of unlimited revolutionary aims on the basis of non-negotiable claims of justice, and it risks dragging the world back into the seemingly boundless destruction of the 19th and 20th centuries.

But insofar as we care about justice and sometimes must go to war to render it, a reconstructed classical Chinese military philosophy offers a particularly sophisticated model from which to draw. Its comprehensive conception of good governance offers insights on standards for action and intervention, and offers viable justicial content for contemporary *jus ad bellum* considerations beyond sovereign self-defense or human rights.

In doing so, it provides a challenging but promising alternative to relying on rights discourse, while simultaneously providing robust reinforcement for rights clams by addressing the circumstances under which those rights can be meaningfully held. Its promotion of political consistency across domestic and international realms should also be an important (even if not the only) consideration for liberal democracies at least.

The empirical connection between good governance and both economic and military achievement only augments the value of thinking about how classical Chinese military philosophy weaves virtue ethics and military ethics into broader questions of good governance and how and why good governance wins wars. Underneath the idealistic focus on virtues lies a nuanced and pragmatic theory of social and global justice.

Notes

1 Views are her own and do not represent the U.S. Naval War College, the U.S. Department of the Navy, or the U.S. government.
2 The UN Charter prohibits "the threat or use of force against the territorial integrity or political independence of any state" (Art. 2.4) and limits the acceptable responses to "threats of the peace, breaches of the peace, and acts of aggression" (Chap. VII).
3 For example, on the separability of domestic from international ethical principles, Confucianism applies to both war and peace situations (Chan 2014, 16), whereas *Daodejing* considers warfighting to be an exceptional situation (§57).
    The prominent and enduring school of realism contrasts with these approaches. Legalist theory, developed by State of Qin's prime minister Shang Yang 商鞅 and philosopher Han Fei 韓非 during the Warring States Period (475–221 BCE), rejected Confucianism in favor of "the autonomy of politics and its independence from morality," as politics should "maintain a viable political order rather than promoting a moral order" (Lo 2015, 251). By the Han dynasty (206 BCE–220 CE), legalism's realist strategic culture was widely practiced.
    Contemporary Chinese just war theory and views of international law remain effectively realist. For example, although Yan Xuetong 閻學通's representative *Ancient Chinese Thought, Modern Chinese Power* (2011) makes much of moral leadership and "humane authority" in international relations, critics consider his "moral realism" to be merely a façade for offensive realism (F. Zhang 2012; Hui

2012; Lo 2016). On historical and contemporary Chinese realist thought, see: Lo 2015; Johnston 1995; Scobell 2003; and Wang 2011.

4 The provenance of some of the works is disputed, and their origins range from late Spring and Autumn Period (5th c. BCE) through the Warring States Period (5th–3rd c. BCE), and possibly the Han Dynasty (2nd c. BCE–early 3rd c. CE) and the Song Dynasty (960–1127 CE). Unlike the six other "military classics," for example, *Six Secret Teachings* was written in contemplation of revolution, by the Zhou against the stronger Shang dynasty (Sawyer 1993, 23).

5 Study of Chinese philosophy often exhibits a Confucian bias, but Confucianism is just one of many broad, rich, and varied traditions in conversation and often at odds with each other, including within the *Seven Military Classics*. Although Confucius spoke sparingly about civility and martiality, he influenced others' military philosophizing, e.g., *Yanzi Chunqiu* (晏子春秋) on virtue's role in securing the state and Mencius 孟子 on the ruler's personal example for conquering the enemy. The moral order and ethical ideal espoused in both *Wuzi* (吳子) and *Sun Tzu's The Art of War* (孫子兵法) resemble that of Confucian philosopher Xunzi 荀子, as does the Legalist-influenced *Si Ma Fa* (司馬法) (Rand 2017, 101, 104, 124–128). See Johnston 1995, 32–108; Sawyer 1993; and Rand 2017, 124–128.

6 Alastair Iain Johnston classifies classical Chinese thought as a "*strategic-culture model*" (versus a realpolitik-dynastic model) that "[reflects] a Confucian-Mencian equation linking moral state government and external security... [such that] even as the empire mobilized resources, strategic culture would dictate policies that manifest the magnanimity of the ruler, his "awesomeness" (威 *wei*) and "virtue" (德 *de*)" (Johnston 1995, 57). See Scobell 2005 and Yuen 2014, 155–174 for other interpretations of Chinese strategic cultures.

7 As early as the Western Zhou period (circa 1100–771 BCE), thinkers wrestled with "the *wen/wu* problem," which addresses the proper uses of and relationship between civility (*wen* 文) and martiality (*wu* 武) in preserving cultural stability. The *Seven Military Classics* take up some aspects of this relationship between good governance and *jus ad bellum* (Rand 2017, 15–18, 124–128).

8 One continuity between classical and contemporary Chinese thought is their shared focus on right as legitimate cause for war, although the similarities largely end there. Following Mao, contemporary writers conceptualize only two kinds of war: just and unjust, i.e., revolutionary and counter-revolutionary (e.g., F. Zhang 1997). For them, *jus ad bellum* is determined entirely by communist revolutionary purpose, to fight "the oppression of a ruling class [and] foreign aggression, and [to] promote social progress" (Wu 1998).

The discontinuities are more striking. The classical literature focuses on military virtues beyond military acumen: courage, wisdom, benevolence, humanity, trustworthiness, loyalty, respect, and dignity. Officers must lead by example and share in their troops' hardships, and possess a sense of justice in order to "judge disputes" and "accept criticism" (*Six Secret Teachings* §19, Sawyer 1993, 62–63; *Sun Tzu's Art of War* III§2, Ames 1993, 225–226; *Huang Shigong* §1, Sawyer 1993, 295–297). (Cf. von Clausewitz on moral courage.) A general's character is considered so essential to military success that these texts largely ignore soldiers and their desired attributes.

In contrast, contemporary Chinese thought's treatment of military virtues is unfortunately thin and largely caricatures self-sacrifice and strength of will, e.g., "resolute and stubborn will to fight, heroic [and] indomitable spirit, and the combat style of not fearing sacrifice and not fearing difficulty, so as to overwhelm and defeat the enemy" (Lectures on the Science of Air Force Campaigns, ed. Dai Jinyu 1990). There is little discussion of why or how these virtues contribute to military success, as if victory will surely follow from their mere exercise, and the deafening silence on military leaders is telling. Their moral and political virtues are simply assumed, by virtue of their equally assumed alignment with communist ideology.

9 Even Mohists seem to leave ajar the door to justifiable punitive intervention (as opposed to military aggression), although their stance is more opaque (Loy 2015).

10 *Si Ma Fa* §3 (Sawyer 1993, 137).

11 *Si Ma Fa* §3 (Sawyer 1993, 137).

12 *Si Ma Fa* §1 (Sawyer 1993, 126).

13 *Si Ma Fa* §1, (Sawyer 1993, 128).

14 *Tang Taizong and Li Weigong* §§1, 3 (Sawyer 1993, 332, 350).

15 *Six Secret Teachings* §7 (Sawyer 1993, 47).

16 *Huang Shigong* §1 (Sawyer 1993, 293).

17 E.g., *Tang Taizong and Li Weigong* §3 (Sawyer 1993, 348), and *Sun Tzu's The Art of War* (Ames 1993, 85).

18 *Huang Shigong* §3: "The Sage King does not take any pleasure in using the army. He mobilizes it to execute the violently perverse and punish the rebellious" (Sawyer 1993, 305). *Six Secret Teachings* §12: "The Sage Kings termed weapons evil implements, but when they had no alternative, they employed them" (Sawyer 1993, 51). *Wei Liaozi* (尉繚子) §2 adds probability of success to the requirement of right intention: "The army cannot be mobilized out of personal anger. If victory can be foreseen, then the troops can be raised. If victory cannot be foreseen, then [the mobilization] should be stopped" (Sawyer 1993, 243).

19 *Huang Shigong* §3 (Sawyer 1993, 306). Daoist ambivalence toward violence goes even further, as "Taoists shun three generations [of a family] serving as generals. [Military teachings] should not be carelessly transmitted, yet should also not be not transmitted. Please pay careful attention to this matter" (*Taizong and Weigong* §3, Sawyer 1993, 360).

20 *Sun Tzu's The Art of War* §4 (Ames 1993, 115).

21 *Sun Tzu's The Art of War* §2 (Ames 1993, 107–108).

22 *Si Ma Fa* §1, Sawyer 1993, 126.

23 *Sun Tzu's The Art of War* I§3 (Ames 1993, 111–112). See also *Sun Tzu's The Art of War* III§2 (Ames 1993, 231). *Tang Taizong and Li Weigong* §3 cites Sun Tzu: "an army which can cause men to submit without fighting is the best; one that wins a hundred victories in a hundred battles is mediocre; and one that uses deep moats and high fortifications for its own defense is the lowest. If we use this as a standard for comparison, all three are fully present in Sun-tzu's writings" (Sawyer 1993, 360). See also: *Six Secret Teachings* §13 (Sawyer 1993, 53) and *Wei Liaozi* §2 (Sawyer 1993, 243, 260–261).

24 *Sun Tzu's The Art of War* II§5 (Ames 1993, 193).

25 Cf. von Clausewitz, Liddell Hart.

26 See also, e.g., *Six Secret Teachings* §§1, 8 (Sawyer 1993, 42, 47) and *Si Ma Fa* §3 (Sawyer 1993, 136).

27 Yuen 2014, 37–38, 70, 115–116.

28 *Huang Shigong* §1 (Sawyer 1993, 294).

29 *Shi shi qishu jiangyi* 施氏七書講義 [1222] (Johnston 1995, 134, 286).

30 For example, the highly influential Confucian view holds that morality is determined by an independent natural law, based in human nature or in Heaven. Its source is heteronomous—located outside the person—and is to be apprehended by the ruler, rather than created through reason or preferences.

31 *Huang Shigong* §1 (Sawyer 1993, 294).

32 *Six Secret Teachings* §2 (Sawyer 1993, 42).

33 *Huang Shigong* §3 (Sawyer 1993, 304).

34 *Six Secret Teachings* §6 (Sawyer 1993, 45).

35 *Huang Shigong* §3 (Sawyer 1993, 304).

36 *Six Secret Teachings* §13 (Sawyer 1993, 53).

37 *Sun Tzu's Art of War* III§2 (Ames 1993, 231).

38 The importance of righteous governance is common across multiple schools of Chinese philosophy, and various aspects of this have been distilled as idioms over time,

among them: "emphasize civility, deemphasize martiality; stress virtue and downplay physical strength" (重文輕武重德不重力 *zhong wen qing wu, zong de bu zhong li*), "if one has virtue, one cannot be matched [by an enemy]" (有德不可有敵 *you de bu ke you di*), and "display virtue and do not flaunt the military instrument" (觀德不耀 兵 *guan de bu yao bing*) (Johnston 1995, 63–64).

39 *Huang Shigong* §1 (Sawyer 1993, 292).

40 He continues, "When their bodies submit the beginning can be planned; when their minds submit the end can be preserved" (*Huang Shigong* §3, Sawyer 1993, 303).

41 *Si Ma Fa* §1 (Sawyer 1993, 127).

42 *Six Secret Teachings* §1 (Sawyer 1993, 42). See also *Wei Liaozi* §2 (Sawyer 1993, 243).

43 The idea that virtue is alluring even to "barbarians" gets reinforced in Chinese literature, none too subtly. For example, in the novel *Romance of the Three Kingdoms* (circa 14th c.): in 225 CE, a general captured then released the enemy tribal king seven times instead of executing him, even though the "barbarian" said he would continue to fight and then did so, to demonstrate his superior morality. After seven iterations, the tribal king whole-heartedly submitted, exclaiming, "Seven times a captive and seven times released. Surely there was never anything like that in the whole world. I know I am a barbarian and beyond the pale, but I am not entirely devoid of a sense of propriety and rectitude. Does he think I feel no shame? O Minister, you are the majesty of Heaven…I and my sons and grandsons are deeply affected by your all-pervading and life-giving mercy. Now how can we not yield? (Chap. 87–90) The fable demonstrates the limited purpose of wars—the general returned the tribal king's lands once the latter embraced righteousness, thus eschewing the material or strategic advantages of military victory—and the proper spirit with which to conduct war in order to evoke submission.

44 Referring to the Sage King's army, *Huang Shigong* §3 proclaims, "Now using the righteous to execute the unrighteous is like releasing the Yangtze and Yellow rivers to douse a torch, or pushing a person tottering at the edge of an abyss. Their success is inevitable!" (Sawyer 1993, 305–306).

45 This intentionally overlooks the responsibility of soldiers who fought an unjust war, willingly or otherwise.

46 *Six Secret Teachings* §40 (Sawyer 1993, 87). *Wei Liaozi* §8 (Sawyer 1993, 254) echoes prohibitions against attacking both innocent cities and men.

47 *Si Ma Fa* §1 (Sawyer 1993, 126–128).

48 *Six Secret Teachings* §40 (Sawyer 1993, 87).

49 Alastair Iain Johnston explains, "This use of benevolence and righteousness as a political tool in a broader offensive policy is a mechanism by which a 'guest' (i.e., invading) army can be turned into a 'host' in enemy territory" (Johnston 1995, 181).

50 *Wuzi* (吳子) §1 (Sawyer 1993, 208).

51 Other major Chinese works show similar concern with *jus post bellum*: for example, the historical *Tso-chuan* (左傳) recounts that after the Battle of Pi, the king of Ch'u quotes *Shih-ching* (詩經 Classic of Poetry, Book of Songs), refuses to build a battle monument on the bodies of the Chin dead, and questions his own virtue and sense of right and wrong (Kierman 1974, 46).

52 Post-war justice entails benevolence as well as righteous punishment, such as executing the guilty as appropriate (*Si Ma Fa* §1, Sawyer 1993, 128).

53 *Six Secret Teachings* §§2, 3, 7 (Sawyer 1993, 42–44, 47). *Huang Shigong* §1 (Sawyer 1993, 294). *Wuzi* §1 (Sawyer 1993, 207–208). *Wei Liaozi* §§10–11 (Sawyer 1993, 259–260).

54 *Wuzi* §1 (Sawyer 1993, 208).

55 Huang Zongxi's "mirror for princes" treatise, *Waiting for the Dawn: A Plan for a Prince* [1663], also tackled the tedium of rulership. An unorthodox Confucian, Huang rejected the paternal analogy of politics. He considered the sage king's position not a great prize but one of great responsibility: to provide economic well-being, customs and ceremonies, education, moral training, and military defense,

and to prevent a festering bureaucracy that "breeds indifference and irresponsibility." Good governance institutions include the roles, powers, selection of, and constraints on the prince (such as a strong prime minister and a cabinet), ministers, administrative departments, and legislature; tax and financial systems; the land system; education; the military; and especially proper laws and the rule of law. "Only if there is governance by law can there be governance by men...unlawful laws fetter men hand and foot, even a man capable of governing well," he cautioned (de Bary 1993, 19, 56, 80, 12–14, 20, 23, 26, 99).

56 *Huang Shigong* §1: "Thus it is said, 'Draw in their men of character and valor and the enemy's state will be impoverished.' These valiant men are the trunk of a state. The common people are its root. If you have the trunk and secure the root, the measures of government will be implemented without resentment" (Sawyer 1993, 294). §III: "When the ruler's munificence extends to the people, Worthy men will give their allegiance. When his munificence reaches the multitudinous insects, then Sages will ally with him. Whomever the Worthy give their allegiance to, his state will be strong. Whomever the Sages support, [under him] the six directions will be unified. One seeks the Worthy through Virtue, one attracts Sages with the Tao" (Sawyer 1993, 303).

57 *Huang Shigong* §1 (Sawyer 1993, 298).

58 *Huang Shigong* §3: "When the power of life and death lies with prominent, powerful families, the state's strategic power is exhausted. If [they] bow their heads in submission, then the state can long endure. When the power of life and death lies with the ruler, then the state can be secure" (Sawyer 1993, 306).

59 *Huang Shigong* §1 (Sawyer 1993, 298).

60 *Huang Shigong* §3: "Employing the discontented to govern the discontented is... 'contrary to Heaven.' Having the vengeful control the vengeful, an irreversible disaster will result. Govern the people by causing them to be peaceful. If one attains peace through purity, then the people will have their places, and the world will be tranquil" (Sawyer 1993, 305).

61 *Huang Shigong* §3: "Thus it is said, 'One who concentrates on broadening his territory will waste his energies; one who concentrates on broadening his Virtue will be strong'" (Sawyer 1993, 304).

62 Michael Walzer's secular domestic theories inform his just war theory, but their connection is looser.

63 Based on Correlates of War Project data (inter-state military conflicts with at least 1,000 battle casualties) for all wars 1816–1990, when democracies initiated wars during that period, they won 93% of the time; as targets of aggression, they still won 63% of wars, compared with dictatorships and oligarchies (Reiter and Stam 2002, 28–29).

64 E.g., Tilly 1975.

65 Correlation between democratic governance and greater military effectiveness means that democracies are comparatively more capable of securing the *underlying conditions* (namely, stability and security) for justice and right. Stability and security are necessary but insufficient conditions for justice and right, however, and it is a separate step to show that just and righteous governments are then necessarily right in violently and coercively imposing like governments on others (Chiu 2019, 224–225).

66 Virtue ethics usually emphasize motive or intent, which can be difficult to reconcile with dominant contemporary ethical theories that prioritize procedure and/or outcome.

67 Other Chinese schools of thought also offer possibilities, such as the language of "social justice" and fairness in Daoism or the permissibility (if not the right) of complaint and rebellion in response to bad rule in Confucianism.

68 See, for example, Bauer 1969; Deaton 2013. There are also the difficult questions of *when* exactly to intervene militarily.

69 See Schwartzman 2012 on comparative political theory methods to "conjecture" across cultures and provide reasons to adherents of other comprehensive doctrines.
70 *Huang Shigong* §§1, 2 (Sawyer 1993, 293, 300–301).
71 *Six Secret Teachings* §6 (Sawyer 1993, 46).
72 *Six Secret Teachings* §3 (Sawyer 1993, 44). *Si Ma Fa* §2 (Sawyer 1993, 129).
73 Some Western schools of thought also advocate a global regime, e.g., Stoic cosmopolitanism (e.g., Epictetus, Marcus Aurelius) or global federation (e.g., Kant, Beitz, Pogge).
74 In fact, this problem is anticipated in *Huang Shigong*, which warns of the dangers of a militarized society, politically governed by the military, as those states will be more prone to conflict.
75 Three features of modern warfare inadvertently generate a "trial by combat" structure that confers only effective right from war: (1) the political nature of war, (2) its limited justicial purposes, and (3) limited qualifications for legitimate participation. This arrangement leaves questions of justicial right problematically unresolved, especially at a time when wars are again becoming increasingly ideological (Chiu 2019, 193–233).
76 The Nuremberg Trials and some other criminal trials notwithstanding, political leaders are generally not held to account for their *jus ad bellum* decisions, which presents a glaring moral gap.
77 McMahan 2009; McMahan 1994; McMahan 2004a; McMahan 2004b.

Bibliography

Ames, Roger T. 1993 [5th c. BCE]. *Sun-Tzu: The Art of Warfare.* Translated by Roger T. Ames. New York: Ballantine Books.

Ames, Roger T. and David L.Hall. 2010 [5th–3rd c. BCE]. *Dao De Jing: A Philosophical Translation.* Translated by Roger T. Ames and David L. Hall. New York: Ballantine Books.

Augustine. 1887 [5th c. CE]. *Contra Faustum.* Translated by Richard Stothert. Buffalo, NY: Christian Literature Publishing, Co. www.newadvent.org/fathers/140601.htm.

Augustine. 1998 [5th c. CE]. *The City of God against the Pagans.* Translated by R. W. Dyson. New York: Cambridge University Press.

Aquinas, Thomas. 1920 [13th c. CE]. *The Summa Theologica.* 2nd ed. Translated by Fathers of the English Dominican Province. www.newadvent.org/summa.

Chan, Joseph. 2014. *Confucian Perfectionism: A Political Philosophy for Modern Times.* Princeton, NJ: Princeton University Press.

Chiu, Yvonne. 2011. "Liberal Lustration." *Journal of Political Philosophy* 19, no. 4: 440–464.

Chiu, Yvonne. 2019. *Conspiring with the Enemy: The Ethic of Cooperation in Warfare.* New York, NY: Columbia University Press.

Choi, Ajin. 2004. "Democratic Synergy and Victory in War, 1816–1992." *International Studies Quarterly* 48, no. 3: 663–682.

von Clausewitz, Carl. 1989 [1832]. *On War.* Edited and translated by Michael Howard and Peter Paret. Princeton, NJ: Princeton University Press.

De Bary, William. 1993. "Introduction to Huang Tsung-hsi." In *Waiting for the Dawn: A Plan for the Prince*, pp. 1–86. New York, NY: Columbia University Press.

Freedom House. 2018. *Freedom in the World 2018: Democracy in Crisis.*

Huang Tsung-hsi (Huang Zongxi 黃宗羲). 1993 [1663]. *Waiting for the Dawn: A Plan for the Prince – Huang Tsung-hsi's Ming-i-tai-fang lu.* Introduction and translation by Wm. Theodore de Bary. New York, NY: Columbia University Press.

Hui, Victoria Tin-bor. 2012. "Building Castles in the Sand: A Review of *Ancient Chinese Thought, Modern Chinese Power*." *Chinese Journal of International Politics* 5, no. 4: 425–449.

Johnston, Alastair Iain. 1995. *Cultural Realism: Strategic Culture and Grand Strategy in Chinese History*. Princeton, NJ: Princeton University Press.

Kierman, Jr., Frank A. 1974. "Phases and Modes of Combat in Early China." In *Chinese Ways of Warfare*, edited by Frank A. Kierman, Jr. and John K. Fairbank, pp. 27–66. Cambridge, MA: Harvard University Press.

Lectures on the Science of Air Force Campaigns, ed. Dai Jinyu. 1990. *Excerpt in Modern Chinese Military Thought on Strategy & Ethics: A selection of authoritative writings 1990–2000*, compendium of translations, edited by Phillip A. Karber (2013): 9–10. McClean, VA: The Potomac Foundation.

Lo, Ping-cheung. 2012. "The *Art of War* Corpus and Chinese Just War Ethics Past and Present." *Journal of Religious Ethics* 40, no. 3: 404–446.

Lo, Ping-cheung. 2015. "Legalism and Offensive Realism in the Chinese Court Debate on Defending National Security 81 BCE." In *Chinese Just War Ethics: Origin, Development, and Dissent*, edited by Ping-cheung Lo and Sumner B. Twiss, pp. 249–280. London: Routledge.

Lo, Ping-cheung. 2016. "Can China Rise Peacefully? A Critical Appraisal of Yan Xuetong's 'Moral Realism'." unpublished manuscript. Presented at Workshop on Military Ethics, Florida State University, November 2016.Luo, Guanzhong. 2002. [1522, written circa 14th c. CE] *Romance of the Three Kingdoms* (三國志平話). Translated by Charles H. Brewitt-Taylor. Singapore: Tuttle Publishing.

Loy, Hui-chieh. 2015. "Mohist arguments on war." In *Chinese Just War Ethics: Origin, development, and dissent*, edited by Ping-cheung Lo and Sumner B. Twiss, pp. 226–248. London: Routledge.

McMahan, Jeff. 1994. "Self-Defense and the Problem of the Innocent Attacker." *Ethics* 104, no. 2: 252–290.

McMahan, Jeff. 2004a. "The Ethics of Killing in War." *Ethics* 114, no. 4: 693–733.

McMahan, Jeff. 2004b. "Innocence, Self-Defense, and Killing in War." *The Journal of Political Philosophy* 2, no. 3: 193–221.

McMahan, Jeff. 2009. *Killing in War* (Uehiro Series in Practical Ethics). Oxford: Oxford University Press.

Neff, Stephen C. 2010. "Prisoners of War in International Law: The Nineteenth Century." In *Prisoners in War*, edited by Sibylle Scheipers, pp. 57–73. Oxford: Oxford University Press.

Rand, Christopher C. 2017. *Military Thought in Early China*. Albany, NY: SUNY Press.

Reiter, Dan and Stam, Allan C. 1997. "*The Soldier's Decision to Surrender: Prisoners of War and World Politics*." Presented at the Annual Meeting of the American Political Science Association, Washington, DC, August 1997.

Reiter, Dan and Stam, Allan C. 2002. *Democracies at War*. Princeton, NJ: Princeton University Press.

Sawyer, Ralph D. 1993. *The Seven Military Classics of Ancient China* (武經七書). Translated by Ralph D. Sawyer. New York: Basic Books.

Schwartzman, Micah. 2012. "The Ethics of Reasoning from Conjecture." *Journal of Moral Philosophy* 9, no. 4: 521–544.

Scobell, Andrew. 2003. *China's Use of Military Force: Beyond the Great Wall and the Long March*. Cambridge: Cambridge University Press.

Scobell, Andrew. 2005. "Strategic Culture and China: IR Theory versus the Fortune Cookie?" *Strategic Insights* IV, no. 10 (October).

World Bank. 2018. Data: GDP per capita (current US$).

Tilly, Charles. 1975. *The Formation of National States in Western Europe*. Princeton, NJ: Princeton University Press.

Twiss, Sumner B. and Chan, Jonathan. 2012. "Classical Confucianism, Punitive Expeditions, and Humanitarian Intervention." *Journal of Military Ethics* 11, no. 2: 81–96.

Twiss, Sumner B. and Chan, Jonathan. 2012. "The Classical Confucian Position on the Legitimate Use of Military Force." *Journal of Religious Ethics* 40, no. 3: 447–472.

Walzer, Michael. 2015 [1977]. *Just and Unjust Wars*. 5th ed. New York: Basic Books.

Wang, Yuan-kang. 2011. *Harmony and War: Confucian Culture and Chinese Power Politics*. New York: Columbia University Press.

Wu, Chunqiu. 1998. "Grand Strategy: A Chinese View." Excerpt in *Modern Chinese Military Thought on Strategy & Ethics: A selection of authoritative writings 1990–2000*, compendium of translations, edited by Phillip A. Karber (2013), pp. 15–18. McClean, VA: The Potomac Foundation.

Yan, Xuetong. 2011. *Ancient Chinese Thought, Modern Chinese Power*. Edited by Daniel Bell and Sun Zhe, translated by Edmund Ryden. Princeton, NJ: Princeton University Press.

Yuen, Derek M.C. 2014. *Deciphering Sun Tzu: How to Read The Art of War*. New York, NY: Oxford University Press.

Zhang, Feng. 2012. "The Tsinghua Approach and the Inception of Chinese Theories of International Relations." *Chinese Journal of International Politics* 5, no. 1: 73–102.

Zhang, Yunxun. 1997. "China's Ancient and Modern Military Dialectical Thought." Excerpt in *Modern Chinese Military Thought on Strategy & Ethics: A selection of authoritative writings 1990–2000*, compendium of translations, edited by Phillip A. Karber (2013), p. 12. McClean, VA: The Potomac Foundation.

7

NORMATIVITY OF WAR AND PEACE

Thoughts from the *Han Feizi*

Erik Lang Harris

The ideas of the political thinkers of early China who have often been labeled "Legalists" vary significantly in many ways but they share an overarching similarity in their approach to politics: as the great Sinologist A. C. Graham (1989, 269) noted, they were the first "to start not from how society ought to be but how it is." What exactly this characterization implies in terms of political and military activities depends in part on how these thinkers understood the world around them and how they understood human nature.[1] Recently, though, there have been scholars who have argued that these Legalist thinkers advocated a statecraft that:

> "was more than realpolitik; it was machtpolitik; that is, a policy of relentless pursuit and use of power in domestic as well as interstate relations. The ultimate goal of statecraft was the creation of a world order under its dominion."
>
> *(Lo 2015a, 6)*

I believe that this analysis hits on something quite important that is found in texts such as the *Book of Lord Shang* and the *Han Feizi*. However, I also wonder whether it may be missing some elements, as I shall try to articulate below. Before beginning this analysis, however, I should note that I am in broad agreement with those who argue that questions of war and violence are not, on the Legalist perspective, questions that can be answered by appeal to morally based criteria. And, while there are numerous texts that one could draw from in pursuing these ideas, I wish to limit myself here to the text of the *Han Feizi*, attributed to the last great philosopher writing before the unification of the empire under the Qin dynasty in 221 BCE, Han Fei.

DOI: 10.4324/9781003336372-9

Throughout the text of the *Han Feizi*, we see opposition to traditional (and often, though not solely, Confucian) perspectives on a wide range of state activities, both internally and externally. This antipathy towards traditional morality-based criteria for justifying state actions extends to the questions of if, when, and how to wage war. In what we may today think of as reasoning akin to Western conceptions of *political realism* or realpolitik, Han Fei clearly argues that considerations of morality have no place, either in questions of war and peace or, indeed, in broader questions of politics more generally (Harris 2013).

This is not, though, to say that Han Fei is uninterested in broader questions of when war and peace are justified. However, such justifications are based not on appeals to moral concerns. Rather, the fundamental question to be answered is how the strength, security, and flourishing of the state would be affected by warfare. By viewing questions of war and peace from the perspective of state consequentialism, Han Fei believes that we can ascertain the answers to questions not only about if and when to engage in warfare, but also to questions of what particular actions can be justified within the context of war.

In what follows, I will explore these issues and focus on the following points: 1) political normativity for Han Fei rests on a hypothetical imperative; 2) Han Fei's fundamental hypothetical imperative leaves no room for moral considerations; and 3) Han Fei's fundamental concern for the state does not *necessarily* lead to an advocacy of "machtpolitik" – the relentless ever expansionary pursuit of more and more power and control over the world.

Political Order – The Fundamental Hypothetical Imperative

Han Fei is famous for his advocacy of clearly promulgated laws and an explicit set of rewards and punishments attached to the adherence or violation of these laws respectively. However, unlike what some have argued, the goal of this system was not to provide the ruler with a totalitarian power enabling him to do as he wished.[2] Rather, Han Fei's fundamental goal in advocating such a system was to ensure the strength, stability, and security of the *state*. Han Fei is a state consequentialist – he is first and foremost concerned with ensuring a well-ordered state.

Unlike the Confucians, the Daoists, or even the Mohists, Han Fei does not tie the importance of a well-ordered (or harmonious) state to some underlying moral goal. It is not that such a state is valued because of the opportunities for human flourishing that it offers. Rather, the well-ordered state stands as a goal in and of itself. It may well be in the interests of those within the state to have such a state, and it may well be the case that such a state allows them to flourish. Indeed, Han Fei does believe that a well-ordered, strong, stable, and flourishing state is one in which the ruler and the people all stand the greatest chance of surviving and living materially flourishing lives, but arguing for this claim is not how Han Fei tries to justify the state. Rather, the normativity that Han Fei offers in support of his political vision is of a non-moral, hypothetical variety. The specific claims that he makes about how the state should be organized and run are all predicated on the assumption that the ruler (the one who is making decisions about how to organize the state) wants his state to be strong, stable, secure, and long-lasting. The normativity here, the force of the "should" or "ought," is on a par with the normativity inherent in the claim, "If you wish to become an underwater welder, you should obtain both a commercial diving certification and an AWS underwater welding certification." The point of such a claim is certainly not that *you* morally ought to pursue these two certifications! Indeed, the vast majority of people in the world live very good lives without ever having done anything that even remotely compares to this. Rather the point here is that if you wish to achieve something X, in this case start a career in the exciting field of underwater welding, then engaging in action Y, which here involves obtaining the appropriate training and certification, is the appropriate action to take. At the fundamental level, the normativity is a normativity that points to success criteria.

So, on Han Fei's account, *if* you wish to create a strong, stable, and prosperous state, then the appropriate (and, indeed, the *essential*) thing to do is to maximize order both within this state and beyond the state. And it may very well be the case that doing this maximizes the collective welfare of those within the state, even though that is not the direct aim. If, on the other hand, you have no desire to have a strong, stable, and prosperous state, then there is no reason to follow any of Han Fei's dictates. For Han Fei, political order is not intrinsically valuable – nothing has intrinsic value on his account. Rather, it is simply a fact that most rulers wish to rule over strong, stable, and prosperous states, and most people wish to dwell in such states, and political order is the tool that allows them to do so most reliably.

Morality in Political and Military Endeavors

Now, developing a political theory that predicates itself on the hypothetical imperative to ensure a strong, stable, and secure state does not in and of itself necessarily set itself in opposition to morality. Indeed, another early Chinese thinker, Xunzi (who on some accounts was Han Fei's teacher), argued that it was simply impossible to establish a strong stable, secure state that was not undergirded with a moral foundation. On his account, political order was predicated on an underlying moral order.[3]

Han Fei, however, does not believe that the political problem is, at heart, a moral problem. That is, on his account those things that lead to a strong, stable, and flourishing state are non-identical to those things that are morally good or right. And insofar as the former is what he advocates pursuing, the latter will, at times at least, need to be discarded. Indeed, not only is it not efficacious to think in terms of morality when deciding how the state should act, it can be downright disastrous.

We can begin to see the dangers that Han Fei saw in according with morality in his discussion of warfare – and, more precisely, in his discussion of one particular famous battle in 638 BCE between the troops of the state of Chu and those of the state of Song, led by Duke Xiang:

> Duke Xiang of Song was fighting the people of Chu at the Zhuogu River. The people of Song had already formed ranks while the people of Chu had not yet crossed the river. The Commander of the Right, Gou Qiang, hastened forward and remonstrated: "The people of Chu are numerous while those of Song are few. Please order the attack while the people of Chu are only halfway across the river and have not yet formed ranks. [In this way] they will certainly be defeated."
>
> Duke Xiang replied, "I have heard a Gentleman say, 'Do not wound someone more than once, do not capture those whose hair is turning grey, do not push people into danger, do not compel people into impossible situations, do not sound the attack drums [when the enemy] has not yet formed ranks.' Now, attacking Chu when they have not yet forded the river is to do harm to moral standards. I request that you wait until the people of Chu have finished fording the river and formed their battle array and only after that sound the attack drums to send the officers forward."
>
> The Commander of the Right said, "My lord does not care about the people of Song or whether their stomachs are cut open and their hearts scooped out. You are only concerned with moral standards and that is all."
>
> The Duke responded, "If you do not return to your ranks, I shall implement the law [i.e., punish you]."
>
> The Commander of the Right returned to the ranks. Only after the people of Chu had formed their ranks and composed their battle array did the duke thereupon drum his troops forward. The people of Song suffered a great defeat and the duke was injured in his thigh, dying three days later. This is the disaster arising from a personal admiration for benevolence and moral standards.
>
> *(Lau and Chen 2000, 32/87/20–25)*[4]

Han Fei's point in recounting this story is quite clear. Whenever moral concerns would lead one to act in ways that are inimical to the strength, security, and stability of the state, they must be discarded. This has vast implications for warfare in particular. On Han Fei's account, moral concerns should never be raised in the context of warfare – whether it be questions of whether or not to engage in warfare or questions of how to act during war. The question of whether to go to war is answered by reference to whether doing so is the most effective, and most efficient, means of ensuring the strength, security, and flourishing of the state. And the particulars about how the war is to be fought are to be answered by reference to those military actions that will most efficiently and effectively allow the state to ensure its strength, security, and flourishing. Indeed, as P. C. Lo has noted, in the *Han Feizi*:

> There was no concern for proper conduct in war other than that which guaranteed military success. There were no scruples aiming to restrict violence and human suffering on both sides, and no mercy was to be shown to enemy soldiers, whether captured or surrendered. There was no self-imposed restraint or limit on the use of lethal violence to serve national interests.
>
> *(Lo 2015a, 6)*

I agree with everything Lo says here. However, we should also keep in mind that this position does not countenance senseless violence, indiscriminate killing, or suffering for suffering's sake. War is not to be fought by reference to what is virtuous; neither, however, is it to be fought with an aim toward viciousness. Indeed, Han Fei would argue that engaging in vicious warfare will be just as detrimental as engaging in virtuous warfare, and for much the same reason. In both cases, one allows emotions to cloud one's judgment – to act based on feelings as opposed to acting based on what will most effectively ensure the survival and thriving of the state itself. And in this light, we do well to keep in mind that Han Fei saw emulating the actions of both the sage kings Yao and Shun and the vicious tyrants Jie and Zhou in the same vein – they are dangerous to the state because their actions to not take the state itself as the reason for action.

I say this not in an effort to paint Han Fei in a better light. If executing thousands of soldiers who have already surrendered or massacring hundreds of innocent non-combatant women and children truly will allow for the greatest security at the least cost, then Han Fei would advocate it. However, if it is calculated that winning over the "hearts and minds" of innocent villagers or drawing surrendered troops to one's own side would better serve the interests of the state, then that is what should be done – and killing them would be wrong. (Not morally wrong, of course, but wrong based on Han Fei's fundamental hypothetical imperative). Indeed, Han Fei would argue that part of why the horrible tyrants Jie and Zhou were ultimately defeated and destroyed was precisely because they were vicious in ways that were politically unjustifiable and that the fear, anger, and horror that their actions wrought led to the downfall of their reigns.

Political Order – is Machtpolitik the Logical Consequence?

While the preceding exposition provides us with a foundation for exploring and evaluating whether one should engage in war or not, given the relevant circumstances, and a set of criteria for determining what particular actions one should engage in if at war, it still leaves open the question of what international relations would look like. That is, what sort of international scenarios are those under which a particular state can best ensure that it not only survive but that it thrive? Can we envision an international system that had numerous states interacting without conflict? Or does Han Fei think that the only way that a state can be secure is to ensure that all potential competitors are overtaken and exterminated

That is, does Han Fei's understanding of realpolitik lead him to advocate machtpolitik – the idea that in any situation in which it is possible to increase one's power, one must do so? P. C. Lo and others have noted that there are aspects of texts such as the *Han Feizi* that do incline us to the belief that they advocated *machtpolitik* (Lo 2015b). However, through the rest of this paper, I wish to suggest that there are aspects of the *Han Feizi* that may allow for, and even encourage, alternate interpretations. In doing so, though I will not be completely denying the *machtpolitik* interpretation. I believe that there are scenarios under which a Han Feizian ruler would engage in *machtpolitik* – circumstances under which it is the appropriate action to take. My claim will merely be that this is not the only viable scenario, that there are circumstances in which such a relentless pursuit of superiority would be inappropriate. As Han Fei is always keen to point out, one should not always follow a particular method or action. Rather, one should always be asking the question of what action is most effective at ensuring a particular result, given the current (and anticipated) circumstances. And what actions are most effective at ensuring a particular result will vary as circumstances vary. As Han Fei says:

> Those who do not understand [the conditions of] order inevitably say, 'Do not vary the old; do not change what has become the norm.' As for changing or not changing, the [true] sage does not listen to that question – he sets straight and orders the state and that is all. As for whether or not to vary the old and whether or not to change the norm, this depends upon whether what is old and what is the norm are acceptable or not.
>
> *(Lau and Chen 2000, 18/31/17–18)*

While this passage is, in its context, an attack on the Confucian appeal to follow the actions of the sages of yore and an exhortation against a particular way of governing, it is equally a piece of positive advice and a recognition that no method, and this would include the method of *machtpolitik*, is eternally applicable. So, while there may very well be scenarios in which the ruler should engage in *machtpolitik*, as Lo and others have noted, this will not always be the case – and, I think, could not be the case, given the premises that Han Fei himself accepts.

In demonstrating this claim, I want to begin by looking at a chapter that is often neglected. The *Han Feizi* is quite a lengthy text and consists of a diverse range of material – from concise essays in support of particular political theories to screeds against its opponents to collections of historical stories and their implications to the earliest extant commentaries on the *Daodejing of Laozi*. And, as we shall see, in its commentary on the *Laozi*, the *Han Feizi* offers an intriguing discussion of warfare that seems to push against the machtpolitik interpretation. There are, it must be admitted, a range of debates about the authenticity of the chapter within the *Han Feizi* that comments on the *Laozi*, and there are those who think that it does not originate from the hand of Han Fei. I do not have the space to delve into questions of authenticity here. However, as my interests are more philosophical than Sinological in nature, part of my investigation is to examine the extent to which this chapter should be read as expressing a political vision that is compatible with the rest of the *Han Feizi*.

With this in mind, let us turn to the *Han Feizi*'s discussion of the *Laozi*. The received version of *Laozi* 46 itself reads as follows:

> When the world has the Way, fleet-footed horses are used to haul dung.
> When the world is without the Way, war horses are raised in the suburbs.
> The greatest misfortune is not to know contentment.
> The worst calamity is the desire to acquire.
> And so those who know the contentment of contentment are always content.
>
> *(Ivanhoe 2003, 49)*

While this chapter has been interpreted in a variety of ways, it is commonly seen as expressing an opposition to at least certain versions of warfare – an opposition that would itself seem to be in tension with ideas of machtpolitik. It is for this reason that it may be valuable to examine Han Fei's commentary to this chapter. And, although it is long, that commentary is worth reproducing at length here:

> The ruler who has the Way harbors no complaints or enmity against his neighbors and rivals outside the state and inside the state he shows potency and kindness to his people. Outside the state, harboring no complaints or enmity against his neighbors and rivals, he greets the feudal lords with the proper rituals and appropriate standards. Inside the state showing potency and kindness to this people, he works at the root of things in his ordering of human affairs. If one greets the feudal lords with the proper ritual and appropriate standards, then the occasions for military drafts will be few. If in ordering human affairs one works at the root, then licentiousness and extravagance will cease. When horses are put to their greatest use, it is for providing armor and weapons outside the state, and supplying licentiousness and extravagance inside the state. Now the ruler who has the Way rarely uses armor and weapons outside the state, and he forbids licentiousness and extravagance inside the state. When the superiors do not employ horses in fighting wars and making expeditions

northwards, and the people do not use horses to transport licentious goods over great distances, then their accumulated strength is devoted to the fields. When they accumulate their strength to work at the fields, then they must, moreover fertilize and irrigate. And so the text says, "When the world has the Way, then fleet-footed horses are used to haul dung."

(Hutton unpublished)[5]

There is a lot that we can draw from this commentary, and unfortunately we cannot delve into everything. However, for our purposes, Han Fei appears to be arguing that it is possible for one state to engage with other states in a non-contentious fashion, and that doing so is preferable (in many instances, at least) to military conflict. Undergirding this assumption seems to be a particular conception of human nature, namely that while people have desires, interests, and preferences, these are not infinite. That is, it is possible for the people to have all that is necessary for them to live good lives without continually trying to wrest more from those around them. Further, it is possible for states to have enough to be strong, stable, and secure without endlessly attacking and consuming other states. And, finally, it is possible for all the relevant states to restrain their offensive actions when such a situation is achieved – equilibrium is possible.

Is this always possible? Well, no – as always, it depends on the particular circumstances. However, Han Fei is clear elsewhere in the text that his conception of human nature is not as pessimistic as that of the Confucian Xunzi, who believed that human beings originally have self-interested desires that are unending and necessarily lead us into conflict with one another. Yes, we have interest sets that are primarily focused on the self, and yes, they can lead us into conflict, but whether they do so or not depends on external circumstances, and in particular whether, as a matter of fact, it is possible to satisfy our interests without forcibly taking from others.

Han Fei continues his commentary on the *Laozi* 46 by arguing that in situations where this course of action is possible a consequentialist calculation should lead us to realize that the costs of continued conflict is much greater than any benefits that we might hope to achieve. As he says:

If the ruler of the people does not have the Way, then inside the state he will tyrannize and abuse his people, and outside the state he will invade and bully his neighboring states. If inside the state he oppresses and abuses his people, the people's livelihood will be depleted. If outside the state he invades and bullies, then troops must often be raised. If the people's livelihood is depleted, then the birth of livestock will decrease. If troops must often be raised, then one will run out of soldiers. If the birth of livestock decreases, then war-horses will be lacking. If one runs out of soldiers, then the army will be in grave danger. If war-horses are lacking, then female horses will be sent out. If the army is in grave danger, then close ministers will be drafted. Horses are most greatly used by the army. 'The suburbs' denotes its closeness. Now in this case what the army is provided with are female horses and close ministers. And so the text says, "When the world is without the Way, war horses are raised in the suburbs."

(Hutton unpublished)

Here again we see the term the Way, or the *dao* (道), invoked. One of the basic meanings of the term is a physical path, but in early Chinese texts its connotation broadened to refer to a way of doing something, and, in many cases, the right way to do them. Thinkers like Laozi and Zhuangzi used it even more broadly, referring to the underlying nature of the universe. In this sense, we can think of the *dao* as the way things are. Indeed, it would not be incorrect to construe much of the debate in early Chinese philosophy as a debate over the right Way to live.[6] For Han Fei, the Way is conceived of as the source of everything that exists and as providing the standard by which everything can be evaluated. It gives rise to the patterns and regulations of the natural world and is simultaneously 'the way things naturally are' and the reason why things are this way. As such, the Way provides a model or set of rules that need to be followed as a precondition for survival or success.[7]

How, though is this relevant to our discussion? Well, someone who understands the Way will only engage in warfare when it is necessary – and will be able to evaluate the conditions that would make it necessary. One who does not grasp the Way will engage in warfare when it is neither necessary nor beneficial. And this is quite problematic. After all, as all within the Warring States period of China were quite aware, war has numerous downsides. If a ruler is constantly battling neighbors, this increases the resources he needs to access. And these resources come from his people – conscription to obtain the troops that he needs and higher taxes to obtain the material resources necessary to wage war among others. Further, when farmers are conscripted and pulled away from their farming duties, food stores necessarily decrease, taking a toll both on the people of the state but also on the overall strength and security of the state. States that continually pursue more power and resources may well find themselves diminished rather than strengthened.

Han Fei's commentary continues:

If people have [excessive] desires, then their plans and calculations will be chaotic. If their plans and calculations become chaotic, then their desires will become even greater. When their desires become even greater, then their perverse inclinations will conquer them. If their perverse inclinations conquer them, then in their affairs they will take short cuts. If they take short cuts then misfortunes and difficulties will arise. Viewing it from this perspective, misfortunes and difficulties are born from perverse inclinations, and perverse inclinations tempt people by approving [excessive]

> desires.... And so the text says, "of crimes, none is greater than approving of [excessive] desires."
>
> *(Hutton unpublished)*

This again implies that a ruler who understands the Way will understand that certain desires must be restrained if one is to successfully rule a state. And again, this is a point that is raised throughout numerous chapters of the *Han Feizi*. The text is rife with examples of rulers who acted on the basis of excessive desires that led to their downfall either from those within their state or from those without. And so offensive wars aimed at gaining more territory, power, or in pursuit of other excessive desires can be quite dangerous.

Now, to be sure, this analysis does not serve to provide a prohibition on expansionary offensive warfare. However, it clearly indicates that there are many bad reasons to go to war and, when read with the earlier commentary, clearly recognizes the possibility of and advocates for the attainment of peaceful co-existence among states. The alternative will often (but of course, not always) lead to disaster. As the commentary on *Laozi* 46 concludes:

> And so if the desire for benefit is strong, then one will feel distress. If one feels distress, then anxiety will arise. Anxiety arises and then one's wisdom declines. When one's wisdom declines, then one will lose the ability to measure and weigh. When one loses the ability the ability to measure and weigh, then one will at rashly. If one acts rashly, then misfortunes will arrive. Misfortune arrives and then anxiety will disturb one's innards If the anxiety disturbs one's innards, then one will feel pain. And if misfortunes press near from outside, then one will feel bitterness. When bitterness and pain are strewn among one's stomach and intestines, then the injury to the person will be excruciatingly painful. If it is extremely painful, then one will retire and fault oneself. This retiring and faulting oneself is born from desiring benefit. And so the text says, "There is no fault more painful than the desire to acquire."
>
> *(Hutton unpublished)*

Implications for our Understanding of the *Han Feizi's* Views on War and Peace

What sort of overall message are we to draw from Han Fei's commentary on the *Laozi*? Well, perhaps the most important message that comes across is that peace among rival states is possible and, in many cases, preferable. That is to say, conflict and war is not inevitable in the international arena even if there are a number of roughly equally powerful states bordering one another. Furthermore, in many instances just such a situation is often going to be desirable. That is to say, simply because expansionary warfare is possible, and even if the ruler has good reason to think that it would be successful, this is not in and of itself a reason to engage in such warfare. Rather, warfare has significant consequentialist costs and in many instances the costs will be far greater than the benefits that may be gained, particularly in the case of expansionary warfare. This calculation, therefore, seems to indicate that Han Fei is not advocating for an unchanging policy of "relentless pursuit and use of power in domestic as well as interstate relations." Nor does he have a static ultimate goal of "the creation of a world order under its dominion."

On the other hand, as we have seen, Han Fei never rules out the possibility that, under certain circumstances, the appropriate thing to do is to continually amass and increase power and perhaps even create a world order under its dominion. Rather, the key to answering this question requires us to return to the fundamental principles undergirding his political system. We need to ask whether, given the particular circumstances under which we are operating, ensuring the strength, security, and prosperity of the state requires the state to go to war, or whether it is better served by peace. And there is no overall, general, good for all situations in answering this question. It can only be answered by an in-depth understanding of both internal and external circumstances and an evaluation of how each of the relevant aspects of these circumstances potentially impact the strength, security, and survival of the state.

Indeed, it may very well have been the case that Han Fei would argue that, during his time period, given the particular circumstances of the Warring States period, the only way for a state to ensure that it survive and thrive was to defeat all comers. But this is not a theoretical precondition upon which he wishes to rest his theory. He can have no theoretical pre-commitments to *machtpolitik*, even if there are scenarios in which its pursuit is appropriate and desirous. In evaluating whether or not to pursue ever increasing power, the question is never whether doing so is the virtuous thing or whether it is in pursuit of a moral good. Further, the fact that hundreds or thousands of soldiers and innocent non-combatants may perish does not in itself provide a moral reason to abstain from war. The negative consequences to state stability and security of such losses must certainly be weighed, but not the moral value of those lives.

Han Fei's overall political theory is much more nuanced than is possible to express here. And a complete understanding of the particular criteria that must be met in order to justify military action must await a fuller analysis. Therefore, I do not take this chapter to provide readers with a set of tools that will allow them to reliably determine whether, in any particular situation, Han Fei would have advocated war or peace. What I do hope to have provided, however, is a better sense of those things that Han Fei saw as relevant to such a decision. And further, I hope to have demonstrated that Han Fei neither saw interstate conflict as necessary nor necessarily desirable.

Notes

1 For a brief overview on 'Legalism' see Harris 2014. For an opposing view, see Goldin 2011.
2 Such a view of Han Fei as an apologist for totalitarianism is seen not only in the work of Western scholars, but also in the writings of Chinese and Japanese scholars. See, for example, Matsudaira 1911, 10; Creel 1953, 113; Rubin 1970; Xiao 1982, 247; Mou 1983, 173; Fu 1996; Song 2010, chapter 2.
3 For a more detailed account of Xunzi's political philosophy, see Harris 2016. For an alternate interpretation, see El Amine 2015.
4 Translations of the *Han Feizi* are my own.
5 Translations of the *Han Feizi*'s commentary on the *Laozi* throughout are adapted from the manuscript translation of (Hutton unpublished).
6 Hence the title of A.C. Graham's famous introduction to Chinese philosophy, *Disputers of the Tao*. See Graham 1989.
7 For a more comprehensive account of the Han Fei's understanding of the Way and the role that it plays in his philosophy, see Harris (forthcoming).

Bibliography

Creel, Herrlee Glessner. 1953. *Chinese Thought from Confucius to Mao Tsê-Tung*. New York: Mentor.

El Amine, Loubna. 2015. *Classical Confucian Political Thought: A New Interpretation*. Princeton, NJ: Princeton University Press.

Fu, Zhengyuan. 1996. *China's Legalists: The Earliest Totalitarians and their Art of Ruling*. Armonk, NY: M.E. Sharpe.

Goldin, Paul R. 2011. "Persistent Misconceptions about Chinese 'Legalism'." *Journal of Chinese Philosophy* 38, no. 1: 88–104.

Graham, A. C. 1989. *Disputers of the Tao: Philosophical Argument in Ancient China*. La Salle, IL: Open Court.

Harris, Eirik Lang. 2013. "Han Fei on the Problem of Morality." In *Dao Companion to the Philosophy of Han Fei*, edited by Paul R. Goldin, 107–131. New York: Springer.

Harris, Eirik Lang. 2014. "Legalism: Introducing a Concept and Analyzing Aspects of Han Fei's Political Philosophy." *Philosophy Compass* 9, no. 3: 155–164.

Harris, Eirik Lang. 2016. "Xunzi's Political Philosophy. " In *The Dao Companion to Xunzi*, edited by Eric L. Hutton, pp. 95–138. New York: Springer.

Harris, Eirik Lang. Forthcoming. "The Dao of Han Fei." In *The Oxford Handbook of Chinese Philosophy*, edited by Justin Tiwald. Oxford: Oxford University Press.

Hutton, Eric L., trans. (unpublished) "Explaining the Laozi." In *Explaining and Illustrating the Laozi: The Earliest Commentaries on the Daodejing*, edited by Eirik Lang Harris and Eric L. Hutton.

Ivanhoe, Philip J., trans. 2003. *The Daodejing of Laozi*. Indianapolis, IN: Hackett Publishing.

Lau, D. C. and Ching Chen, Fong eds. 2000. *A Concordance to the Hanfeizi*. Hong Kong: The Commercial Press. [Published in Chinese: 劉殿爵、陳方正《韓非子逐字索引》香港: 商務印書館].

Lo, Ping-cheung. 2015a. "Varieties of Statecraft and Warfare Ethics in Early China: An Overview." In *Chinese Just War Ethics: Origin, Development, and Dissent*, edited by Ping-cheung Lo and Sumner B. Twiss, 3–24. New York: Routledge.

Lo, Ping-cheung. 2015b. "Legalism and Offensive Realism in the Chinese Court Debate on Defending National Security 81 BCE." In *Chinese Just War Ethics:*

Origin, Development, and Dissent, edited by Ping-cheung Lo and Sumner B. Twiss, pp. 249–280. New York: Routledge.

Matsudaira, Yasukuni. 1911. *Han Feizi, with Explications in Modern Published Japanese Kanpishi kokujikai.* Tokyo: Waseda Daigaku Shuppanbu. [In Japanese 松平康國《韓非子國字解》東京: 早稻田大學出版部。].

Mou, Zongsan. 1983. *Nineteen Lectures on Chinese Philosophy: A Brief Introduction to Chinese Philosophy and its Implications.* Taibei: Taiwan Xuesheng Shuju. [Published in Chinese: 牟宗三《中國哲學十九講: 中國哲學之簡述及其所涵蘊之問題》台北: 台灣學生書局。].

Rubin, Vitaliĭ Aronovich. 1970. *Ideologiia i kul'tura drevnego Kitaia (Chetyre silueta) [Ideology and Culture of Ancient China (Four Sketches)].* Moscow: Nauka.

Song, Hongbing. 2010. *A Reevaluation of Han Feizi's Political Thought.* Beijing: China Renmin University Press. [Published in Chinese: 宋洪兵《韓非子政治思想再研究》北京: 中國人民大學出版社。].

Xiao, Gongquan. 1982. *A History of Chinese Political Thought.* Taibei: Linking Publishing. [Published in Chinese: 蕭公權《中國政治思想史》台北: 聯經出版公司].

8

WAR AND PEACE ACCORDING TO HUANG-LAO PHILOSOPHY

Based on the *Huangdi sijing*

Ellen Y. Zhang

In this chapter, I intend to explicate the political thought of Huang-Lao in conjunction with its views on war and peace in the Pre-Han and Han periods of ancient China. Compared with other early Chinese philosophical schools and their influences, Huang-Lao was relatively unknown in the English-speaking world before an archaeological discovery in southern China in1973. Its relation to Daoism (both philosophical Daoism and religious Daoism) and other schools of thought in ancient China is still under heated debate. My analyses will be focused on the *Huangdi sijing*《黃帝四經》(*Four Books the Yellow Emperor*), one of the most representative texts of the Huang-Lao tradition, supplemented by certain other texts related to Huang-Lao. The discussion consists of four parts: (1) The definition of Huang-Lao; (2) Huang-Lao's political philosophy; (3) The conception of war in *Huangdi sijing*; and (4) A comparative perspective in light of the contemporary discourse on just war. The chapter aims to offer a unique Huang-Lao's views on warfare in its own terms while at the same time bringing it into a conversation with the current debate on war and peace.

I. Defining Huang-Lao

In the *Shiji* 史記 (*Records of the Grand Historian*) where historian Sima Qian 司馬遷 (145–90 BCE) of the Han dynasty classifies the intellectual trends in China during the Warring State period and the early Han, a school of thought is identified as "Huang-Lao 黃老." However this term is never systematically explained, nor does it gain a widespread consensus on its precise definition.[1] Nevertheless, Huang-Lao is understood by many scholars as a branch of Daoism that employs Laozi's doctrines combined with Legalistic elements to address socio-political issues. The term "Huang-Lao" is a reference to ideas

DOI: 10.4324/9781003336372-10

deriving from the legendary Huangdi 黃帝 (the Yellow Emperor) and Laozi 老子, although the thought of Huang-Lao is not confined to the teachings of Laozi's Daoism.[2] Some Chinese Scholars such as Xiong Tieji 熊鐵基and Wu Zegang 吳則光call Huang-Lao "Neo-Daoism 新道家" (Huang 2000, 27), arguing that the "School of Daoism" (*daojia* 道家) mentioned by the Han historian Sima Tan 司馬談 (d. 110 BCE), the father of Sima Qian and the "School of Syncretism" (*zajia* 雜家) mentioned in the *Han shu* 《漢書》 (*History of the Former Han*) both refer to Huang Lao.

Apart from Laozi, some other key philosophical thinkers generally associated with the Huang-Lao thought are not viewed as Daoists, such as Tian Pain 田駢 (370–291 BCE), Shen Dao慎到 (395–315 BCE), Shen Buhai 申不害 (420–337 BCE), and Han Fei 韓非 (295–233 BCE). All of them are closely associated with the Legalist tradition.[3] What is intriguing is that these thinkers integrate a technique of institutions of centralized government with the Daoist notion of sagely government that lends to a form of "instrumental Daoism."[4] Based upon Laozi's concepts of *dao* (way), *de* (potency or efficacy)[5] and *wuwei* (non-action), Huang-Lao explicates the effectiveness of governmental administration, political leadership, and warfare.

One of the major problems in studying Huang-Lao is to identify a definitive source text for the tradition. Usually, scholars examine the tradition from the texts of other schools since most texts that are taken as Huang-Lao are deeply syncretic, drawing together selected ideas from many different schools in an attempt to present them in a harmonious way.[6] Among these schools, Laozi's Daoism is clearly foremost and the School of Legalism (*fajia* 法家), the School of Military (*Bingjia* 兵家), and the School of the Yin-Yang (*Yinyang jia* 陰陽家) also contribute a very significant portion of these ideas as well.[7] In this sense, Huang-Lao is not simply an amalgam of Laozi and Huangdi but a complicated fusion of eclectic thought (Csikszentmihalyi 1994, 7–14).[8]

In 1973, archaeologists working near the city of Changsha in southern China uncovered a tomb known as Mawangdui馬王堆that held the bones and the goods of the ruler of one of the early Han kingdoms that had been located in that area. Among the grave goods were found a set of silk scrolls with texts written on them, known as the *Boshu* 帛書 (*Silk Manuscripts*). One of them is the long-lost *Huangdi sijing* 《黃帝四經》 (*Four Books of the Yellow Emperor*, hereafter abbreviated as *HDSJ*).[9] The name of this lost set of texts is recorded in the bibliographic monograph of the *Han shu* and the collection is viewed as representative works of Huang-Lao thought. Other texts with definitive Huang-Lao orientation include the *Wenzi*《文子》,[10] the *Guanzi* 《管子》 (*Book of Master Guan*),[11] the *Heguanzi* 《鶡冠子》 (*The Master Pheasant-Cap*),[12] the *Huananzi* 《淮南子》 (*Book of the Prince of Huainan*),[13] the *Lüshi Chunqiu* 《呂氏春秋》 (*Master Lü's Spring and Autumn Annals*)[14] and the military *Huang Shigong Sanlüe*《黃石公三略》 (*Three Strategies of Huang Shigong*).[15]

The archeological discovery of Mawangdui has initiated a renewed scholarly interest in the study of the long-standing puzzle concerning what has been known as "the Huang-Lao branch" of Daoism in past decades. The *HDSJ* has also been examined as a body of writing to represent Huang-Lao political viewpoint.[16]Since Huang-Lao thought dominated the intellectual life of late Warring States and early Han China, its influence was pervasive across a broad spectrum of intellectual and cultural endeavors. Therefore, we cannot ignore Huang-Lao thought if we want to have a better picture of debates on war and peace in ancient China.

II. Major Themes in Huang-Lao's Political Philosophy

As a complex political philosophy, Huang-Lao draws its thought from a broader politico-philosophical spectrum, looking for solutions to incessant wars, political chaos, and strengthen the order of the feudal state. Sima Tan (190–110 BCE), the first historian classifying different philosophical schools during Pre-Han and Han periods of ancient China, defines Daoism as a school which values non-action, tranquility, softness, spontaneity, and emptiness (Chen 2013, 16–20; Fischer 2015, 8–9). Sima particularly mentions that Daoism postulates the idea of change and the action in response to the natural transformation of things. In fact those values are not only argued in the *Laozi* 《老子》 (aka. the *Daodejing* 《道德經》; hereafter abbreviated as *DDJ*), but also contended in the Huang-Lao thought. No wonder some scholars insist that Daoism, in the eyes of Sima Tan, *de facto* means Huang-Lao rather than Lao-Zhuang as conventionally perceived.[17] At the same time, Sima Tan recognizes the syncretic nature of this stream of thought, pointing out, "In the practice of their [political] techniques, they accord with the great sequence of the *Yingyang*, select the good of the *Ru* and *Mo*, and adopt the essentials of the *Ming* and *Fa*."[18] I will first highlight several concepts that influence the way Huang-Lao approaches to relevant issues before discussing Huang-Lao's view on war.

1 The Conception of the Cosmic Dao

Like Laozi's Daoism, the Dao is the key concept in Huang-Lao. It not only produces everything but also represents a cosmic order, which determines the social structure, political order, and human life. Upon this cosmic Dao, Huang-Lao develops its triad of *Dao-Li-Fa* 道-理-法 (Dao-Principle-Law). For Huang-Lao, both *fa* and *li* serve as administrative method. In the *HDSJ*, one reads:

> The Dao gives birth to laws. Laws are the measuring-line of gain and loss and what illuminates straight and crooked. Who holds to the Dao gives birth to laws and dares not contravene them; once laws are set up, he dares not discard them. After one can stretch plumb spontaneously, one can know the world and not be of two minds.
>
> *(HDSJ 1.1.1)*[19]

...It is the principle of heaven and earth that the four seasons have their regulations... As for the principles of human affairs, they depend on whether one complies with the way of heaven or go against it.

(HDSJ 1.8.1 & 2)

As the citations above indicate, Huang-Lao speaks of a law-based model of governing on the basis of the onto-cosmic Dao.[20] Randall P. Peerenboom sees Huang-Lao's interpretation of the Dao as "foundational naturalism", meaning naturalism based upon a cosmic natural order that includes both the *rendao* 人道 (the way of humans) and *tiandao* 天道 (the way of heaven). To be more specific, the Dao represents the onto-cosmic level, the Principle represents the natural level, and the Law represents the political level (Peerenboom 1993, 27–31).[21] The same argument is found in the *HDSJ* in which human social order is viewed as being based upon and in harmony with the cosmic order. It is also in consistence with Sima Tan's description of Huang-Lao Daoism which maintains that one's actions should harmonize with the Dao that is formless yet powerful.

For Huang-Lao, the *Dao* is the cosmic order while the *De* 德 (Potency) denotes a particular manifestation or power of the cosmic order. John S. Major puts it well when he points out that for Huang-Lao thinkers, the Dao is the "highest and most primary expression of universal potentiality, order, and potency", and "is expressed in cosmic order, which embraces both the world of nature and the human world" (Major 1983, 12). Although the triad of heaven-earth-human was employed by most other philosophical schools, especially Confucianism, Huang-Lao is more explicit in using an ontological argument to argue for the social order and the necessity of having laws to maintain the order.

2 The *Wuwei* Method of Ruling

The art of governance (*zhiguo zhishu* 治國之術) is one of the major concerns of Huang-Lao's political philosophy which, according to *HDSJ*, can be explained in three principles: (1) the principle of the cosmic Dao, (2) the principle of *De* or Potency, and (3) the principle of human society. It argues that the ruler should be "sincere" inside and "correct' outside. In the section of "Correct Kingship of *HDSJ*," it states:

In the first year of governing, the king should follow the customs of his people. In the second year, the king should utilize his people's potency (*de*). In the third year, the king enables his people to gain the profit. In the fourth year, the king implements laws and decrees. In the fifth year, the king should make his people proper in behavior. In the sixth year, he makes his people fearful and respectful. In the seventh year, he should be able to organize his people for military expeditions (*zheng* 正/征).

(HDSJ 1.3.1.)

The statement clearly shows a combination of Daoism and Legalism with the latter being more conspicuous. At noted above, Huang-Lao represents a syncretic thought of Daoism and Legalism, and this syncretism lies in the use of the concept *wuwei* 無為 (non-action) as a model for the art of ruling. *Wuwei* (non-action) is one of the key concepts of Daoism, linking directly with its political philosophy. When Sima Tan defines Daoism, he makes it clear that Daoists advocate *wuwei*, but they also claim that by the practice of *wuwei* nothing is left undone. This *wuwei* method is called "quiescence and non-action" (*qingjing wuwei* 清靜無為) by Feng Youlan, a well-known contemporary Chinese philosopher, referring to the political practice of the early Han Emperors who "were following the tenets of Laozi" (Feng 1983, 175).

As a technique of governance, *wuwei* indicates that the ruler should act without coercion and contrivance but rather follow the spontaneous order of things. To be more specific, the ruler should leave more space for his people to be self-ordering, self-managing, and self-transforming. In terms of Daoist thought, Laozi is critical about a strong and manipulative bureaucracy that extends its control over every aspect of the people's lives. For example, in the chapter 17 of the *DDJ*, there is a passage speaking of different models of rulership in which we read:

> With the most excellent rulers, people only know they have them,
> The next best are the rulers whom people love and praise,
> Next are the rulers whom people fear,
> And the worst are the rulers whom people despise
>
> (DDJ 17)[22]

The passage reflects the perennial debate on good government in ancient China. For Laozi, a minimal government is the highest form. A humane government is the next highest. A law-centered government is next to the worst one. Nevertheless, Laozi's Daoism still assumes the need for a hierarchical political structure instead of advocating for a pure form of political anarchism as observed by Roger Ames and David Hall (Ames and Hall 2003, 102). In the same vein, the argument of the ruler's non-intervention is also made in Huang-Lao thought. Thus, we read:

> If a ruler really does not actively attend to affairs, the myriad things (including people) will find their respective places. They will, on their own accord, behave according to their designated station in life, and they will not contend with one another.
>
> (HDSJ 4.1.7)

The passage reminds us a statement made by Shen Buhai, a Legalist philosopher and politician whose thought influences Han Fan's Legalist thought significantly. Yet as a socio-political form of thought, Shen's notion of *wuwei* diverges from Laozi's Daoism in that functions more as strategic means to achieve specific political goal:

> The ruler is like a mirror, which reflects the light, doing nothing, yet because of its mere presence, beauty and ugliness present themselves. ... The ruler's method is that of complete acquiescence... He does not act, yet as the result of his non-action, the world brings itself to a state of complete order.
>
> *(Creel 1970, 64)*

Japanese scholar Sokichi Tsuda 津田左右吉 has also observed that the Huang-Lao thought incorporated into the Legalist thought can be traced back to the thought of early Legalists such as Shen Buhai and Shen Dao (Goldin 2012, 274–275; Hsiao 1979, 385–386). For example, Shen, argues in terms of *wuwei*:

> The ruler is like a mirror, which reflects the light, doing nothing, yet because of its mere presence, beauty and ugliness present themselves. ... The ruler's method is that of complete acquiescence... He does not act, yet as the result of his non-action, the world brings itself to a state of complete order,
>
> *(Creel 1970, 64)*[23]

Hsiao-Po Wang and Leo S. Chang make the argument about the connection of a proto-Legalist thought of the *HDSJ* to the Daoist thought reflected in the *DLJ*. They also point out a striking similarity between the content of the *HDSJ* and the Legalist thought.[24] In the *Hanfeizi* 《韓非子》 there are five chapters that either deal with the *DDJ* directly or indirectly, including "An Expository Commentary on the Teachings of Laozi" ("Jie Lao" 〈解老〉) and "An Illustration of the Teachings of Laozi "("Yu Lao" 〈喻老〉).[25] Nevertheless, Legalists (as well as Huang-Lao to a certain extent) sees the *wuwei* method as a form of "political technique" or statecraft (*shu* 術). The politicization of *wuwei* is also applied to Huang-Lao's argument of strategies and tactics in warfare. It should be noted also that Han Fei dismisses Laozi's "philosophy of peace and quietude" (*qingtan zhi xue* 恬淡之学)" as "useless" and criticizes Laozi's "vague and illusive language" (*huangwu zhi yan* 恍惚之言) as "lawless." It is quite clear that Laozi's Daoism does not advocate strict laws and regulations, and that Laozi's idea of peace is too "passive" for Han Fei and Legalism.

III Correlativity of *Wen* and *Wu*

There are a series of antithetical concepts in Huang-Lao's political thought, such as order and chaos (*zhi-luan* 治亂), survival and destruction (*cun-wang* 存亡), strong and weak (*qiang-ruo* 強弱), (rest and movement (*jing-dong* 靜動),

and compliance and rebellion (*ni-shun* 逆順). Among these concepts, there is an antithetical pair that appears quite often, namely, *wen* 文 and *wu* 武. These two words are normally understood as "cultured-ness" and "martiality" respectively, and the interdependence of which is intrinsically associated with the idea of social stability and political order in ancient China.[26] Meanwhile, Huang-Lao uses pair to denote a naturalistic circle of life and death:

> To begin in *wen* [life] and to end in *wu* [death] is the Dao of heaven and earth. That the four seasons have regularities is the principle of heaven and earth. ...That three seasons are for coming to harvest and one season for dying is the Dao of heaven and earth.
>
> *(HDSJ 1.8.1)*

Then Huang-Lao links this life-qua-*wen* vis-à-vis death-qua-*wu* to human actions, contending that "moving and resting in accord with heaven and earth is called *wen*" (*HDSJ* 1.5.4).[27] Peerenboom calls this form of naturalism "correspondence naturalism (Peerenboom 1993, 32). Roger T. Ames and David L. Hall call it "processual and relational cosmology" (Ames and Hall 2003, 167).

At the same time, *wen* and *wu* are linked to Huang-Lao's correspondence epistemology. For example, the observation of the movement of *wen* and *wu* in the *HDSJ* is elucidated in light of penetrating insight (*guan* 觀) and dynamic balancing (*cheng* 稱), and both of which are crucial capacities for good leadership and governance.[28] While *wen* indicates the civil and refined side of governance, *wu* indicates the military and authoritative side of governance.

The section "Four Principles" (*Sidu* 四度) of the *HDSJ* speaks of four basic principles to rule the state: culturedness (*wen*) and *wu* (matiality), *jing* 靜 (quietude) and *zheng* 正 (rectitude)."

> Quietude results in peace and stability; Rectitude results in good governing; culturedness results in insightfulness; Martiality results in strength. Peace and stability lead to the basis; Good governance wins the support of people. ...By thoroughly knowing these four principles, all-under-heaven can be well-ordered and the state can be peaceful.
>
> *(HDSJ 1 5.4)*

It follows with a statement that "culturedness results in insightfulness and martiality results in strength (*wen ze ming; wu ze qiang* 文則明；武則強). Obviously, this statement is different from that in the *DDJ* where Laozi claims that one should drop "sharp and military instruments" and "civilized and cultured skillfulness" (*DDJ* 57), i.e., one should drop both military coercion and moral edification, since according to Laozi, both *wen* and *wu* can lead to war and violence (Zhang 2015, 493–497). In contrast, Huang-Lao contends for both soft and hard powers as correlative strategies for politicians or statesmen-thinkers of the time. Thus, it claims, "If both *wen* and *wu* function well, then all-under-heaven will obey the ruler" (*HDSJ*, 1.3.5).

Clearly, we see a syncretic understanding of the *wen/wu* tradition in Huang-Lao which, according to Christopher C. Rand, "attempted to refocus the *wen/wu* debate on the need for balance and reciprocity between martiality and civility" (Rand 2017, 26–27). Huang-Lao thus takes civil administration as the foundation to prepare military expedients if necessary.[29] It follows that domestic peace relies on sound governance, and a powerful army (for the sake of self-defense and punitive expeditions) relies on a strong economy.[30] The legalist orientation is quite obvious here, particularly in its support of the Legalist idea of "wealthy nation and strong military" (*fubing qiangguo* 富國強兵). For Huang-Lao, wealthy nation and strong military are the foundation of peace (*an* 安) and social order (*zhi* 治). Chaos or disorder (*luan* 亂) which includes rebellion and war is a sign of failure to properly govern. Therefore, it states in the *HDSJ*, "When the foundation of the state is attacked and its achievements are destroyed, a chaotic situation increases and the state is lost" (*HDSJ* 1.5.5). In order to "rectification of disorder" the ruler needs to pay attention to three elements: the foundation of governance, the responsibility of the ruler, and the support of common people:

> An act of rebellion results in the loss of the foundation. Disorder results in the loss of responsibility. Rebelliousness results in the loss of heaven. Brutality results in the loss of people.
>
> *(HDSJ 1.5.2)*

James D. Sellman has an insightful observation when he argues that the emphasis on interdependency of military and government affairs is *de facto* a Legalist argument (Sellman 2002, 57). Laozi's Daoism is relatively vague on this issue except the saying that "One governs a kingdom by normal rules (*zheng* 正) and fights a war by exceptional moves (*qi* 奇)" [*DDJ* 57], which suggests that different methods should be adopted when one is engaging in civil and military affairs. This is in contrast to the Confucian thought that ethical principles apply to both peace and war situations. As Joseph Chan puts it, "Martial arts and the use of force are supposed to the very antithesis of the Confucian ideal of benevolence and harmony, yet they are necessary in the non-ideal world and can, and should, be practiced with a view to the ideal" (Chan 2014, 16).

IV. Huang-Lao's Position on War

1 Warfare

Warfare was one of the most urgent issues in ancient China, and many wars were supported by the need of territorial expansion among competing states. In Laozi's Daoism, we find a strong sentiment against the violence of wars, as it says:

> Weapons are ominous instruments (*buxiang zhi qi* 不祥之器),
> Not the instruments of the cultured and refined
> They are used only when there is no other alternative.
>
> (DDJ *31*)

The phrase "ominous instruments" is adopted by later military texts after the *DDJ* and is often referred to as "instruments of evil" (*xiongqi* 凶器; Ryden 1998, 55). In the *HDSJ*, war is linked to three inauspicious signs" (*sanxiong* 三兇): (1) to be fond of ominous instruments, (2) to act in a manner against virtue, and (3) to indulge in desires (*HDSJ* 1.7.11; 138).[31] The third point, being indulging in desires, is worthy of attention. Huang-Lao is sided with Laozi, maintaining that the excessive desires of the ruler is one of the major reasons for conflict and violence. Therefore, we read in the *HDSJ*:

> He who is greedy for profit from the world will suffer calamity from it. He who is greedy for the profit from one state will receive that state's misfortune.
>
> (HDSJ 1.7.11)

> He never put his people in a situation of poverty. He never incited wars. He never initiated riots. …He never resorted in intrigues and conspiracies. He never resolved a knotty problem presumptuously without consulting others. He never intended to plunder the territory of others. He never intended to seize houses belonging to others by force. He carefully pacifies his people in order to comply with submitting oneself to heaven and earth.
>
> (HDSJ 2.14.3)

Following to Laozi's Daoism, Huang-Lao connects violent behaviors directly the proliferation of desires (*yu* 欲) of the ruler and too much strife and competition (*zheng* 爭) in society. Therefore, Huang-Lao contends:

> Void without form, its axis all dark; it is what the world of things is born from. In their birth-nature there is that which harms, called desire, called not knowing what is enough…
>
> (HDSJ 1.1.2)

War or warfare (*zhanzheng* 戰爭) in modern Chinese means "contentiousness or competition through fighting or force of arms." As it is well known, one of the persistent arguments in Laozi's *DDJ* is that problems of various kinds arise when rulers are too contentious and too aggressive. We find a similar argument in the *HDSJ* when it says, "It is true that those who initiate contentions suffer from misfortune" (*HDSJ* 2.3.4). At the same time, Huang-Lao employs the Daoist language of softness and femininity to speak of a good ruler:

He modeled himself after feminine conduct so that the things produced were soft. ... upright and impartial, he was fond of bestowing benefits and did not contend with others.

(HDSJ 2.14.3)

We can see that Huang-Lao distinctly adopts the Daoist concepts of the *yin* or feminine and the *yang* or masculine. In a Daoist manner, it promotes the "strategy" of femininity characterized by softness, humility, and accommodation. Huang-Lao also advocates the "strategy" of femininity in terms of military non-aggressiveness:

Masculine conduct is characteristically arrogant and proud while feminine conduct is yielding, modest, respectful and frugal. ... As regards disasters and suffering, [generally speaking], those who act first will suffer misfortune whereas those who act last will secure good fortune.

(HDSJ 2.7.2–3)[32]

Whoever is fond of applying masculine conduct should be called: "one who is harmful to life." ... If he fights, he will not overcome his enemy.

(HDSJ 2.7.4)[33]

Nevertheless, Huang-Lao's submission to Laozi's teachings takes a turn when it speaks of the advantage of being militarily strong:

When a state is strong, it has the ability to issue orders to other states. When a state is weak, it will have to obey orders issued by other states. When there is a balance of power among the states, they will compete according to certain regulations.

(HDSJ 3.1.32)

Here Huang-Lao acknowledges the importance of military power in interstate affairs. It also takes practical position on war, claiming that while it not good to be engaged in contention, "those who do not engaged in contention [when they have to] have no achievement" (*HDSJ* 2.3.4). Accordingly, rather than taking a pacifist position on war, Huang-Lao admits that some wars are necessary: It states, "A sage-ruler persists in neither disarmament nor warfare. Only in a case of necessity will he undertake military actions" (*HDSJ* 3.1.13), and "He never engaged in military action without waiting for the deterioration of those who had assumed rebellious attitudes (*HDSJ* 2.14.3). Like most other philosophical thinkers at the time, Huang-Lao speaks of the engagement in war in terms of an act of "being compelled to" or having "no other alternative" (*budeyi* 不得已). That is, War is a last resort to maintain or restore peace and order.

The same expression is also employed in the *DDJ,* according to which, "last resort" refers to a situation when a state is attacked by others and self-defense is the only way out.

According to Huang-Lao, therefore, warfare or the martial (*wu*) should be utilized when the civil cannot do its job. That is, the king can use *wen* or the civil as a major means of governing and supplement *wen* with *wu* when it deems necessary. After the calamity of war and violence, the king should use *wen* again for consolation. In Confucianism, there is a saying that if remote people (e.g., the people of the conquered state) refuse to submit, it is better to use "cultivated virtues" (*wende* 文德) to attract them.[34] There is no such statement in the *HDSJ,* but a similar idea is implied.

It is also interesting to see that Huang-Lao appeals to its triad of heaven-earth-man for the justification of war, upon which, three elements are listed:

> If warfare does not take it form from heaven, warfare cannot be initiated. If it does not take earth as its model, warfare cannot be managed. If its performance and law do not rely on man, warfare cannot be brought to a successful conclusion.
>
> *(HDSJ 2.8.1; Yates 2005, 16)*[35]

As a matter of fact, using the triad of heaven-earth-man to make political plans, including war preparation, is a common practice not only in Huang-Lao writings, but also in other syncretic texts. For instance, in the *Guanzi* it says, "To climax and to return, to flourish and to wane is the Dao of heaven and earth as well as the principle of human affairs" (Chang and Feng 1998, 34). The triad suggests a normative claim that if one fails to comply with the Dao, there is punishment:

> As for the principle of human affairs, it depends on whether one complies with the Dao of heaven or rebels against it. If one's achievement transgresses the Dao of heaven, there is punishment by death.
>
> *(HDSJ 1.8.2)*[36]

Meanwhile, the idea of the Dao of heaven is also employed in Huang-Lao thought to explicate a strategic logic and military rationality, as expressed in the notion of timing or the timeliness of heaven (*tianshi* 天時):

> According to the timeliness of heaven, initiating a punitive expedition against states that are suffering destruction ordained by heaven is called *wu*. If the military sword of control is followed by *wen*, then success will be achieved. He who utilized two measures of *wen* and one measure of *wu* will become a king.
>
> *(HDSJ 1.5.3)*

> If one rests and acts in accordance with right timing, heaven and earth will be with one.
>
> (HDSJ *2.6.5*)

The idea of the "timeliness of heaven" is a crucial term in Laozi's Daoism, denoting that one's action should be timely in being fittingly attuned with the world.[37] In the *Sunzi bingfa* 《孫子兵法》 (*Master Sun's The Art of Warfare*), we also see sections that discuss the utilization of the timeliness of heaven and "the natural advantage of the earth" (*dili* 地利).[38] According to the *Sunzi bingfa*, important technique includes observing the enemy's army for changes in strategic circumstances and preparation for quick troop movement. Similar to the *The Art of War*, the idea of "strategic potentials" or "strategic advantages" (*shi* 勢) in Huang-Lao refers to a hidden structure and pattern, or power within certain situation or condition which can be employed to achieve a particular outcome. Accordingly, strategic potentials emphasize what is dynamic and appropriative timing (Ivanhoe 2011, 30).

Apart from the timeliness of heaven and the natural advantage of the earth, Huang-Lao speaks of the "unification of the people" (*renhe* 人和), that is, winning the minds of the people and mobilizing the public support for war. It emphasizes the idea that *wen* as a domain of civil officialdom is more important than *wu* as a means of military operation since *wen* is the way to win the people's support before using the army. As a matter of fact, the idea of the unification of the people is not something only discussed by Huang-Lao. Xunzi 荀子 (third century BCE), a Confucian who influenced the thought of Legalism, also considers the unification of the people more important than the timeliness of heaven and the natural advantage of the earth. In a passage of the *Xunzi* (*The Book of Master Xun*) when Xunzi has a conversation with the Lord of Linwu, one reads:

> …It was a general principle that fundamental requirements to be met before using the army in attacks and campaigns was the unification of the people… Hence to be good at winning the support of the people is also to be expert in the use of the army. Thus, the essential principle of warfare consists in nothing more than being good at gaining the support of the people.
>
> (Xunzi *15:1*) (*Knoblock 1990, 219*)

It seems that Xunzi grants a higher authority to *wen*, or civilian leadership whereas Huang-Lao gives the same attention to timeliness of heaven and winning the support of the people.

2 Right to war: punitive interventions

Huang-Lao in general tends to take a realist position on warfare in that it rarely engages in an explicit ethical discourse. For example, the concept of *de* is understood as "potency" or "power" rather than "virtue" as one sees in the Confucian tradition. Nevertheless, we can still find passages in the *HLSJ* where normative claims linked to the idea of "just war" are made, at least from an instrumentalist perspective. For Huang-Lao, the justification of going to war can be established in two situations: (1) self-defense and (2) punitive interventions. In Laozi's Daoism, it endorses (1) but not (2).[39]

Instead of strenuous opposition to military aggression, Huang-Lao sees military power as a necessary means of self-defence and keeping the peace. Meanwhile, the idea of war as means to other ends is recognized in Huang-Lao's acceptance of punitive expeditions not only permissible but also necessary in some cases. In other words, a punitive intervention as an offensive campaign against a state can in principle be justified as a response to its aggression against other states or its own people. In addition, Huang-Lao seems to more concerned with the right cause (expulsion of tyrants, for instance) rather than specific right authorities. The Confucian notion of the "son of heaven" (i.e., sage-king) as sovereign authority is not mentioned in the *HDSJ*. [40]

According to the ancient Chinese literature, the origin of recorded punitive expeditions starts with Huangdi, the Yellow Emperor, who was the inventor of weapons and the first one to use weapons, yet with a noble cause. As Mark E. Lewis points out, in ancient China "war was justifiable primarily as the highest form of punishment through which the ruler could suppress large-scale deviance or criminality and thereby bring peace and harmony to the people and to the world" (Lewis 2006, 185). For example, the *HDSJ* identifies "punishment and attack" (*zhujin* 誅禁) as effective means to "stop war" or "halt the violation." The notion of using war and violence to stop war and violence is well expressed in the image of Huangdi during the tribal wars of antiquity, where we find one of the earliest *jus ad bellum* arguments in ancient China:

> Thereupon, Huangdi brought out his great axe, bestirred his armed soldiers, raised himself the drum and baton, went to meet Chi You, and captured him. The Emperor authored a covenant, which said: "He who reverses righteousness and acts contrary to the times, shall be punished as Chi You; reversing right and turning one's back on the exalted, the law for him shall be to perish and die exhausted.
>
> *(HDSJ 2.3.4)*[41]

Based on the traditional narrative, Huangdi is portrayed as the first sage-ruler who brought peace to the world by a war against his enemy Chi You who was viewed as a "public evil" and dangerous to "public order." The image of Huangdi employing military force to ensure peace and stability has been well illustrated in many other Chinese ancient texts. A punitive expedition as such aims to restore social order. It clearly points to the idea that a punitive expedition is a legitimate and viable means to stop evil doings and injustice. The

Guanzi states that Huangdi's warfare aimed to "establish the law and not change it, thus people would be secure under his law" (Ryden 1998, 54). Many other literatures also point out that Huangdi was a paragon of virtue even though he used weapons to defeat his enemy.

In the same vein, the *HDSJ* notes the importance of punitive interventions when the purpose of the military action is to remove tyranny and save the people. In the section "Bases for Military Expeditions" (*Benfa*本伐) of the *HDSJ*, it states: There are three reasons to wage war: (1) waging war for profit (*li* 利); (2) waging war for righteousness/justice (*yi* 義); and (3) waging war for anger and resentment (*fen* 忿) [*HDSJ* 2.11.1]. It warns against warfare out of gaining profit or out of revenge, maintaining that only warfare out of justice has a hope to be successful. It then explains that waging war for righteousness means a military expedition is used to stop disorder and prohibit atrocities, to promote the worthy and to dismiss those without integrity.[42] Moreover, Huang-Lao contends that it is only for a right cause that the multitude will fight hard even at the cost of their lives.[43]

The Chinese word *yi* has the meaning of "justifiable," "righteous" or "just." It also contains the meaning of "duty" or "obligation." Therefore, unlike improper wars for profit or out of resentment, the war with *yi* indicates a right cause and proper warfare. According to Huang-Lao, the war is justifiable not only when it has a right purpose but also when it corresponds to the pattern of the cosmic Dao. If the word "ethics" is used here to refer to "norms of behavior", whether they are formulated as laws or as moral principles, Huang-Lao has a kind of ethics, that is, an appeal to cosmological justification in the sense of correlation between the cosmic law and the human actions.[44] To apply this cosmological justification to punitive intervention, it means that only military interventions which have a good cause and conducted at the right time are likely to be successful.[45]

Along with the *HDSJ*, the *Lushi Chunqiu* 呂氏春秋 (*Master Lü's Spring and Autumn Annals*, hereafter abbreviated as *LSCQ*), another representative work of Huang-Lao, is more explicit in arguing for the *jus bellum* dimension of a punitive expedition when the author speaks of "punishing the tyrant" (*zhu baojun* 誅暴君) and "rescuing the suffering people" (*zhen kumin* 振苦民). It is exactly against these two notions that the author of *LSCQ* speaks of a debate between "keeping the army" or "using weapons" (*jubing* 舉兵) and "disbanding the army" or "abolishing the use of weaponry" (*yanbing* 偃兵) in light of the "just war" or "just army" (*yibing* 義兵). According to *LSCQ*, one should not talk about the evil of warfare without making a distinction between war for selfish material gains and war for rescuing people from suffering. When mentioning the role of the *yibing* in the chapter called "On Mobilizing the Military" (*Dangbin* 蕩兵), the *LSCQ* states:

> Nowadays, there is no greater act of injustice and no greater harm to the
> people of the world than failing to distinguish between just (*yi*) and

unjust wars (not-*yi*) by rejecting rescuing and protecting. It follows that it is not allowed to either adopt a policy of punitive warfare or to condemn it, to either adopt a policy of defensive warfare or to condemn it. It is only the use of force in a just/righteous cause that can be considered proper. If the war is just, both punitive and protective attacks are allowed; if the war is unjust, neither a punitive nor protective attack is allowed.

(LSCQ: Dangbing)[46]

Basically, Huang-Lao shares a similar view to the *LSCQ*. Nevertheless, the *HDSJ* cautions against the war even if it may look "just" in the beginning, because things "could develop in an opposite direction when pushed to extreme," and thus those who launch the war will suffer (Chang and Feng 1998, 172). On this note, Huang-Lao's view resembles the one expressed in *DDJ*, although the latter was more reserved on military intervention. At the practical level, Huang-Lao acknowledges the danger in military operation since warfare is full of uncertainty, friction, and chance. Thus, it says, "If a state loses its order in a punitive expedition, then the state will fall into great chaos. ... When the objective of the punitive expedition is not properly executed, the state that launched the punitive expedition will in turn be confronted with disasters" (*HDSJ* 1.2.1).

Since Huang-Lao does not emphasize the role of sage-king as the legitimate authority for carrying out just military action as one sees in the Confucian argument, the right cause becomes the most important argument. I think that one of the key reasons is that Huang-Lao, like most politicians of the time, looks for the plausibility of a "unitary empire under the heaven" (*tianxia*天下) to replace the chaotic structures of interstate relationship. As for the question regarding who could carry out this mission, the "son of heaven" or a hegemon, it does not matter. The idea of unification through military power can be traced back to the image of Huangdi, the legendary Yellow Emperor who in the *HDSJ* is depicted as being the first ruler to unify an empire.

Another idea mentioned in the *HDSJ* that fits one of the contemporary *jus ad bellum* argument is the principle of probable success. That is, the military action must have a good chance of achieving its aims. According to Huang-Lao, even if a war, like a punitive expedition, can be justified, one should not engage in it when one is not ready to do so. *HDSJ* mentions that there are three situations which lead to one's death: (1) one acts out of irritation without considering his strength; (2) one overindulges in desires without limit; and (3) one fights against enemy forces far from stronger than one's own (*HDSJ* 3.1.36). Here, Huang-Lao emphasized the ability of the ruler to govern and discipline himself by prudential judgment, since it would be unwise to initiate a military operation when the possibility of winning is slim. A similar idea is maintained in the *Sunzi bingfa* or *Master Sun's Art of War* where it describes five conditions to determine who will be victorious, and one of them is to know when to fight and when not to fight.

II. Right conduct in war

In his essay "The Just War in Early China," Lewis contends that the early Chinese doctrine of was focuses largely on right to war (*jus ad bellum*) rather than right in war (*jus in bello*). In terms of just war in ancient China, Lewis points out that "Having justified the use of violence as a method of securing order, little attention was paid to restraints on the form or degree of violence (Lewis 2006, 97). Lewis's observation is challenged by scholars who have offered some strong cases for *jus in bello* arguments in early Chinese classics (Twiss and Chan 2012, 81–96). We also find similar arguments in the *HDSJ* and some of which reflect the Daoist position given in Laozi's *DDJ*. Let me give some examples.

Firstly, do not kill the captured enemy. The *HDSJ* mentions three kinds of killing that are viewed as "unjust" (*bugu* 不辜): (1) impulsive killing of the worthy; (2) killing those who have already surrendered; and (3) punishing those who are innocent. Among these three unjust actions, not killing those who have already surrendered points to right conduct in war.

Secondly, proportionality in using force. Huang-Lao maintains that one should know when to stop "committing unnecessary violence" (*shidang* 失當). It is also called following the "ultimate of heaven" (*tianji* 天極) in the *HDSJ*. It should be noted that in the contemporary just war discourse, the principle of proportionality concerns how much force is morally appropriate. In fact, the idea of "being appropriate" (*dang* 當) in military force is a repeated argument in the *HDSJ* which emphasizes that the military action deployed should fit the situation, not too much, nor too little. For example, it states, "According to the rectification of names, carry out punitive military expeditions against enemy states. Stop when what is desired has been achieved" (*HDSJ* 3.1.25). It also says, "If one transgresses the ultimate and loses what is fitting, heaven's punishment will be avoidable" (*HDSJ* 1.2.4).

The notion of avoiding an extreme is one of the key ideas in Daoist philosophy, and Huang-Lao follows this rule when it emphasizes the idea of "necessity," i.e., one should not act until he has to act; when he acts, he acts with prudence. Therefore, we read in the *HDSJ*, "The Realization of the Dao follows the tendency of necessity. When the tendency of necessity is followed, his action will not be pushed to the extremes" (*HDSJ* 2.11.3). The preceding passage also explains that sometimes one fails to achieve the goal of the war because "things develop in the opposite direction when pushed to extreme" (*HDSJ* 2.11.2).

However, there are passages in the *HDSJ* showing a seemingly Legalist position, arguing that in executing a punitive operation against a state which "should be punished and ruined," i.e., to punish the wrongdoers in a way deserved by them (*HDSJ* 1.2.1). Such kind of punitive justice in the sense that "one should be eliminated due to one's crime" (*dangzui dangwang* 當罪當亡) is called "the function of heaven" by Huang-Lao. Thus, we read:

> When engaging in war, do not think from the perspective of the *yang* (nourishing life); When engaging in farming (in a peaceful time), do not think from the perspective of the *yin* (killing). Do not exhaust the land and do not press people into labor in a coercive manner.
>
> *(HDSJ 1.27)*

The distinction is also made between conducts in warfare and manners adopted in a peaceful time. The passage further explains that during the war, if fighting with too much kindness, the conquering state will be punished. Another passage offers a specific example:

> When a sage-ruler launches a punitive intervention, he will annex another state by destroying its capital city and the outer walls, by burning its bells and drums, by disposing its fortune, by dispersing its sons and daughters, and by dismembering the conquered territory into feudal estate of the worthies.
>
> *(HDSJ 1.26)*

According to Chen Guying, a contemporary Daoist scholar and commentator of the *HDSJ,* this passage represents the Huang-Lao's response to "five things to do" in a punitive expedition, namely, the city, the walls, the fortune, the people, and the land. However, Chen offers a Daoist interpretation by arguing that the aim of the destructive action described in the passage above is to prevent the conquered state from being held in the hand of the conquering ruler since the purpose of the ruler of the conquering state should not be to meet his own desire (Chen 2007, 36–52). However, one can ask if destroying its capital city and the outer walls, and burning its bells and drums are out of proportionality. The violent actions permitted in the *HDSJ* are different from what is promoted in other Daoist and military texts. In the *Liu Tao* 六韜, a military classic for example, it states clearly, "Do not set fire to what the people have accumulated; do not destroy their palaces or houses, nor cut down the trees at gravesides or altars" (Ryden 1998, 60). It seems that Huang-Lao is concerned about the operational strategy, like what Sunzi does in his method of warfare, when he discusses the plan for attacking the enemy's fortified cities, yet even for Sunzi, such method of attack is the least desirable one (Lo 2015, 74).

Contrary to Huang-Lao, Laozi's Daoism calls for the idea of "compassion" (*ci* 慈) even in the battlefield, claiming that "Those who are compassionate will win the war" (*DDJ* 69). However, it is not the case in the war between Huang-Lao and his enemies. The following passage is a detailed account of violent actions during the tribal wars between the Huangdi and Chi You 蚩尤 recorded in the *HDSJ*:

> Huangdi lifted his halberd, snatched up his weaponry and personally confronted Chi You. Capturing him in due course, Huangdi peeled off

his skin to make a target and ordered people to shoot at it. Those who most often hit were rewarded. He then cut off Chi You's hair, hung it above the heavenly gate and called it the banner of Chi You. Stuffing his stomach in order to make a ball, he ordered the people to kick it. Those who most often kicked it were rewarded. Huangdi then fermented his bones and flesh and threw them into the bitter broth and had the people drink it. In this way, the emperor established order.

(HDSJ 2.5.5; Peerenboom 1993, 89)

It is a graphic picture of the war recorded in the chapter entitled "Rectification of Discord" (*Zhengluan* 正亂) in the *HDSJ*. The infliction of suffering for the sake of suffering or revenge is obvious here. Nevertheless, Huangdi is praised the *HDSJ* as well as by some other Pre-Han texts as a "righteous warrior," and the story of Huangdi's war against Chi You (the symbol of "public evil") is seen as an example of deploying war to stop war. The destruction of Chi You's body signifies the idea of a "total destruction of military arms" (*xiaobing* 銷兵).[47]

Additionally, the discrimination between the combatants and civilians, or what is called the "distinction principle" today is not especially mentioned by Huang-Lao.[48] It should be noted, however, that a similar doctrine in bellicose operations did exist prior to the Warring States period as implied in the military rituals mentioned in Confucius' *Lunyu* 《論語》 (*Analects*), yet the noble chivalric class collapsed with the destruction of Zhou Dynasty and the practice of relevant rituals or chivalric rules had been long abandoned.[49] Nevertheless, we can find relevant points in some military texts during the Warring States period that speak of avoiding unintended civilian casualties. For example, in the section on the:

> "Foundation of Humanness (*Renben* 人本) of the *Sima fa* 司馬法 (*The Military Methods of Sima*, also known as *The Marshal's Art of War*), there is a specific discussion regarding the rule of protecting elder people and children in war.[50]

(Lo 2015, 39).

According to Qian Mu 錢穆, a well-known scholar in Chinese thought, early military texts like the *Sima fa* have preserved the element of chivalric codes of the ancient noble class (Qian 1996, 68). Yet we do not see this code brought up anywhere in the *HDSJ*.

4 Right conduct after war

Several passages in the *HDSJ* deal with specific right conduct after a punitive intervention or what may be called *jus post bellum*. For example, in the section entitled "The Order of a State" (*Guoci* 國次), it states:

> When a state is annexed, and the conquering state rebuilds the capital city and the outer walls, settles therein, delights in its music, enjoys its property, plunders its beauties, this is called going against the Dao of heaven and acting crudely. Such a conquering state will face danger that will finally lead to its own destruction.
>
> *(HDSJ 1.2.5)*

The passage above does not indicate directly what a conquering state should do, but what they should not do, since post-war conduct also influences the moral justification for war. In a way, the passage in *HDSJ* reminds us of what Laozi says, "The sage rears things but lay no claim to them. He does his work but sets no store by it. He accomplishes his task but does not dwell in it" (*DDJ* 2). The passage in *HDSJ* is followed by the statement that "only the sage is able fully to follow the ultimate of heaven and to administrate his state according to what is appropriate for the ways of heaven" (*HDSJ* 1.2.5). For Huang-Lao, since the purpose of a punitive expedition is not to take profit from the conquered state, the army should not kill civilians for the sake of taking their properties. So it says, "If the ruler of the conquering state tries to take without giving, the conquered people will not remain in subjugation for long" (*HDSJ* 1.2.1). Here, we see that the argument is given from a consequentialist position. In another passage, it says, "When one achieves military success and yet one does not know how to make proper use of the fruit of his victory, one will suffer disasters caused by oneself" (*HDSJ* 2.8.2).

Concurrently, the idea of "being appropriate" denotes Huang-Lao's principle of following the pattern or spontaneous order of nature (such as the transformation of four seasons) instead of coercion and overdoing. Huang-Lao insists that to be over-anxious for quick results may on the contrary bring about the loss of one's achievement, a principle emphasized in war conduct, and this principle also applies to the post-war rebuilding and winning the hearts of the people of the conquered state.

In the *DDJ*, Laozi contends that those who win a war are required to go the battleground to self-reflect and should feel grief and sorrow. The event is treated as a funeral since military victory is not a thing of beauty. The *DDJ* states that winning the war is no different than attending a funeral, for the occasion entails more sadness than joy, more regret than glory (*DDJ* 31). Huang-Lao does not offer a similar position explicitly except to say that one should hold a modest posture both before and after military action. The reason given is strategic rather than ethical: it is better not to show off since the strong and the weak always exchange places in a cyclical process.

5 Other related issues

What strikes me is that in Huang-Lao's political and military arguments, there is little discussion of the virtues of a sage-ruler to vindicate a military operation except for a couple of occasions where we encounter a few moral concepts from other schools, such as "inclusive love" (*jianai* 兼愛) from Mohists and "humaneness" (*ren* 仁) from Confucians. Very often, the focus of discussion is to seek what is fitting and to be responsive to the fluid and processional nature of things. Or to put it differently, we see more strategic concerns than ethical considerations.

Despite the concept of *yi* (or just) is used in Huang-Lao, it is not a key one; instead, we find the concept of *zheng* 正, denoting at least five meanings used in the *HDSJ*: (1) rectitude, (2) impartiality, (3) correction or rectification, (4) military expeditions (*zheng* 征), and (5) governance (*zheng* 政). While the concept of rectitude implies the idea of righteousness and goodness, the concept of impartiality focuses on the idea of getting rid of personal biases. In terms of (3) and (4), it suggests that a punitive operation is seen as a military action aiming at the "correction of wrongdoings." Therefore, in the *HDSJ*, we find following statements:

> Those who are the most *zheng* (righteous) abide in quietude; the utmost quietude induces sagacity.
>
> *(HDSJ 1.1.6)*

> *Zheng* (Rectitude) results in good governing.
>
> *(HDSJ 1.5.4)*

> If a ruler and chief ministers follow the Dao, they are able to *zheng* (rectify) all-under-Heaven.
>
> *(HDSJ 1.5.12)*

> The movement of heaven is always *zheng* (correct) and trustworthy.
>
> *(HDSJ 2.5.1)*

> Conscientiously observe *zheng ming* (the correct name) of mine [referring to Huangdi], and do not lose the constant forms of mine....
>
> *(HDSJ 2.5.6)*

> Those who are *zheng* (righteous), though lowly, will make progress.
>
> *(HDSJ 3.1.35)*

Accordingly, it would simplify Huang-Lao's position on war if we hold that Huang-Lao is completely amoral. Huang-Lao is not exactly the same like those proto-Legalists who do not attempt to articulate natural or ethical foundations for social and political order, nor do they provide any metaphysical grounds for human conduct.[51] In addition, since Huang-Lao incorporates concepts from other traditions of the time, its ethical principles with regard to warfare are eclectic in nature, which reflect how Huang-Lao deploys a syncretic response to external threats.

Concluding Remarks from a Comparative Perspective

In this chapter, I have talked about the Huang-Lao thought from its political philosophy to its position on war. I have also discussed the syncretic nature of the Huang-Lao tradition, especially its similarities to and differences from Laozi's Daoism. In short, the Huang-Lao thought can be summarized as follows:

First, since Huang-Lao itself is a diverse system of thought coming from a wide range of the intellectual tradition of the ancient China, its views as well as its practices of warfare are by no means coherent and systematic. Like many other texts attributed to philosophical masters of the Warring States period, the *HDSJ* should not be considered as a "book" but rather a collection of essays authored by different people at different times. For instance, it believes in punitive justice and austere law on the one hand and speaks of Laozi's compassion, Mozi's inclusive love, and Confucius' humaneness on the other.[52] The lack of coherence appears to be not only in the difference in historical settings but also in whether a Huang-Lao thinker is more Daoism-oriented or more Legalism-oriented, the former would adopt more a pacifist position whereas the latter more a realist and pragmatic one.

Second, Huang-Lao's military position is more dependent on and responsive to the context and to the particulars that collectively constitute the totality of the reality. In addition, Huang-Lao regards war as a means for China's unification in order to achieve peace in a long run. As a result of these two reasons, Huang-Lao is more willing to be engaged in punitive expeditions, a position divergent from the Laozi's Daoist one that holds a skeptical view about any wars except for self-defense.

Third, as a manual for the art of governing, the discussion on warfare in the *HDSJ* is, like Laozi's *DDJ*, more from the perspective of the ruler or the commander rather than specific tactics in warfare as we see in most military books such as the *Master Sun's Art of Warfare*. Thus, the discussion of warfare is based upon the framework of its political philosophy or statecraft in general. The exposition on warfare, however, remains fragmentary.

Fourth, while different from many other competing schools of the time, Huang-Lao seems to be quite reserved on military codes of virtue such as those by the military leaders mentioned in other military texts. As a result, it requires further study to see whether the notion of weakness and softness promoted by Huang-Lao is simply a military way of deception (*gui* 詭) and deceit (*zha* 詐). It remains a question insofar as how we see the relationship between the *do*-able and the *dao*-able, or between the "winning way" (*dao* as skillful tactics) and the 'ethical way' (*dao* as moral guidance).

From a comparative perspective, some of Huang-Lao's views are relevant to the contemporary debate on war and peace. Although very often Huang-Lao does not stick to fixed principles but seek the efficacy of *de*, rendering in English as "excellence" or "potency." It suggests that Huang-Lao is more concerned with strategies in warfare. Nevertheless, the ethical ideas are not completely absent when Huang-Lao speaks of the principle of the cosmic order, the patterns of heaven and earth, and the principles of nature, all of which serve as the model of the ultimate way for humans to emulate and practice. Meanwhile, we find passages in the *HDSJ* that bear a resemblance to the just war discourse in the Western tradition, such as right to war, right conduct in war, and right conduct after war, as noted above.

Although Huang-Lao does not spell out explicitly why war needs to be morally justified, its insistence on the Dao as a cosmic order as well as humans' necessity to follow the pattern of the Dao implies ethical considerations. In other words, Huang-Lao's view on war is not merely regarding war as a regrettable and inevitable fact of life or taking a realist position in that what matters is nothing but winning the war. Huang-Lao rejects the idea we cannot stop wars since they are natural consequence of human society. Like Laozi's Daoism, Huang-Lao criticizes those rulers who seek to aggrandize their desires and self-interests by waging wars. Nevertheless, Huang-Lao does not take a pacifist position in that the use of killing force against other human beings can never be justified.

In contemporary context, the idea of peace is very often associated with "pacifism," a concept meaning an opposition to war, militarism, or violence. Etymologically, the word "pacifism" is derived from the word *pax* (pacific), meaning "peace making" between conflicting states. Huang-Lao's definition of peace is simply an "absence of chaos and war." In other words, its position on peace is *ipso facto* a perspective on violence or chaos. Therefore, Huang-Lao maintains to have a "wealthy nation and strong military" for the sake of keeping peace both in domestic and interstate affairs. For Huang-Lao, a strong military force is an important means to achieve peace. Yet it is not in the sense a state with military power could take on "policing" functions as we see under the UN peace-making operation today; rather a state with military power would intimidate the aggressive action of other states. We see a Legalist bent in this argument. Meanwhile, we see a similar argument during the cold war between the U.S. and the Soviet Union.

In ancient history of China, Daoist philosophy is well-known for its various descriptions of "peaceful societies." To an extent, it resembles the Hellenistic concept of "Irene" (a state of non-war) although in the case of Daoism peaceful societies are often taken as a utopian imagination instead of historical reality. For Laozi's Daoism, the Dao means giving birth to life, and war, in contrast, means killing and death. Despite that Huang-Lao shares Laozi's pacifist sentiment yet it maintains that the use of killing force against a wrongdoer (like tyrannical rulers) can be justified. The concept of *zheng* (rectification) also means "to correct that past mistake" through setting proper behavioral rules for those whose moral characters are lacking or inferior. By so doing, Huang-Lao's position is closer to just war theory today. Meanwhile, it is interesting to see that Huang-Lao attempts not be polarize peace (*wen*) and violence (*wu*) by contending that the former is used to win the hearts and the minds of the people but the latter can stop the violence in a more effective way.

The contemporary argument on "right to war" (*jus ad bellum*) includes just cause, legitimate authority, right intention, necessity or last resort, and reasonable hope of success. As abovementioned, Huang-Lao mentions all the items except for legitimate authority. The war waged by Huangdi is considered just in the *HDSJ*, not because of Huangdi, the most important symbols for Huang-Lao, but because of the cause and his intention, that is, fighting with the evil. The issue concerning legitimate authority as a just war condition is a disputable one during the warring period of ancient China when a normative power to govern is in question.[53] Meanwhile, no concept like "rights" is used as a foundational principle for determining the morality of war for Huang-Lao since there is no such concept in the Chinese tradition, especially when the concept of rights is based on individualism.

As for "right in war" (*jus in bello*), Huang-Lao mentions military necessity and proportionality, contending that the army using excessive force leads to its own destruction.[54] Like Laozi, Huang-Lao also argues for keeping a humble attitude in the battle field. But the principle of distinction between combatants and civilians is not raised as an ethical concern.[55] In today's just war theory, however, the distinction is viewed as a cornerstone of international humanitarian law (IHL) that includes a set of rules which seek to limit the effects of armed conflict.

How about the relationship between "right to war" (*jus ad bellum*) and "right conduct in war" (*jus in bello*)? Military ethics, including IHL, normally separates these two principles. Nevertheless, some scholars question such kind of separation, since it is "perfectly possible for a just war to be fought unjustly and for an unjust war to be fought in strict accordance with the rules" (Walzer 2000, 21). Feminist scholars tend to be more concerned with the separation with "right to war" to "right conduct in war." For example, Lucinda J. Peace contends that "right to war" denotes a priority given to state authority makes decisions to wage war instead of the individual who goes to war, yet this state-based argument which has often ignored an individual soldier in war who may lack the ability to understand the moral issues involved in going to war (Peach 1994, 155).[56] J. B. Elshtain the separation may fail to address the problem when the subject of war becomes a means of transgression of what is called "the classic levels-of-analysis" in terms of the individual, the state, and the [anarchic] international arena (Elshtain 1985, 44).[57] Zhou Liang, from a perspective of the Chinese tradition, also challenges the separation embedded in the just war argument, contending that the "right to war"/"right conduct in war" relationship should be construed as "essentially holistic and harmonized" rather than having "right conduct in war" subordinated to "right to war" (Liang 2021, 319). Liang argues that "the first legal implication of the proposed constructive *ad bellum/in bello* relationship would be judging the legality of the use of force through the lens of the conduct during warfare" (Liang 2021, 318). Nevertheless, we often find it difficult to reconcile the need to mobilizing public support for war on the one hand, and the moral duty to conduct war and conclude peace in a just manner on the other.

Another issue concerning Huang-Lao's argument of "all under the heaven." In his study of the *Art of Warfare*, Ames has observed:

> [W]hat makes any military action "appropriate" and "proper" (*yi*) as opposed to "self-seeking" (*li*) is the claim that it serves the quality of the sociopolitical order as a whole rather than any particular interest group within it. Those persons promoting military engagement must make their argument on the necessity of such action to revive and reshape the shared world order.

The idea of "the shared world order" suggested by Ames refers to the ancient Chinese concept "all under the heaven" (*tianxi*天下), denoting the idea of "one unitary empire under the heaven." It often implies that warfare is not just for the peace of a particular state, but for all states involved. As Liang puts it, national sentiment in its modern meaning was absent in pre-modern China (Liang 2021, 313). This notion is also implied in the *HDSJ* when it speaks of peace and order for all as an outcome of warfare. The idea of humanitarian intervention or R2P (responsibility to protect) in contemporary global context has a similar intention as wars like humanitarian intervention is not simply for the purpose of releasing the people from the suffering (owing to situations like genocide, war crimes, crimes against humanity, and ethnic cleansing) but also for the purpose of global peace.[58] Nevertheless, concepts such as global peace, global governance, and cosmopolitanism are contested ones in contemporary political and ethical debates.

In sum, Huang-Lao's position on warfare reflected in the *HDSJ* entails elements of the just war argument despite its emphasis on efficiency and utility. Huang-Lao's justification for war is an exercise of practical reasoning, which suggests reason in relation to human action. This involves "not only reasoning about appropriate means to ends but also understanding the appropriateness of ends themselves". We study the Huang-Lao thought and the *HDSJ* not just as a way to the understand the ancient thought of China but also a means to make sense of the mindset of China and her relations to other states today, for much of Chinese history holds instruction and explanation for the actions of modern and contemporary China.

Notes

1 The term "Huang-Lao" is not clearly defined in any of the classical texts. It is a diversified philosophical system that has no unified doctrine. Harold D. Roth contends that the original meaning of Chinese *Daojia* 道家 (the Daoist School) refers to Huang-Lao rather than the traditional understanding of the philosophies of "Lao-Zhuang" 老莊 of Daoism. See Sima Qian, 1985; Roth 1991, 599–650.

2 Unlike the "Lao" portion which is based on the Laozi (also known as *Daodejing* 《道德經》) as its source thought, there is no defining text for "Huangdi" or the Yellow Emperor in the classical corpus.

3 In the *Shiji*, it mentions that both Shen Buhan and Han Fei based their ideas on Huang-Lao thought. Some modern shcolars regard Laozi's Daoism as the basis on which Legalist political ideas are built. For example, Feng Youlan points out, the Legalist school has been greatly influenced by Daoism". See Creel 1970, 49.

4 Because of its specific agenda to deal with the socio-political problem of the time, some scholars such as Herrlee Creel and Benjamin Schwartz have adopted terms "purposive" or "instrumental" Daoism to characterize the Huang-Lao tradition. Other scholars, especially scholars from Mainland China tend to examine Huang-Lao with the Legalist framework or in the context of the *Jixia* Academy 稷下之學 during the Warring States Period where the tradition was originated. See Liu 1995.

5 Literally speaking, *de*, 德 in ancient Chinese also means "gain" (*de*, 得). Rather than translating as "virtue" *de* in Daoism and Huang-Lao is understood as potency or efficacy. For instance, Ames and Hall translate "The highest virtue or good (*de*) like water" in the Daodejing as "The highest efficacy like water." See Ames and Hall 2003.

6 As a matter of fact, as Mark Csikszentminhalyi has pointed out, the Huang-Lao thought is by no means a coherent system as its writing are quite diverse. He thus defines Huang-Lao as "a complex or group traditions" with different dimensions (Csikszentminhalyi 1994, 9, 53).

7 Nevertheless, some scholars hold a different view on the nature of Huang-Lao. Tu Weiming, for example, argues that "The Huang-Lao doctrine is neither Daoist nor Legalist in the conventional sense, nor is it, strictly speaking, a form of Legalized Daoism. It is rather, a unique system of thought" (Tu 1979, 107).

8 According to Mark Csikszentmihalyi, Huang-Lao would be more suitably defined as "a complex or group of traditions" with different dimensions (Csikszentmihalyi 1994, 9). W. Allyn Rickett, a scholar of the *Guanzi*, contends that the most Huang-Lao text, like text of the *Guanzi*, is a mix of Daoist, Yin-Yang Correlativist, Huang-Lao, and Legalist writings authored from the fifth to first centuries BCE (Rickett 1985, 3, 37).

9 The text of the *HDSJ* is dated to the mid-late Warring State Period and prior to the unification of the Qin (Yates 1997, 4). It is considered the earliest known Huang-Lao text representing its cosmological thought along with political philosophy (Wu 1985).

10 The *Wenzi* is a Daoist text allegedly written by a disciple of Laozi. See Yoshinobu, 2007, 1041–1042.

11 The *Guanzi* is a political and philosophical text attributed to Guan Zhong 管仲 (720–645 BCE), the Primary Minister of the State of Qi. Part of the *Guanzi* has long been recognized as an important source for Huang-Lao thought. For the historical relationship of the *Guaizi* to Huang-Lao and the Jixia Academy, see Chen 1998; Rickett, 1984.

12 The *Heguanzi* (*Master Pheasant Cap*) is a collection of philosophical treatises from the Warring State period. Its content covers a wide range of topics with diverse philosophical sources. In the bibliography monograph of the Han shu, it was listed as a Daoist writing, but it was also seen as a military classic. See Peerenboom 1991, 169–186

13 The *Huainanzi* is a collection of philosophical and political treatises compiled by scholars at the court of Liu An, king of Huainan, in the second century BCE. It is "a tightly organized, sophisticated articulation of Western Han philosophy and statecraft" See Major 2010. There is one chapter in the text that exclusively deals with military strategies. See Ryden 1998.

14 The *Lüshi Chunqiu* (*Master Lü's Spring and Autumn Annals*) is a collection of philosophical treatises complied by the end of the Warring States period (241 BCE) under the leadership of Lü Buwei, the state of Qin. Similar to the *Huangdi sijing* of the Huang-Lao school, the text has been viewed by most scholars as

eclectic in nature where one can find ideas from Daoism, Confucianism, Legalism and etc. Yet it should be noted that the strict sectarian divisions in terms of different philosophical schools (*jia* 家)is a kind of *a posteriori* reconstruction, so it is quite a common practice for the Warring States thinkers to borrow ideas from each other without fearing across the scholastic line. For the English translation of the text, see Knoblock and Riegel, 2000.

15 The *Huang Shigong Sanlüe* (*Three Strategies of Huang Shigong*) is a military text during the first or second century CE. It summarizes the basic teachings of the Military School but with strong Daoist underpinnings, especially when it uses the Daoist concepts of weakness, humility and non-action to talk about military operation (Lo 2015, 3–26).

16 There are several translations of the *Huangdi sijing* into English. For this chapter, the translation of the text is based on Leo S. Chang and Yu Feng (1998), *The Four Political Treatises of the Yellow Emperor. Original Mawangdui Texts with Complete English Translations*, Honolulu: University of Hawaii Press. I have also consulted Yates' *Five Lost Classics: Tao, Huang-Lao, and Yin-Yang in Han China*. See Yates 1997.

17 Historians have acknowledged that Sima Tan is familiar with Huang-Lao master so his description of the Daoist school may reflect his association with his Huang-Lao background. See Schwartz 1985, 23–45; Chen 2007, 29.

18 English translation from Kidder Smith (Smith 2003, 129–156).

19 The English quotation of the *HDSJ* used in this chapter is based on Yates's translation (1997), supplemented by the translation by Chang and Feng (1998). Some modifications have been made by me for the sake of consistency.

20 Chad Hansen observes that Huang-Lao's Dao is more metaphysical than that expressed in Daoist philosophy especially one by Zhuangzi, for the latter refers to multitude of perspectival discourses rather than an absolute and singular metaphysical entity (Hansen 1983, 24–55).

21 Nevertheless, Peerenboom maintains that Legalism, different from Huang-Lao, tends to be a legal positivist rather than a natural law theorist (Peerenboom 1993, 140).

22 The English translation of the *DDJ* chapters used in this essay is my own. I have consulted the translation by Ames and Hall 2003.

23 For a detailed study of Shen Buhai, see Creel 1974; Chen 2007, 10.

24 Wang and Chang contend that in the Legalist writings of the *Hanfeizi*, Laozi's views are "more pervasive and more widely interspersed than is generally believed" and that Han Fei utilizes Laozi's "dialectical logic" in that the "lord of men" must lie low in order to control from above" (Wang and Chang 1986, 106).

25 Scholars studying Han Fan have recognized that the comments and interpretations on the *Laozi* in the *Han Feizi* reinforces the argument of the possible connection of Legalism and Huang-Lao (Schwartz 1985, 343).

26 For a more detailed discussion on this issue, see Rand 2017.

27 For Huang-Lao, the concept of wen 文follows its original etymological connotation, meaning, "pattern" (*wen* 紋), i.e., the pattern of the natural world. In contrast with Confucian *wen* with its explicit ethical implications, Huang-Lao's *wen* suggests more naturalist and epistemological implications.

28 To be more specific, a sage-ruler should be able to use *guan* to observe, contemplate, and penetrating insight of the inner workings of the universe, and use *cheng* to balance which enables him to have timely responses to the challenges of the situation around him.

29 In *HDSJ* 1.4.7, it states, "the government of virtuosity by *wen* reaches down to the most minute and meticulous; and the government of martiality by *wu* reaches up to… (In fact, two characters are missing here in the original text, and some scholars have suggested that these two missing characters in parallel to the two characters in the antecedent sentence should mean "the most comprehensive and exhaustive" (Chang and Feng 1998, 118).

30 Yet such a syncretic view declined gradually in the Han period, as Rand points out, for it was rejected by prominent literati, and this decline "paralleled chronologically the weakening of Huang-Lao influence in government affairs" (Rand 2017, 151).

31 The notion of being against virtue (*nide* 逆德) here refers to the act of "recklessness" or an "absence of prudence." Yates translates *nide* as "oppositional virtue," meaning "courage." That is, "one should let things follow their course, not try to force changes through acts of courage contrary to the natural process" (Yates 1997, 237).

32 In this context, Huang-Lao takes the Daoist position that it is better for one "not to precede over others" which resembles Laozi's idea of "not daring to come to be first in the world." See chapter 67 in the *DDJ* where one reads, "A leader without first wish to follow is only courting death."

33 The *yin* or the feminine maintained by Huang-Lao can be viewed as an administrative strategy as well as a military strategy, well expressed in a saying that "You should pretend that you are weak so that your enemy will be fooled." On this note, Huang-Lao's strategy is close to that adopted by Legalists.

34 See the *Analects* where Confucius says, "if there are distant subjects who do not submit, you can attract them by cultivating refinement and virtue" (*Analects* 16:1).

35 See the section in the *HDSJ* entitled "The Feature of an Army" (*Bingrong* 兵容). Yates also interprets this passage in terms of divination which was a popular practice in Huang-Lao (Yates 2005, 15–43).

36 No doubt, the Dao of heaven's sanction does not correspond directly to any condition in just war theory today. However, in the Huang-Lao philosophical and political context, the Dao of heaven and earth can be interpreted as fulfilling the condition of proper authority for warfare.

37 This idea of timeliness corresponds to the description of Daoism in the *Shiji*, which is also links to Huang-Lao's technique (*shu*) of governance. Thus, timeliness means one should shrift with times and changes in response to the natural order of things, or one should act based on following and compliance (*yinxun* 因循).

38 To an extent, timeliness also entails the strategy of military surprise, which is argued in both the *Sunzi bingfa* and the *DDJ*. In the latter, Laozi contends that "One governs a kingdom by normal rules and fights a war by exceptional moves (*DDJ* 57). Exception moves (*qi*) in this context means "attacking by surprise." Also see Ames and Hall's translation of this statement: "Bring proper order to the state by being straightforward/And deploy the military with strategies that take the enemy by surprise" (Ames and Hall 2003, 165).

39 For a detailed discussion, see Zhang 2015, 209–25.

40 For the Confucian argument for right authority, see Twiss and Chan 2012.

41 In fact, the Chinese word *wu* 武, the martial, force of arms, entails two components, that is, *zhi* 止 (meaning stop) and *ge* 戈 (meaning weapons). The word thus means "stopping war or weapons through war and weapons."

42 The word *fa* 伐 is used here to indicate the ides of attacking or punishing.

43 Laozi's Daoism may have reservation on this point, for it would challenge that what really matters was the power of the ruler and its right to impose his political agenda by force.

44 According to Yates, the triad of heaven-earth-man in Huang-Lao refers to military prognostication or divination without any ethical implications, but I think that the triad can be interpreted in an ethical way as well.

45 According to Wu, Huang-Lao responded to the social issues in the transition between the Warring States period and the twilight of China's unification it expresses a hope of a peaceful world after the unification (Wu 1985, 119). On this note, Huang-Lao's position is analogous to that in the *LSCQ*.

46 The translation from the Dangbing chapter is mine with the consultation of the translation of the *LSCQ* by Knoblock and Riegel. See Knoblock and Riegel 2000. The author of the *LSCQ* argues that waging war can be medicines to cure as well as poisons to kill. It all depends on how warfare is employed.

47 In his commentary on this passage, Chen Guying offers another explanation, contending that the destruction of Chi You's body has something to do with the ancient belief in *guisui*鬼祟, that is, the evil spirit of the ghost that haunts the people when he is in the underworld (Chen 2007, 260).

48 In fact, the distinction principle was almost absent during the military confrontations in the Warring States period, as opposed to early European culture in which the principle of distinction derived from the context of "chivalric society" (Lewis 2006, 186). On warrior's aristocracy and its disappearance in ancient China, see Lewis 1990. In his Just and Unjust Wars, Michael Walzer defends the distinction as the central part of jus in bello (Walzer 2000, 144–46).

49 For instance, we read in the *Analects*, "The Master said: "When the Way (just government) prevails in the realm, then ritual, music and military campaigns are all initiated by the emperor. When the Way declines in the realm, then ritual, music and military campaigns are initiated by the nobles (*Analects* 16:2). See A. Charles Muller's translation at www.acmuller.net/con-dao/analects.html (cited on 20 August 2022).

50 The idea of humaneness (*ren*) during the war is emphasized in the *Sima fa*. As it puts in the military orders given to the army by the Prime Minister: "When you enter the offender's territory, do not do violence to his gods; do not hunt his wild animals; do not destroy earthworks; do not set fire to building; do not cut down forests; do not take the six domesticated animals, grains, or implements. When you see their early and young, return them without harming them".

51 Scholars like Edmond Ryden holds the view that Huang-Lao is one of "the most realistic and least idealistic ones" responding to the issues of warfare (Ryden 1998, 13–14). I think that Ryden gives this conclusion because he focuses on the Legalistic dimension in the Huang-Lao thought.

52 For example, according to the *Shiji*, some members who belonged to the Huang-Lao camp during the early Han period insisted on a political policy opposing the military campaigns in the north against the Xiongnu, nomadic and militant tribes. Yet in the *HDSJ*, we see a very different position.

53 Confucianism was the only school during the warring period that retain the legitimacy requirement. In the West, the argument of legitimate authority can be traced back to the early just war theories of Augustine and Aquinas, who used the criterion to distinguish war as a lawful activity from criminal violence. In a contemporary context, some scholar such as James T. Johnson argues that the head of the state using military force can be the legitimate authority (Johnson 2005, 14–24).

54 The war of Huangdi against Chi You is the only exception.

55 Xunzi makes a distinction between the guilty and the innocent in war, contending that "in punitive expeditions, punishment is not extended to the [populace], but rather only to those who have caused anarchy among them" (Knoblock 1988, 227; Liang 2021, 314).

56 In fact, Watzer has acknowledged the problem of "the moral equality of combatants" when he discusses specific principles in *jus in bello*.

57 The "the classic levels-of-analysis" refers to the approach designed by Kenneth Waltz in his book *Man, the State, and War* (1959) in which he explains the causes of war by distinguishing three levels (or "images"), i.e., the individual, the state, and the international system.

58 For discussions on humanitarian intervention, see Twiss and Chan 2012, 81–96.

Bibliography

Ames, Roger T. 1993. *Sun-Tzu: The Art of Warfare*. New York: The Ballantine Publishing Group.

Ames, Roger T. and Hall, David L. trans. 2003. *Daodejing: Making This Life Significant*. New York: The Ballantine Publishing Group.

Bai, Tongdong. 2009. "How to Rule without Taking Unnatural Actions (無為而治): A Comparative Study of the Political Philosophy of the *Laozi*." *Philosophy East & West Philosophy East & West* 59, no. 4: 481–502.

Ban, Gu. 1983. *History of the Han Dynasty*. Beijing: Zhonghua shuju. [Published in Chinese: 班固《漢書》北京　中華書局].

Chan, Joseph. 2014. *Confucian Perfectionism: A Political Philosophy for Modern Times*. Princeton, NJ: Princeton University Press.

Chang, Leo S. and Yu, Feng. 1998. *The Four Political Treatises of the Yellow Emperor Original Mawangdui Texts with Complete English Translations*. Honolulu, HI: University of Hawaii Press.

Chen, Guying. 2007. *Notes and Commentaries on the Huangdi sijing*. Beijing: Shangwu Publication House. [Published in Chinese: 陳鼓應《黃帝四經今注今譯》北京: 商務印書館].

Chen, L. K. and Winnie Sung, H. C. 2014. "The Doctrines and Transformation of the Huang-Lao Tradition." In *Dao: Companion to Daoist Philosophy*, edited by Xiaogan Liu. Springer.

Chen, Li-kuei. 1991. *Huang-Lao Thought of the Warring States Period*. Taipei: Lianjing chuban gongsi. [Published in Chinese: 陳麗桂《戰國時期的黃老思想》台北: 聯經出版公司].

Chen, Li-kuei. 1998. *Bibliography of research on Han philosophers, 1912–1996*. Taipei: Center for Chinese Studies. [Published in Chinese: 陳麗桂《兩漢諸子研究論著目錄 1912–1996》台北: 漢學研究中心].

Chen, Li-kuei. 2013. *The Daoist Thought in the Han Dynasty*. Taipei: Wunan tushu. [Published in Chinese: 陳麗桂《漢代道家思想》台北: 五南圖書。]

Chen, Qiyou. 2000. *Han Feizi, with new collations and commentary*. Shanghai: Gujichban. [Published in Chinese: 陳奇猷《韓非子新校注》上海: 古籍出版].

Chen, Qiyou. 2002. *Springs and Autumns of Mr. Lü, with new collations and explanations*. Shanghai: Gujichban. [Published in Chinese: 陳奇猷《呂氏春秋新校釋》上海: 古籍出版].

Creel, Herrlee G. 1970. *What is Taoism? And other studies in Chinese cultural history*. Chicago, IL: University of Chicago Press.

Creel, Herrlee G. 1974. *Shen Pu-hai: A Chinese political philosopher of the fourth century B.C.* Chicago, IL: University of Chicago Press.

Csikszentmihalyi, Mark. 1994. *Emulating Huangdi: The theory and practice of Huang-Lao 180–141 BCE*. Stanford, CA: Stanford University. ProQuest Dissertation Publishing.

Elshtain, J. B. 1985. "Reflections on war and political discourse: Realism, just war and feminism in a nuclear age." *Political Theory* 13, no. 1: 39–57.

Feng, Youlan. 1983. *A History of Chinese Philosophy*. Princeton, NJ: Princeton University Press.

Filipiak, Kai, ed. 2017. *Civil-Military Relations in Chinese History: From Ancient China to The Communist Takeover*. London and New York: Routledge.

Fischer, Paul. 2015. "The Creation of Daoism." *Journal of Daoist Studies*, no. 8: 1–23.

Goldin, Paul, ed. 2012. *Dao Companion to the Philosophy of Han Fei*. New York and London: Springer.

Hansen, Chad. 1983. "A Tao of 'Tao' in Chuang Tzu." In *Experimental Essays on Chuang-tzu*, edited by Victor H. Mair. Honolulu, HI: University of Hawaii Press, pp. 24–55.

Hsiao, Kung-chuan (蕭公權). 1979. *A History of Chinese Political Thought, Vol 1: Beginning to the Sixth Century A. D.* Translated by E. W. Mote. Princeton, NJ: Princeton University Press.

Huang, Hanguang. 2000. *The Explication of Huang-Lao Thought.* Taibei: Ehu Publishing. [Published in Chinese: 黃漢光《黃老之學析論》台北: 鵝湖出版].

Ivanhoe, P. J. 2011. *Master Sun's Art of War.* New York: Hackett Publishing Company.

Johnson, James. 2005. "Just War, As it Was and Is." *First Things* 149(January): 14–24.

Knoblock, John, trans. 1990. *Xunzi: A Compete Translation and Study.* Vol. II. Stanford, CA: Stanford University Press.

Knoblock, John. 1988. *Xunzi: A Translation and Study of the Complete Works.* Stanford, CA: Stanford University Press.

Knoblock, John and Riegel, Jeffrey. 2000. *The Annals of Lü Buwe.* Redwood City, CA: Stanford University Press.

Lewis, Mark Edward. 1990. *Sanctioned Violence in Early China.* Albany: State University of New York Press.

Lewis, Mark Edward. 1999. *Writing and authority in early China.* Albany: State University of New York Press.

Lewis, Mark Edward. 2006. "The Just War in Early China." In *The Ethics of War in Asian Civilizations.* London and New York: Routledge.

Liang, Zhou. 2021. "Chinese perspectives on the *ad bellum/in bello* relationship and a cultural critique of the *ad bellum/in bello* separation in international humanitarian law." *Leiden Journal of International Law* 34, no. 2: 291–320.

Liu, An. 1966. *The Book of the Prince of Huainan.* Taipei: Zhonghua shuju. [Published in Chinese: 劉安《淮南子》台北: 中華書局].

Liu, Xiaogan. 1985. "The Huang-Lao School in Post-Zhuangzi." *Zhexue Yanjiu* 6. [Published in Chinese: 劉笑敢〈莊子後學中的黃老派〉《哲學研究》: 第6期].

Liu, Xiaogan, ed. 2014. *Dao: Companion to Daoist Philosophy.* London and New York: Springer.

Lo, Ping Cheung. 2015. "The Art of War corpus and Chinese just war ethics past and present." In *Chinese Just War Ethics: Origin, development, and dissent,* edited by P. C. Lo and Sumner B. Twiss. London and New York: Routledge.

Major, John S. 1993. *Heaven and earth in early Han thought: Chapters three, four and five of the Huainanzi,* SUNY series in Chinese philosophy and culture. Albany, NY: State University of New York Press.

Major, John S. 2010. *The Huainanzi: A Guide to the Theory and Practice of Government in Early Han China.* New York: Columbia University Press.

Peach, Lucinda J. 1994. "An Alternative to Pacifism? Feminism and Just-War Theory." *Hypatia* 9, no. 2: 152–172.

Peerenboom, Randall P. 1990. "Natural Law in the *Huang-Lao Boshu.*" *Philosophy East and West* 40, no. 3: 309–329.

Peerenboom, Randall P. 1991. "Heguanzi and Huang-Lao thought." *Early China* 16: 169–186.

Peerenboom, Randall P. 1993. *Law and morality in ancient China: The silk manuscripts of HuangLao.* Albany, NY: State University of New York Press.

Qian, Mu. 1996. *The Outline of Chinese History.* Beijing: Shangwu yishuguan. [Published in Chinese: 錢穆《國史大綱》北京: 商務印書館。]

Rand, Christopher C. 2017. *Military Thought in Early China.* Albany, NY: State University of New York.

Rickett, W. A. 1985. *Guanzi: Political, Economic, and Philosophical Essay from Early China*. Princeton, NJ: Princeton University Press.

Roth, Harold D. 1991. "Psychology and Self-Cultivation in Early Taoistic Thought." *Harvard Journal of Asiatic Studies* 51, no. 2: 599–650.

Ryden, Edmund. 1998. *Philosophy of Peace in Han China: A Study of the Huainanzi* (Ch. 15, On Military Strategy). Taipei: Ricci Institute.

Sellmann, James D. 2002. *Timing and Rulership in Master Lü's Spring and Autumn Annals (Lüshi hunqiu)*. Albany, NJ: State University of New York Press.

Schwartz, Benjamin. 1985. *The world of thought in ancient China*. Boston, MA: Belknap Press of Harvard University Press.

Sima, Qian. 1985. *Records of the Grand Historian*. Beijing: Zhonghua shuju. [Published in Chinese: 司馬遷《史記》北京：中華書局。]

Smith. Kidder. 2003. "Sima Tan and the Invention of Daoism, 'Legalism,' 'et cetera'." *The Journal of Asian Studies* 62, no. 1: 129–156.

Twiss, Sumner B. and Chan, Jonathan K. L. 2012. "Classical Confucianism, Punitive Expeditions, and Humanitarian Intervention." *Journal of Military Ethics* 12, no. 2: 81–96.

Tu, Weiming. 1979. "The 'thought of Huang-Lao': A reflection on the Lao Tzu and Huang Ti texts in the silk manuscripts of Ma-wang-tui." *Journal of Asian Studies* 39, no. 1: 95–110.

Twiss, Sumner B. and Chan, Jonathan. 2012. "Classical Confucianism, Punitive Expeditions, and Humanitarian Intervention." *Journal of Military Ethics* 12, no. 2.

Walzer, Michael. 2000. *Just and Unjust Wars*. 3rd ed. New York: Basic Books.

Wang, Hsiao-Po and Chang, Leo S. 1986. *The Philosophical Foundations of Han Fei's Political Theory*. Honolulu, HI: University of Hawaii.

Wei, Qipeng. 2004. *Textual Examination of the Yellow Emperor's Letter according to Mawangdui Silk Manuscripts from the Han Tomb*. Beijing: Zhonghuashuju. [Published in Chinese: 魏啟鵬《馬王堆漢墓帛書黃帝書箋證》北京：中華書局。]

Wu, Guang. 1985. *A Comprehensive study of Huang-Lao*. Hangzhou: Zhejiang renmin chubanshe. [Published in Chinese: 吳光《黃老之學通論》杭州：浙江人民出版社: 1985。]

Yan, Changyao, ed. 1996. *The Note and Interpretation of the Guanzi*. Changsha: Yuelu. [Published in Chinese: 顏昌嶢編《管子校釋》長沙：岳麓。]

Yates, Robin. 1997. *Five lost classics: Tao, Huang-Lao, and Yin-Yang in Han China*. New York: The Ballantine Publishing Group.

Yates, Robin. 2005. "The History of Military Divination in China." *EASTM* 24: 15–43.

Yoshinobu, Sakade. 2007. "Wenzi." In *The Encyclopedia of Taoism*, edited by Fabrizio Pregadio, 1041–1042. Abingdon: Routledge.

Zhang, Ellen Y. 2015. "*Zheng* as *Zheng*? A Daoist Challenge to Punitive Expeditions." In *Chinese Just War Ethics: Origin, Development, and Dissent*, edited by P. C. Lo and Sumner B. Twiss. London and New York: Routledge.

9

ZENG GUOFAN'S MILITARY ETHICS[*]

Jonathan Chan

Introduction

In this chapter, I shall discuss Zeng Guofan's military ethics which I take to be the most important Confucian view of military ethics in the late Qing dynasty. Zeng Guofan (1811–72), who played an important role in launching China on the path of modernization, was regarded as a great politician and strategist, a prominent Confucian scholar and a great general. Among his other prominent achievements, Zeng is well-known for creating the *Xiang* (Hunan province) Army which was built on the basis of Confucian ethical doctrines and defeated the Taiping Rebellion Army after a nine-year battle with the Taiping rebels.[1] Zeng was also well-known for his military writings as they had significant influence on the military thoughts of the then contemporary military leaders such as Zuo Zongtang and Li Hongzhang as well as the military leaders of the Republic of China (1912–49) such as Cai E[2] and Jiang Jieshi (Chiang Kai-shek),[3] and even Mao Zedong[4]. As a military leader, Zeng was a controversial figure. Some historians suggest that following the defeat of the Taiping army, Zeng's troops stormed Nanjing and slaughtered much of the city's population, for which Zeng was held responsible. On the other hand, Zeng was highly regarded by historians for his effort in rebuilding that city. In this paper, I shall discuss Zeng's view of military ethics by doing the following things: (1) discussing his reasons for forming the Hunan Militia Army and launching civil war against the Taiping Army; (2) discussing the moral principles which Zeng advocated for governing the build-up of the army; and (3) giving a critical analysis of Zeng's military ethics.

[*] The author is thankful to Prof. Barney Twiss for reading the draft of the paper and his valuable suggestions.

DOI: 10.4324/9781003336372-11

To understand Zeng Guofan's military ethics, we need to bear in mind two things. First, Zeng was not only a theorist but also a practitioner of his own military doctrines. Prior to Zeng, there were very few people who were both theorists and practitioners of military doctrine, not to mention having the opportunity to create, train, and command a huge and powerful army based on self-taught or self-synthesized principles of war. Thus, it would help us to understand Zeng's military ethics to not only analyze his ethical doctrines but also examine his practices in warfare. Second, central to Zeng Guofan's military ethics is the idea of "ruling the army according to rituals", which is in turn derived from the core idea of his Confucian philosophy of "ruling according to rituals". However, this philosophical doctrine of rituals is merely one part of his Cheng-Zhu philosophical outlook, albeit one of the important streams of Neo-Confucianism developed in the Song dynasty. Thus, it would also help us to understand Zeng's military ethics by tracing it back to his Neo-Confucian philosophy.

Before going into details, it might be useful to have a brief sketch of the development of Zeng's political career. As mentioned earlier, Zeng Guofan was an eminent statesman, military general, and prominent Confucian scholar. Historians often attributed the Tongzhi Resurgence (the revival of the Qing dynasty at mid-course) to Zeng's efforts to arrest the decline of the Qing dynasty. Zeng came from a Hunan farming family. His father was a local teacher. He had studied with his father since he was five years old. He passed the prefectural examination in 1833. He then went to study at the famous Yuelu Academy in Zhangsha Prefecture in 1834 and passed the provincial examination in the same year. It is noteworthy that the academy had been a scholarly center since the time of Song dynasty. Zhu Xi and Wang Yangming, two great masters of Song-Ming Neo-Confucianism, had given lectures there. Wang Fuzhi, a great master of Confucianism in the period from late Ming to early Qing (17th century), had studied there. The academy was famous for encouraging the application of the Confucian ideals to practical issues, which in turn had a great influence on Zeng Guofan's intellectual development: for example, taking a balanced approach to elaborating Confucianism by emphasizing both the metaphysical and the practical aspects of Confucianism. In 1838, Zeng passed the imperial examination granting him the title of *jinzhi*, a scholar who had successfully passed the highest level of the imperial examination in Imperial China, and he joined the Hanlin Academy, the highest academic institution in imperial China, two years later. He served in Beijing for more than 13 years, during which time he worked hard in interpreting the Confucian classics. During this period, he visited the Neo-Confucian master *Tang Jian* and studied with him about the metaphysical principles of morals. He then focused his study on Cheng-Zhu philosophy and put this philosophy into practice including, for example, rising at dawn, meditating, maintaining a respectful mindset, reviewing classics and history, regularly practicing calligraphy, and keeping an introspective diary. During the period in Beijing, he was appointed Junior Deputy Director of the Board of Rituals, and later served as Associate Director of the boards of Defense, Works, Justice, and Finance. He was an outspoken and conscientious statesman, repeatedly criticizing the policies of the Qing court.[5]

With respect to his military career, Zeng was not only a military general but also the founder of a powerful army, the *Xiang* Army, which became well known for defeating the Taiping rebellion armies. In 1852, Zeng Guofan returned to his hometown, Xiangxiang in Hunan, to observe a three-year mourning ritual for his mother's death. However, the Taiping rebels who initiated the famous Taiping rebellion in 1850 became a serious threat to the Qing court when they entered into the region of south-central China in 1852 reaching the shores of the Yangzi River. The Qing imperial troops were apparently unable to stop the Taiping rebels' advancement. Zeng was then called into service and appointed Commissioner for Local Defence for Hunan province, responsible for organizing local militias to fight against the Taiping rebels. In the process of organizing the local military force, Zeng soon found that the military and civil officials were so corrupt and incompetent that he had to build an entirely new army. He carefully recruited soldiers for the new army, particularly commanders and officers of senior rank. His recruiting standard included the candidates' evidence of high quality of morals. To use his phrase, the commanders and officers must be persons of *zhong yi xue xing*, i.e., loyal, righteous, brave and determined. His ideal commanders were Confucian scholars with excellent practical military skills. For lower ranking officers and soldiers, the ideal candidates were simple and untarnished men of peasant stock. Zeng chose the top commanders; they in turn chose their subordinate officers, who in their turn recruited the rank and file largely from their own villages. The commanders Zeng chose – such as Yang Yuebin, Peng Yulin, Li Xubin and Ta Qibu – turned out to be the most capable generals in the *Xiang* Army.[6] The *Xiang* Army proved to be the most powerful military force of the Qing court by defeating the Taiping Army in many important battles. The *Xiang* Army under Zeng recovered Zhangsha, Wuchang, and Hanyang in 1854. Zeng was then appointed Vice Minister of Board of War. In 1860, he was given the brevet title of Minister of War and appointed Governor-General of Liang Jiang with the title of Imperial Commissioner. In 1864, Zeng reached the peak of his political career by finally destroying the Taiping Army and recovering the Taiping capital of Nanjing.[7] He was then rewarded with the noble title "First Class Marquis Yiyong" and the entitlement to wear double-eyed peacock's feather (a great honour).

Reasons for undertaking military action against the Taiping militia

In the mid-19th century the Qing court was seen by most people as a corrupt and incompetent regime. It was deemed ineffective and incompetent in responding to a series of natural disasters, economic problems, and military defeats by the Western powers. The incompetence of the Qing court was clearly revealed in the First Opium War with the British Empire in 1842, in which China was defeated in a very humiliating way. As a result of this defeat, Hong Kong was ceded to the British Empire under the Nanking (i.e., Nanjing) Treaty. We might wonder, as an authentic Confucian who was earnest about putting his Confucian ideals into practice, why Zeng Guofan wanted to defend such a corrupt and incompetent government by taking up the leadership in fighting against the Taiping rebels, which were comprising mainly of Han Chinese. A clue to the answer can perhaps be found in Zeng's famous 1854 official proclamation against the Taiping rebels "A Proclamation of Calling for Arms against the Bandits of Guangdong and Guangxi" (Zeng 1854).

This proclamation conveyed three messages. The first message concerned the cruelty of the Taiping militia. Zeng pointed out that the Taiping militia had, for example, killed millions of people and had robbed them of their properties including money, even clothes, they enslaved men to build city walls and dug trenches, and they forced women to watch out for enemies on the parapets and transport food and coal. The second message concerned the Taiping rebels' destruction of the ethical order established by the Confucian "saints" and their oppression of Confucian intellectuals by forbidding Confucian practices and teaching. For Zeng, the Confucian saints in the past established the ethical order of society, defining the proper relation between the emperor and his subjects, father and son, the senior and the junior as well as the proper relation between the noble and the commoner. This ethical order was the basis of society that no one should defy. However, the Taiping rebels promoted brotherhood and sisterhood among all people, referring to all men as brothers and all women as sisters and declaring that "we are all God's children, created equal by God". What even worse were the Taiping regime's oppressive measures disallowing Confucian intellectuals to study and teach the Confucian classics. This oppression, according to Zeng, constituted not only a repulsive attack on peace and order in the society but also a repudiation of the values that lay at the core of Chinese civilization itself. Zeng called on "all those who can read" to heed and fight against the Taiping rebellion, arguing that no educated person should fold his hands in his sleeves and do nothing. The third message concerned the Taiping rebels' violence against other religions including destruction of temples of other religions. Zeng pointed out that the violence done by the Taiping rebels included burning the Confucian academy, damaging temples of the sages, destroying memorial tablets inscribed with the words of the saints, and burning the statues of the deities, which were things that no bandit had done before.

From "A Proclamation of Calling for Arms against the Bandits of Guangdong and Guangxi", we can identify three types of causes that Zeng Guofan took to be justifications for undertaking military action. They are protection of the civilian population against tyranny, preservation of the core values of a nation's civilization, and ending oppression of a nation's traditional religious practices. The first category is in line with the intent of using military force affirmed by the Confucian righteous war tradition. The other two categories, however, are quite novel. No previous writers in the history of Chinese military thought, to my knowledge, have ever identified preserving Chinese civilization and stopping the oppression of traditional religious practices as justifiable causes for using military force. This *ad bellum* thinking when generalized, especially the former category, has important implications for the justification of using military force. For instance, this *ad bellum* thinking might allow using military force to interdict social transformation which might result in the disappearance of a nation's civilization.

Ruling the army according to rituals

In addition to the *ad bellum* thinking, central to Zeng Guofan's military ethics is the idea of "ruling the army according to rituals." Zeng, following Xunzi, takes rituals to be a set of moral rules as well as a system of decrees and regulations. Ritual defines the 'proper place' for a person and the corresponding duties he has. "Ruling according to rituals" is an idea central not only to Zeng's military thought but also his method of governance. He once said that military matters, economics and miscellaneous things are all things that should be covered by the subject of ritual. He also said, "From inside, without rituals, there is no morality. From outside, without rituals, there is no state governance."[8] "From self-cultivation, family maintenance, state governance to world peace, one thing is needed, which is ritual."[9] He gives himself the following principle "Cultivate oneself according to rituals, manage others by using rituals."[10] Thus, "ruling the army according to rituals" is just the application of the general idea of "ruling according to rituals" to the military realm. In his diary, Zeng once wrote:

> Mencius said, "The Confucian gentleman bears in mind the virtue of *ren* and the ritual." One can live even in a country of barbarians by observing these two things. Then, why can we not rule the army properly?[11]

Ren is regarded as the core of humanity by the Confucians such as Confucius and Mencius and is sometimes translated as "humaneness" because a person of *ren* acts humanely towards others. *Ren* and ritual are two important moral values of Confucianism. For Zeng Guofan, these two Confucian values are inseparable since ritual is the embodiment of *ren*.[12] *Ren* is the inner core whereas ritual is the external expression of the inner. That explains why Zeng said that "without rituals, there is no morality." Thus, for Zeng, ruling the army according to rituals and ruling the army according to *ren* are the two sides of the same coin. *Ren* and ritual are the two inseparable fundamental values constituting Zeng's military ethics. Zeng explicitly stated in his diary that in governing the army, both *ren* and rituals should be used.[13]

In addition to the two fundamental ethical values mentioned above, Zeng's theory of military ethics consists of two other important components, namely, moral leadership in the military and soldiers' virtues.

Moral leadership in the military

Moral leadership is a central theme of Zeng's theory of state governance. He once wrote:

> The principle of governance consists of two maxims, namely, recruiting persons both virtuous and wise and nurturing the people. Whether the social mores and customs are good or not depends on the mind and body, every move, every speech of each person. Thus, social leaders should focus on self-cultivation because so doing will have a great and widespread effect on the people of lower position such that they will follow suit and will do so quickly.[14]

According to Zeng Guofan, the social leaders should be selected from the earnest and solid Confucian scholars who would act upon their ideas. For him, even a few such persons could transform the morale and customs of an entire generation and only these Confucian scholars could do so. Zeng extended this principle of leadership to his military thought. In his view, the commanders or generals of an army should be Confucian scholars as well. These commanders or generals must possess two kinds of virtues – excellent military skills and moral virtues. For him the latter is even more important. He said:

> Therefore, I said, the commanders and generals must be persons both wise and brave and excel in both military skills and literary…Perhaps only persons of *zhong yi xue xing*, i.e., persons who are loyal, righteous, brave and determined, would have the qualities mentioned above.[15]
>
> As for the military leaders, four things are of utmost importance: diligent, empathetic, incorruptible, and wise.[16]

Zeng Guofan put this principle of military leadership into practice when he recruited the commanders and officers of senior ranking for the *Xiang* Army. Most of the commanders of the *Xiang* Army were Confucian scholars with excellent military skills. Some of them proved to be the most capable generals in the military history of late Qing dynasty.

Soldiers' virtues

Cultivation of Confucian virtues in all military personnel is another important theme of Zeng's theory of military ethics. Given the importance of moral leadership in Zeng's military ethics, military leaders should possess virtues such as *ren* or humaneness, wisdom, courage, trustworthiness, loyalty, and righteousness so that they could serve as the role model for lower ranking soldiers: if the military leaders take seriously the cultivation of these virtues, the lower ranking soldiers will follow suit. Among the various strands of Confucian virtues, the most important to Zeng was loving people. Zeng wrote:

> Loving people is the most important thing in ruling an army. All military personnel should bear in mind the maxim of loving people. And this maxim should be proclaimed several times every day and taken to be matters of life and death.[17]

Zeng had even written a lyric poem entitled "Song of loving people" for the *Xiang* Army, which conveys the message of not interfering with the civilian population such as not trampling down growing crops, not taking away common people's properties, and not destructing their home. Zeng also had issued a military regulation entitled "Prohibition of interference of civilian population" to the *Xiang* Army:

> As for the principle of using military force, the first important thing is protect common people…To recruit soldiers to exterminate bandits is to love the people. So if interference of civilian population is not prohibited, there would be no difference between the armies and the bandits…[18]

Within the military, the military personnel of different ranking should possess virtues which match their ranking. The superior officer is expected to act like a father to his soldiers. And the subordinate is expected to act like a son to his father. In other words, the relationship among the military personnel is as akin to a kind of family relation. Thus, the superior is expected to have the virtues akin to those of a father and the subordinate the virtues akin to those of a son.

Critical analysis of Zeng's theory of military ethics

In this section, I shall provide a critical analysis of Zeng Guofan's theory of military ethics. Let us begin by focusing on Zeng's *ad bellum* perspective in his "A Proclamation of Calling for Arms against the Bandits of Guangdong and Guangxi."

Zeng's ad bellum *perspective*

In military ethics, there are different *ad bellum* paradigms that tell us under what conditions undertaking military action can be justified. A prominent one is the national defense paradigm (Lee 2012, 73–97). This paradigm allows a state to go to war only if it is in self-defense against another invading state. It is obvious that Zeng's *ad bellum* perspective does not fit the national defense paradigm. And this is natural given the fact that the military conflict between Zeng's militia and the Taiping rebels is not a war between sovereign states, but rather a military conflict between political communities within a state which could be better described as a civil war. There are different types of civil war or military conflict. The civil military conflict between Zeng's militia and the Taiping rebels is a conflict between a counterinsurgent group and an insurgent group seeking to overturn the political status quo, including governmental power and the social order. So the fact that Zeng's *ad bellum* perspective does not fit the national defense paradigm does not mean that the perspective is defective. On the contrary, one might argue that this is due to the failure of the national defense paradigm to provide an adequate explanation of the moral dimension of civil wars. There are, indeed, other *ad bellum* paradigms that can better explain the moral dimension of civil wars. The human rights paradigm is one of those (Lee 2012, 121–133). The *ad bellum* perspective of classical Confucianism is another (Twiss and Chan 2012a; Twiss and Chan 2012b). According to the human rights paradigm, all things being equal, a state or political community may act to aid victims of human rights violations by using military force. The human rights paradigm, then, can provide good reason for taking the first category and the third category of cause for undertaking military action as justifiable in Zeng's official proclamation against the Taiping rebels. To recall, the two categories are protection of the civilian population against tyranny and ending religious oppression. These two causes are deemed to be justifiable for using military force from the perspective of the Confucian righteous war tradition as well. However, it is not clear whether the second category of cause for undertaking military action articulated in Zeng's official proclamation can also be subsumed under the human rights paradigm. To recall, the second category is preservation of the core values of a nation's civilization. One implication of the *ad bellum* thinking deployed in Zeng's official proclamation is that it might allow using military force to interdict social transformation that will result in the disappearance of a nation's civilization. It seems to me that this *ad bellum* thinking not only cannot be explained by the human rights paradigm and on some occasions might contradict the principle of this paradigm. Consider a massive social movement which occurs in a nation and has support from foreign states attempting to revolutionize the nation by promoting the abandonment of the nation's civilization. Also suppose that the movement have many followers, and if the movement is not being interdicted, owing to the support from foreign states, the nation's civilization will disappear in the next one or two generations. Does the nation have the right to stop this movement? Is it wrong for the nation to use military force to interdict the movement? This poses a challenge to both the military ethicist who upholds the human rights paradigm and the military ethicist who adopts the military principle that military force is permitted to be used to preserve a nation's civilization. For the human rights military ethicists, they need to explain why the intervention is wrong while many others have very different moral intuitions in the above imagined situation. And for the military ethicists who adopts the above military principle, they need to respond to the criticism of human rights violation.

Virtue ethics in the military context

In what follows, I want to focus my discussion on the contemporary significance of Zeng's ritual theory of military ethics. In the last two decades, more and more attention has been given to virtue ethics in thinking about the morality of war or using military force. This is because moral abuses committed by military forces in military operations have been increased greatly. The human cost of these abuses is high. Many lives of non-combatants were lost. The number of military suicides that are a by-product of traumatic combat experiences and the subsequent violation of moral norms is high as well. This calls for rethinking the role of virtues in training and governing the military. One might ask why focusing on virtues instead of rules and codes of conduct. The reasons are twofold: first, rules and codes are often not flexible enough to deal with the complex military operations nowadays. Second, soldiers lack the motivation to observe the rules and codes when not being watched or found (Olsthoorn 2011, 4). There are, however, some problems with using the approaches of virtues ethics to reduce the occurrence of moral abuses by military personnel. First, the existing literature on virtues in the military context is not overwhelming. The available literature often focuses only on individual soldierly virtues, such as honor, courage, discipline, or loyalty, while broader approaches that attempt to connect different virtues are relatively rare. Second, evidently the conventional soldierly virtues still play some role in the military. However, they are mainly used for furthering military effectiveness. Instrumental in attaining the objectives of the military, they are not particularly helpful to reduce the occurrence of moral abuses by the military personnel. Third, rules and codes of conduct provide objective moral standards to deal with complex operations. Without these rules and codes, the military personnel would have to rely on their subjective moral intuitions to deal complex operations. It seems to me that Zeng Guofan's ritual theory of military ethics can surmount the above difficulties that virtues ethicists encounter. Zeng's theory enriches our understanding of virtue ethics in the military context by being able not only to connect different military virtues under a unified theory but also to explain why the virtues in question are important from the ethical point of view. As we have seen above, for Zeng Guofan, ritual defines the 'proper place' for a person and the corresponding duties he has. It provides us not only a set of moral rules and a system of decrees and regulations. It also informs us what sorts of moral character one should have. As we have seen, according to Zeng, ritual and the virtue of *ren* are inseparable. This understanding of ritual has significant implications for military ethics. Ritual not only provides military personnel a set of moral rules which they must observe but also what characters military personnel should have. In my view, an adequate military ethics needs to tell us both these two things, namely, the rules that military personnel have to observe and the moral character they should have. The military itself is a dangerous weapon. From the Confucian perspective, the persons who have the power to use this weapon must be persons who have important virtues such as humaneness, wisdom, courage, trustworthy, loyalty and righteousness. It is just hard to conceive how one can justify giving such a power to persons who do not have such virtues.

On the other hand, Zeng's military ethics emphasizes not only virtues but also rituals. To understand the importance of the notion of rituals assumed in Zeng's military ethics, we need to consider the notion from a broader Confucian perspective. For many people, ritual is no more than a set of formal rules or procedures that people observe in celebrations or ceremonies. These rules or procedures are social conventions that have a role perceived as far less important than other legal or constitutional bases of society. Confucians hold a very different view on ritual, however. Ritual has a far more important role to play and is not merely a set of formal rules or procedures used in celebrations or ceremonies. Ritual consists of a body of rules or norms of proper behaviour of various levels of specificity which governs action in every aspect of life. It also embodies a fundamental set of moral values, gives structure and coherence to human institutions, and provides a total cultural context in which human needs can be satisfied (Chan 2012). Thus, for the Confucian, rituals are the means of cultivation of virtues and at the same time set moral standards which the behavior of virtuous persons must satisfy. This is exactly why Zeng's military ethics puts emphasis not only on virtues but also on rituals. Zeng's military ethics has both the merits of virtue morality and rule morality.

Notes

1 For a brief biography of Zeng Guofan, see Liu 1994. For a detailed account of Zeng Guofan's life, see Zhao 2011.

2 Cai E (蔡鍔) was one of the most prominent military leaders in the early Republic of China. He went to Japan to study military at the Japanese Military Academy in 1903. After returning to China, he had been appointed to various important positions in the military force of late Qing Dynasty, including the Chief Training Officer of different provincial military academies, the Chief of Staff of the army of Guangxi province, Brigade Commander of Yuennan province, etc. He was elected as the Dudu (the military governor) of Yuennan after the famous Xinhai Revolution (辛亥革命) taking place in 1911. What made Cai an important figure in the history of Modern China was his taking up the leadership of defending the Republic of China, which was founded after the Revolution, by declaring war

against Yuan Shikai (袁世凱). Yuan, the then President of the Republic of China, attempted to restore the monarchy in China in 1915 and was forced to step down in 1916, owing to Cai's punitive expedition. Cai E was an important military theorist in early Republic of China. His military thoughts had a strong influence on both the theoretical development and the practice of the military in modern China. For instance, Cai's military writings largely shaped the military thoughts of Jiang Baili (蔣百里) who was another prominent military theorist with strong military background in the Republican period.

    Zeng Guofan's strong influence on Cai E can be traced back to one of Cai's writings, *Quotations of Ceng (Guofan) and Hu (Linyi) Regarding the Administration of the Armed Forces with Preface and Comments* (Cai 1911), which was reprinted and distributed to the officers and cadets of the Huangpu Military Academy by Jiang Jieshi (蔣介石).

3 Zeng's influence on Jiang can be seen from the fact that Jiang reprinted Cai's *Quotations of Ceng (Guofan) and Hu (Linyi) Regarding the Administration of the Armed Forces with Preface and Comments* and used it as important references in training officers and cadets of the Huangpu Military Academy. He even translated the book into modern Chinese.

4 For a detailed account of Zeng's influence on Mao's military thought, see Xue 2007.

5 For a detailed account of Zeng Guofan's earlier career, see Zhao 2011, 21–48.

6 For the formation of the Xiang Army, see Shi Duqiao et al. 1998, 307–28; Jen 1973, 216–38.

7 For the major battles between the Taiping Army and the Xiang Army, see Shi Duqiao et al. 1998, 150–260; Jen 1973; Gray 1990, 52–76.

8 Zeng Guofan wrote a set of notes on different topics which were collected in his miscellaneous works. Ritual is among one of those topics. The quoted sentence is taken from the note entitled "Ritual" ‹礼› in the second volume of Zeng's miscellaneous works (Zeng 1876b).

9 Ibid.

10 The sentence is taken from the passage 117 of the note entitled "Learning" ‹問學› in the first volume of Zeng's diary (Zeng 1876a).

11 The passage is taken from the passage 23 of the note entitled "Military Strategies" ‹軍謀› in the first volume of Zeng's diary (Zeng 1876a).

12 Zeng Guofan stated clearly the relationship between ren and ritual in "Preface for the Works of Master Wang Chuanshan" ‹王船山遺書序›, which is included in the first volume of Zeng's diary (Zeng 1876a). Zeng also discussed the relationship between the two Confucian values in his "Postscript of *Discussions on the State Officials in the Book of Etiquettes and Ceremonial*" ‹書　儀禮・釋官　後›, which is included in the fourth volume of *Collected Essays of Lord Zeng Wenzheng* (Zeng 1876c).

13 Ibid. 11, passage 24.

14 The passage is taken from the passage 22 of the note entitled "The Way of Governance" ‹治道› in the first volume of *The Diary of Lord Zeng Wenzheng* (Zeng 1876a).

15 The passage is taken from a letter of Zeng's with the title "To Peng Xiaofang and Zeng Xianghai" ‹與彭筱房、曾香海›, which is included in the third volume of *Letters and Correspondence of Lord Zeng Wenzheng* (Zeng 1876d).

16 Ibid. 11, passage 28.

17 See Cai E, "On the Virtue of Ren" ‹仁愛›, *Quotations of Ceng (Guofan) and Hu (Linyi) Regarding the Administration of the Armed Forces with preface and comments* 曾胡治兵語錄, Ch. 8. (This chapter can be found online: www.chinakongzi.org/guoxue/zzbj/bj/lsjd/200708/t20070821_41753.htm).

18 This passage is taken from an official announcement entitled "Prohibition of interference of civilian population" ‹禁擾民之規›, *Miscellaneous Works of Lord Zeng Wenzheng*, Vol. 2 (Zeng 1876b).

Bibliography

Chan, Jonathan. 2012. "Ritual, Harmony, and Peace and Order: A Confucian Conception of Ritual." In *Ritual and the Moral Life: Reclaiming the Tradition*, edited by D. Solomon, R. Fan, and P. Lo, pp. 195–208. New York: Springer.

Chen, Guoqing and Zhang, Keping, eds. 1994. *An Essential Edition of the Complete Work of Zeng Guofan*. Xi'an: Northwest University Press. [Published in Chinese: 陳國慶、張克平編《曾國藩全集精華本》西安；西北大學出版社].

Cai E, ed. 1911. *Quotations of Ceng (Guofan) and Hu (Linyi) Regarding the Administration of the Armed Forces with preface and comments*, Ch. 8. Reprint in 1970. Taipei: Wenhai Press. [Published in Chinese: 蔡鍔編《曾胡治兵語錄》台北　文海出版社].

Gray, Jack. 1990. *Rebellions and revolutions: China from the 1800s to the 1980s*. Oxford: Oxford University Press.

Gu, Zhiming, et al. 2013. *The History of Moral Development of Chinese People's Liberation Army*. Beijing: PLA Publishing House. [Published in Chinese: 顧智明《中國人民解放軍道德建設史》北京；解放軍出版社].

Jen, Yu-wen. 1973. *The Taiping revolutionary movement*. New Haven, CT: Yale University Press.

Lee, Steven P. 2012. *Ethics and War: An Introduction*. Cambridge: Cambridge University Press.

Li, Yumin. 2011. "A discussion on Zeng Guofan's military thought". In *The Positive and Negative Sides of Zeng Guofan*. Edited by Dong Conglin, 45–55. Beijing: Dongfang Press. [Published in Chinese: 李育民；〈略論曾國藩治軍思想〉；《正反曾國藩》董叢林編；北京；東方出版社].

Liu, Mengxue. 1994. "Zeng Guofan: The Man and His Rights and Wrongs (In Lieu of a Preface)." In *An Essential Edition of the Complete Work of Zeng Guofan*. Edited by Chen Guoqing and Zhang Keping. Xi'an: Northwest University Press. [Published in Chinese: 劉孟學；〈曾國藩其人及其是非功過之評說；代序；〉；《曾國藩全集精華本》；陳國慶、張克平；西安；西北大學出版社].

Luo, Ergang. 1984. *The History of Xiang Army*. Beijing: Zhonghua Book Company. [Published in Chinese: 羅爾綱《湘軍兵志》北京；中華書局].

Olsthoorn, Peter. 2011. *Military ethics and virtues: an interdisciplinary approach for the 21st century*. London: Routledge.

Shi, Duqiao, et al. 1998. *A Military History of the Late Qing Dynasty (I). A General Military History of China*, General Editor, The PLA Academy of Military Science, Vol 19. Beijing: Military Science Publishing House. [Published in Chinese: 施渡橋《清代後期軍事史(上)》《中國軍事通史》北京；軍事科學出版社].

Twiss, Sumner B. and Chan, Jonathan. 2012a. "The Classical Confucian Position on the Legitimate Use of Military Force." *Journal of Religious Ethics* 40. no. 3: 447–472

Twiss, Sumner B. and Chan, Jonathan. 2012b. "Classical Confucianism, punitive expeditions, and humanitarian intervention." *Journal of Military Ethics* 11 no. 2, 81–96.

Xue, Xuegong. 2007. *A Study of Huxiang Culture and Mao Zedong's Military Thought*. Hunan: Hunan Normal University Press [Published in Chinese: 薛學共《湖湘文化與毛澤東軍事思想研究》湖南；湖南師範大學出版社].

Zeng, Guofan. 1854. "A Proclamation of Calling for Arms against the Bandits of Guangdong and Guangxi". In *An Essential Edition of the Complete Work of Zeng Guofan*. Edited by Chen Guoqing and Zhang Keping (1994), pp. 640–641. Xi'an: Northwest University Press. [Published in Chinese: 曾國藩；〈討粵匪檄〉；《曾國藩全集精華本》；陳國慶、張克平編；西安；西北大學出版社].

Zeng, Guofan. 1876. *The Complete Work of Lord Zeng Wenzheng*, Vols. 1–12. Edited by Li Hanzhang and Li Hongzhang. Hunan: Chuanzhong Book Company. Reprint in 2011, Beijing: China Bookstore Publishing House. [Published in Chinese: 曾國藩;《曾文正公全集》（十二冊）；李瀚章、李鴻章編；湖南；中國書店出版社] (The reprint version can be found here: www.zhonghuadiancang.com/leishuwenji/12760)

Zeng, Guofan. 1876a. The Diary of Lord Zeng Wenzheng, Vols. 1–2. *The Complete Work of Lord Zeng Wenzheng*, Vol. 1. Edited by Li Hanzhang and Li Hongzhang. Hunan: Chuanzhong Book Company. Reprinted in 2011, Beijing: China Bookstore Publishing House. [Published in Chinese: 曾國藩;《求闕齋日記類鈔》（二冊）《曾文正公全集》（第1冊）; 李瀚章、李鴻章編; 北京; 中國書店出版社].

Zeng, Guofan. 1876b. Miscellaneous Works of Lord Zeng Wenzheng, Vols. 1–4. *The Complete Work of Lord Zeng Wenzheng*, Vol. 7. Edited by Li Hanzhang and Li Hongzhang. Hunan: Chuanzhong Book Company. Reprinted in 2011, Beijing: China Bookstore Publishing House. [Published in Chinese: 曾國藩;《曾文正公雜著》（四冊）《曾文正公全集》（第7冊）; 李瀚章、李鴻章編; 北京; 中國書店出版社].

Zeng, Guofan. 1876c. Collected Essays of Lord Zeng Wenzheng, Vols. 1–4. *The Complete Work of Lord Zeng Wenzheng*, Vol. 7. Edited by Li Hanzhang and Li Hongzhang. Hunan: Chuanzhong Book Company. Reprinted in 2011, Beijing: China Bookstore Publishing House. [Published in Chinese: 曾國藩;《曾文正公文集》（四冊）《曾文正公全集》（第7冊）; 李瀚章、李鴻章編; 北京; 中國書店出版社].

Zeng, Guofan. 1876d. Letters and Correspondence of Lord Zeng Wenzheng, Vols 1–13. *The Complete Work of Lord Zeng Wenzheng*, Vol. 5. Edited by Li Hanzhang and Li Hongzhang. Hunan: Chuanzhong Book Company. Reprinted in 2011, Beijing: China Bookstore Publishing House. [Published in Chinese: 曾國藩;《曾文正公書劄》（第一卷至第十三卷）《曾文正公全集》（第5冊）; 李瀚章、李鴻章編; 北京; 中國書店出版社].

Zhao, Yan. 2011. *A Trilogy of Late Qing Dynasty, Vol. 1: Zeng Guofan*. Hong Kong: Zhonghua Book Company. [Published in Chinese: 趙焰《晚清三部曲之一; 曾國藩》香港; 中華書局].

10

MAO ZEDONG'S ETHICS OF WAR (1927–49)?

Sumner B. Twiss

Introduction

The title of this chapter poses a question that is prompted by a number of considerations. For example, Mao and his thought are often characterized in somewhat contrasting ways, morally speaking: on the one hand, as having an end-justifies-means philosophy without scruples, morals, or ethics; and, on the other, as one of the 20th-century Chinese leaders who displays sustained concern with the hardships, brutality, and grinding want of the poorest peasants.[1] In addition, one encounters Mao's legacy for the People's Liberation Army (PLA) that appears somewhat ambiguous: on the one hand, employing the language of just war, but, on the other, apparently confining this language to "just cause" and avoiding *jus in bello* concerns.[2] Furthermore, in the background lies the broader challenge of how to think about the ethics of revolutionary war (or rebellion or civil war): for example, issues of legitimate authority, last resort, and reasonable success, on the one hand, and non-combatant immunity, proportionate means, and humane treatment of POWs, on the other. In brief, can there plausibly be an ethics of such irregular war? Considerations such these motivated me to wonder whether one could find an ethics of war (or military ethics) – or even a semblance thereof – in Mao's own military writings which might establish a historically rooted baseline of sorts for understanding current PLA military ethics. In what follows, I self-consciously confine my attention to Mao's military writings from roughly 1928 to 1949, supplemented a bit by his earlier diagnostic social analyses (e. g., a 1927 report on the Hunan Peasant Movement as well as much earlier writings).[3]

Mao's explicit military writings span three periods: (1) the internal struggle and conflict after the open split in 1928–35 between the Chinese Communist

DOI: 10.4324/9781003336372-12

Party (CCP) and the Kuomintang (KMT) following their united front (1923–27) against the warlords; (2) the 1937–45 Sino-Japanese War and continuing tensions between the CCP and the KMT during their second united front against the Japanese invasion; and (3) the resumption of open conflict between the CCP and the KMT after the end of the Sino-Japanese war. I want to examine Mao's writings with respect to each of these three phases and ask whether and how they address *ad bellum* and *in bello* issues. Although my interest is not in technical military strategy and tactics per se, I do need to be alert to the distinction between guerrilla war and regular war (whether mobile or positional), and I also think it relevant to discuss some of the virtues Mao associated with the proper exercise of command responsibility. Also, since Mao himself was well educated in Chinese classics and history (as well as other works), it is of interest to know about those thinkers and episodes he evoked and why.[4] I should also point out that, operating on a principle of charitable interpretation, I am explicitly trying to see things from what I imagine might be Mao's point of view. I could be faulted for adopting such a "conceit," but for the sake of ferreting out and understanding Mao's military ethics (if there is such), I think that this strategy is worth is risk. There will be time enough for criticism and correction in the future. I should also mention that the text of this chapter is an integral whole, and so all quotations, references, and footnotes should be taken into account while reading it.

My initial analytical template for examining each of the three phases of Mao's military writings is the standard one used in contemporary just war thinking, supplemented by a recent discussion of it by Anna Scheid in her *Just Revolution*, which focuses on the ethics of political resistance and social transformation in 20th-century South Africa.[5] In her study, she adjusts standard just war categories a bit in light of the South African case – inductively, as she says – to come up with the following categories for examining just resistance against a manifestly oppressive and unjust regime:

Jus ad bellum

Just cause: severe injustice, self-defense of a people, daily attacks on human dignity, structural violence, direct repressive violence;
Legitimate authority which encompasses two things: (1) the governing authority's forfeit of its legitimacy in failing to safeguard justice, in attacking the common good, in disregarding the needs of the community, and in favoring the interests of the few; (2) the revolutionary agent's legitimacy as displayed by its encouraging political participation of all, enjoying the support of the broader population, and controlling and limiting violence in its efforts to institute social and political reform;
Right intention: aim to achieve a just and stable peace, including eventual reconciliation with the oppressor;

Last resort: action begins with nonviolent methods but reaches the point where forms of nonviolent resistance are met with being outlawed and with violent repression with no evidence of the prospect of de-escalation;

Reasonable hope of success: obviously desirable, but wars of self-defense are less subject to this criterion; that is, defense of key values acceptable even against great odds as a form of proportionate "witness" to their importance.

Jus in bello

Noncombatant immunity: no intentional targeting of civilians, taking explicit measures to prevent civilian abuse, and investigating and policing cases of civilian abuse;

Proportionate means: graduated resistance from lesser to greater degrees of violence, combined with offers of negotiated settlement on just terms;

Humane treatment of POWs and all that this entails according to humanitarian law.

And to this template, which I have adjusted slightly (and may do more in what follows), I add identification and explication of, and training in (so far as possible) "command virtues" such as wisdom, courage, prudence, respect, and the like.

Before turning to the three phases of Mao's military writings, I need to be up front about two things that bear on applying the criteria of last resort and legitimate authority to Mao's thought. First, Mao holds to a modified Marxist-Leninist principle of revolution that envisages no real prospect of social and political change being achieved without violence at some stage. In capitalist countries which practice "bourgeois democracy," the task of the party of the proletariat is "to educate the workers and build up strength through a long period of legal struggle...of utilizing parliament as a platform, of economic and political strikes, of organizing trade unions" before taking up arms. "There the form of organization is legal and the form of struggle bloodless... until the bourgeoisie becomes really helpless" and then comes the time "to launch insurrection and war." In China, by contrast, which is "semi-colonial and semi-feudal," and which has "no democracy but is under feudal oppression...it follows that we have no parliament to make use of and no legal right to organize the workers to strike...the task of the Communist Party...is not to go through a long period of legal struggle before launching insurrection and war."[6] Such a political philosophy seems to undermine the applicability of the criterion of last resort.[7] Nonetheless, it is (or will be) interesting to see that Mao still appears to use it in some of his thinking – or so I want to suggest.

Second, while it is the case that in 1928 the KMT established what it regarded as a central government headquartered in Nanjing and received international recognition as the legitimate government of China, it is very unclear that the Chinese population itself regarded KMT governance as fully legitimate. Indeed, at one point during this period it is reported that there were no fewer than three capitals in China (Beijing, Wuhan, and Nanjing) each representing a different political party or splinters thereof.[8] The point here is that legitimate political authority in China was greatly contested, a fact that helps to account for why Mao was so concerned (as we shall see) to argue for the KMT's lack of legitimacy in moral-political terms and for the CCP's ever growing legitimacy in the eyes of the people.

Phase One

Prior to his first set of military writings, Mao published two analytical essays on, respectively, the different classes in Chinese society and the peasant movement in Hunan, both of which reflect his political thought regarding the need for a revolution in China.[9] The first essay, among other things, singles out the big landlords (many of whom had small armies) and the compradors (agents of foreign capital), along with the warlords (with big armies) as oppressors of the common people who comprise a large majority of the Chinese population. The second essay focuses attention on the Hunan peasant association's uprising against the landlords and their armed forces which results in popularly led progressive social, economic, and political reforms. These reforms include the abolition of exorbitant levies, repair of the material infrastructure, provision of expanded educational opportunities, reform of the judiciary, elimination of local government corruption, and the institution of communal decision-making (presaging Communist reform). At this time, the KMT and the CCP had forged an alliance – a united front – to confront militarily the warlords then in control of much of the country. The united front's National Revolutionary Army was headed by Chiang Kai-shek who, in 1927 under the influence of financial interests and alarmed by increasing radical political agitation for social reform, turned against his Communist allies, resulting in their purge and persecution. This split led to the attempted formation of a CCP Red Army against which in the succeeding years Chiang launched no fewer than five military campaigns.

This brief background, however minimal, provides a context for Mao's opening rhetorical salvo in his first military writing, "Why Is It That Red Political Power Can Exist in China?" (October 5, 1928) in SMW, pp. 11–12:

> The present regime of the new warlords of the Kuomintang remains a regime of the comprador class in the cities and the landlord class in the countryside... subjecting the working class and the peasantry to even more ruthless economic exploitation and political oppression...Yet in the last few months both in the north and the south, there has been a growth of organized strikes by workers in the cities and of insurrections by the peasants in the countryside under the leadership of the Communist Party....so as to complete the national revolution, and in carrying out the agrarian revolution so as to eliminate the feudal exploitation of the peasants by the landlord class.

I interpret this passage (and there are many others to the same effect) to be Mao's attempt to justify his revolutionary struggle along the following lines. He clearly identifies a just cause – severe oppression of the many by the few – as well as indicates that the KMT as the oppressor has no real political legitimacy, while his own party (the CCP) has such legitimacy on the basis of representing the people's interests and having their support as revealed in the growth of organized strikes and insurrections. Although it is not said so here, the KMT's ruthless repression of such activities (which had been ongoing for years) arguably comes close to indicating some sort of last resort claim hovering in the background (though see my caveat above about the applicability of this criterion). Mao's intention is clearly to end the people's oppression, and he later goes on to indicate his aim of establishing the people's political power in the form of a "democratic" government along the lines of the Hunan peasant association reforms. In this first writing, Mao represents the "Red Army [as] a split-off from the National Revolutionary Army which underwent democratic political training and came under the influence of the masses of workers and peasants" (SMW, p. 14). What exactly this means and how this growth is to be accommodated is addressed in his subsequent writings during this phase to which I now turn.

In his report of November 25, 1928 "The Struggle in the Chingkang Mountains" in SMW pp. 21–52, Mao explicitly addresses military issues that bear on the composition, training, and discipline of the Red Army. This army, he says, is drawn from other defeated Communist forces, troops defecting from Chiang's Nationalist Army and from warlord armies encountered as well as peasants from various areas and even captured soldiers who are won over by Mao's internal military policies. Mao candidly admits that such a hodgepodge presents problems of discipline – particularly so by the *lumpen*-proletarians (the rootless, vagabonds, outcasts) – and Mao's solution is "to intensify political training" and to institute progressive policies and strict discipline within the army and in its relations with civilians. The internal progressive policies include these, for example: abolition of the mercenary system (employed by the Nationalists and the warlords) so that soldiers "feel they are fighting for themselves and for the people"; officers and soldiers are to share the same hardships; officers are prohibited from abusing their subordinates; officers and their men are to receive equal treatment; ordinary soldiers are given the freedom to hold meetings and speak out; and financial accounts are open for inspection by all. Mao enumerates all of these policies under the rubric of "the practice of democracy" within the army. In addition, military policy regarding captured soldiers includes, for example: providing medical treatment to their wounded and their being given the option to join the Red Army or not, with those choosing to leave being provided with travelling expenses and then set free. With respect to those choosing to join the Red Army, Mao claims that they feel "spiritually liberated" and that the army is "like a furnace in which all captured soldiers are transmuted the moment they come over."[10]

The army's practices and policies with respect to civilians are frankly differential, depending on class background and current behavior (i.e., complicity in oppression), for example: money is explicitly expropriated from local tyrants to cover military expenses; land is confiscated from rich landlords and redistributed to the poorer class according to principles of equality and equity; land taxes are abolished for the poverty-stricken and reduced for those who are better off; representative councils are established for local self-governance; and strict policies are established regarding the prohibition of reckless burning and killing and regarding the protection of the interests of small merchants.[11] Except for the treatment of tyrants and rich landlords, it is hard not to interpret these policies as being aimed at minimizing civilian harm as well as civilian abuses by the military. As well, it is hard not to credit that the internal military democratic practices have a moral side to them, and the same applies to the apparently honorable treatment of captured soldiers. There is no doubt that prudential considerations are also at work here: namely, winning over new soldiers and winning the support of the people for social and political reform. But to fail to recognize that most of these practices and policies fall within what could be charitably called *jus in bello* norms would be myopic, or so I would argue.[12]

On the eve of the Sino-Japanese war in December 1936, Mao returned in a systematic way to what he called "Problems of Strategy in China's Revolutionary War" (SMW, pp. 77–152). Here, beyond recounting the history of the revolutionary war in its two sub-stages (1924–27 and 1927–36), with a nod to the impending war with Japan (characterized as "a third stage that will soon commence"), reiterating the just cause of the people's oppression, emphasizing the Red Army"s *jus in bello* norms of "opposing bandit and warlord ways," and upholding both democracy and strict discipline within the army's ranks, Mao explicitly invokes the language of just war in the following terms (SMW, pp. 30–31):

> War, this monster of mutual slaughter among men, will be finally eliminated by the progress of human society...But there is only one way to eliminate it and that is to oppose war with war...History knows only two kinds of war, just and unjust. We support just wars and oppose unjust wars. All counter-revolutionary wars are unjust, all revolutionary wars are just...But also beyond doubt the war we face will be part of the biggest and most ruthless of all wars...[which] is hanging over us, and the vast majority of mankind will be ravaged unless we raise the banner of a just war...the banner of mankind's salvation...and a bridge to a new era in world history...the era of perpetual peace for mankind.[13]

A statement such as this appears to address only *ad bellum* concerns and includes among these not only internal oppression but also the imminent project of resisting external aggression by an imperialist power – Japan. At the same time, however, Mao's subsequent reflections in this essay appear to break new ground – at least from my point of view – in his extensive discussion of the virtues and characteristics of proper military command. Here I cannot resist introducing some rather extensive quotations that address this matter head-on (SMW, pp. 85, 86, and 88):

> In real life we cannot ask for "ever-victorious generals," who are few and far between in history. What we can ask for is generals who are brave and sagacious and who normally win their battles...generals who combine wisdom with courage. To become both wise and courageous one must acquire a method...What method? The method is to familiarize ourselves with all aspects of the enemy situation and our own...and the necessity of a "flexible application of principles according to circumstances."
>
> A commander's correct dispositions stem from his correct decisions, his correct decisions stem from his correct judgments, and his correct judgments stem from a thorough and necessary reconnaissance and from pondering on and piecing together the data of various kinds...then, he takes the conditions on his own side into account, and makes a study of both sides and their interrelations, thereby forming his judgments...Such is the complete process of knowing a situation which a military man goes through...But instead of doing this, a careless military man bases his military plans on his own wishful thinking, and hence his plans are fanciful and do not correspond with reality. A rash military man relying solely upon enthusiasm is bound to be tricked by the enemy, or lured on by some superficial or partial aspect of the enemy's situation, or swayed by irresponsible suggestions from subordinates that are not based on real knowledge or deep insight.
>
> Some people are good at knowing themselves and poor at knowing their enemy, and some are the other way round; neither can solve the problem...There is a saying in the book of Sun Wu Tzu, the great military scientist of ancient China, "Know the enemy and know yourself, and you can fight a hundred battles with no danger of defeat," which refers both to the stage of learning and to the stage of application...We should not take this saying lightly...We do not permit any of our Red Army commanders to become a blundering hothead; we decidedly want every Red Army commander to become a hero who is both brave and sagacious, who possesses...the ability to remain master of the situation through the changes and vicissitudes of the entire war. Swimming in the ocean of war, he not only must not flounder but must make sure of reaching the opposite shore with measured strokes...[in] the art of swimming in the ocean of war.

The knowledge of self and enemy obviously include comparative insight into and assessment of material conditions, but they also include and depend on crucial subjective factors such as diligence, courage, temperance, prudence, and psychological insight. And Mao drives home this latter point by subsequently invoking an example from Chinese history (SMW, pp. 109–110):

> During the Spring and Autumn Era, when the states of Lu and Chi were at war, Duke Chuang of Lu wanted to attack before the Chi troops had tired themselves out, but Tsao Kuei prevented him. When instead he adopted the tactic of "the enemy tires, we attack," he defeated the Chi army: The battle was joined at Changshao. When the Duke was about to sound the drum for attack, Tsao said, "Not yet." When the men of Chi had drummed twice, Tsao said, "Now we can drum." The army of Chi was routed. The Duke wanted to pursue. Again Tsao said, "Not yet." He got down from the chariot to examine the enemy's wheel tracks, then mounted the arm-rest of the chariot to look afar. He said, "Now we can pursue!" So began the pursuit of the Chi troops. After the victory the Duke asked Tsao why he had given such advice. Tsao replied, "A battle depends upon courage. At the first drum courage is aroused, at the second it flags, and with the third it runs out. When the enemy's courage ran out, ours was still high and so we won. It is difficult to fathom the moves of a great state, and I feared an ambush. But when I examined the enemy's wheel-tracks and found them criss-crossing and looked afar and saw his banners drooping, I advised pursuit."

Although I have shortened the story – leaving out details prior to the actual battle – Mao's own observations on the whole episode brings out his central points about the range of command responsibility in terms of preparation, careful examination, and good judgment expected of a sagacious commander (SMW, pp. 110–11):

> That was a case of a weak state resisting a strong state. The story speaks of the political preparations before a battle – winning the confidence of the people; it speaks of a battlefield favorable for switching over to the counter-offensive…it indicates the favorable time for starting the counter-offensive – when the enemy's courage is running out and one's own high; and it points to moment for starting the pursuit – when the enemy's tracks are criss-crossed and his banners drooping…China's military history contains numerous instances of victories won on these principles.

All of these quotations – taken from many others amid Mao's detailed discussion of guerrilla strategy and tactics- – appear to indicate his deep appreciation for the moral qualities of proper leadership and decision-making on the field of battle, and they appear to reflect Sun Tzu's great emphasis on such qualities as laid out in the opening chapter of *The Art of War* and reiterated throughout that work.[14] As P.J. Ivanhoe once put it in a footnote to his translation of *The Art of War,* specifically referring to Sun Tzu's reference to ethics and leadership in the book's first chapter:

> "Here and in numerous places throughout the text, Sunzi recognizes and values ethical aspects of war, especially, though not exclusively, in regard to leadership. This is important to keep in mind as he is often depicted as a cool and calculating amoralist."
>
> *(Sun 2011, 95)*

To some extent, I think that Ivanhoe's comment, though not intentionally, also applies to at least some aspects of Mao's military writings.

Phase Two

With this said, I now turn to the second phase of Mao's military writings specifically dealing with the War of Resistance against Japan. There are two pointed writings in this period: "Problems of Strategy in Guerrilla War Against Japan" and "On Protracted War" both of which are dated May 1938 (SMW, pp. 153–86 and 186–267, respectively). The first essay yields little new content, and Mao himself admits at the outset of the essay that "regular warfare is primary and guerrilla warfare is secondary" (SMW, p. 153). So I will focus largely on the second essay, but before turning to it, I do want to say that the guerrilla strategy piece throughout, assumptively and without deeper discussion, characterizes Japan's actions against China as "imperialist and barbarous" and Japan itself as "an alien invader...pursuing a most barbarous policy," indicating that China possesses a just cause of self-defense and also indicating awareness that Japan is pursuing its aggression in an unjust manner (I will say more about both points in connection with the second essay) (SMW, p. 160). There are two additional points of interest in this piece. The first is that Mao institutes an economic policy for the guerrilla base areas that mandates equitable distribution of the financial burden for military expenses among the civilians living there: those with money should contribute money, and the peasants should supply within limits grain to the troops. Moreover, this policy mandates the "protection of commerce," including a strict prohibition of the confiscation of shops" (SMW, p. 174). Clearly, the policy displays some concern about civilian welfare in the context of warfare. The second point is that Mao returns to the theme of the wise commander, clearly showing continuing concern about what I have called military virtues. Following extensive discussion of the attributes of seizing and maintaining the initiative and being flexible in adjusting military operations, Mao writes (SMW, p. 163):

> [A] commander proves himself wise not just by recognition of the importance of employing his forces flexibly but by skill in dispersing,

concentrating or shifting them in good time according to specific circumstances. The wisdom in sensing changes and choosing the right moment to act is not easily acquired; it can be gained only by those who study with a receptive mind and investigate and ponder diligently.

While "Problems of Strategy in Guerrilla War Against Japan" says little about military discipline and treatment of prisoners of war, two other of Mao's works written in this period do, and they appear to indicate two different (even contradictory) policies in the latter regard. On the one hand, "On Guerrilla Warfare" (1937) contains a rather large section on establishing military discipline "on a limited democratic basis" (e.g., no beating by superior officers, political liberty and free discussion, equality in hardship among officers and soldiers) that is aimed at creating a discipline of "individual conscience" (Mao's phrase") that is self –imposed (GW, pp. 90–91). This work also reiterates a set of rules regarding relations with civilians (e.g., prohibition of stealing, injunctions to restore or replace what is borrowed, to be honest in transactions, and the like), and it specifically addresses relations with the enemy in the following terms: "We further our mission of destroying the enemy by propagandizing his troops, by treating his captured soldiers with consideration, and by caring for those of his wounded who fall into our hands" – a seemingly enlightened policy regarding POWs (GW, p. 93).

On the other hand, in a series of lectures entitled "Basic Tactics" (1938) delivered by Mao at a military training academy to its guerrilla warfare section, Mao explicitly says in a section on "Surprise Attacks":

"It is not appropriate for a guerrilla unit to take along prisoners, or to acquire large amounts of booty, which hinder our movement. It is best to require the prisoners first to hand over their weapons, and then to disperse them, or to execute them."

(BT, p. 98)

By the same token later in these lectures, Mao also enjoins guerrilla units to "capture [the enemy's] commander in chief alive" and encourages them to "see who can hand in the most arms, and who can take the most prisoners," subsequently indicating that "[they] should bring together the prisoners and booty in the sight of the masses" for the purpose of increasing the latter's morale (BT, pp. 140 and 143, respectively). One could arguably square these two policies regarding POW treatment by regarding the first enlightened policy as the preferable one, where militarily feasible, and the second less-enlightened one (dispersal or execution) as operative when absolutely necessary for military reasons (e.g., in the midst of an ongoing guerrilla action).[15] Let me now turn to the essay on protracted war.

The first thing to be said about this essay is that toward the beginning Mao appears to follow Sun Tzu's principle of examining the enemy and oneself – in this case Japan and China, respectively – on a grand scale. And what he has to say about each is cast in starkly moralistic terms, which I will highlight in bracketed comments (all the following quotations are from SMW, pp. 196–97). Japan, says Mao, is pursuing a "war of aggression" [so China has a just cause of self-defense], and it is conducting its "adventurist war" in a "peculiarly barbarous" way in part due to the "military-feudal character of Japanese imperialism" [unjust conduct]. From Mao's point of view, despite Japan's technological prowess, its aggression and barbarity will arouse such opposition both from with China and from without (internationally) that it cannot succeed. Additionally, given its "inadequate natural endowment" (manpower and material resources) and comparatively small size, Japan will eventually be defeated. By contrast, China's war against Japan is progressive and just, arousing the nation to unity – hence the re-alliance of the CCP and the KMT – and is gaining not only the sympathy and support of most countries in the world but also at least the sympathy of ordinary Japanese citizens. Note that this assessment uses moral terms, and China's just cause, supplemented by its "great size and abundant international support," will result in its eventual success, however protracted it might be in coming.

This comparative assessment and projection of (dare I say it?) reasonable hope of success is preceded by these two extraordinary statements, in the second of which Mao quotes himself from an earlier interview with Edgar Snow (quotations from SMW, pp. 188 and 194, respectively):

> Our perseverance in the War of Resistance...has been possible because of many factors. Internally, they comprise all the political parties in the country from the Communist Party to the Kuomintang, all the people from workers and peasants to the bourgeoisie, and all the armed forces from the regular forces to the guerrilla; internationally, they range from the land of socialism to justice-loving people in all countries; in the camp of the enemy, they range from those people in Japan who are against the war to those Japanese soldiers at the front who are against the war... Every man with a conscience should salute them.
>
> [In response to Snow's question about China's main strategy and tactics] ...In the course of the war, China will be able to capture many Japanese soldiers and seize many weapons and munitions...at the same time China will win foreign aid to reinforce the equipment of her troops...China will be able to conduct positional warfare in the latter period of the war... Japan's economy will crack under the strain of China's long resistance and the morale of the Japanese troops will break under the trial of innumerable battles...The combination of all these and other factors will enable us to... drive the Japanese forces of aggression out of China.

This latter quotation from the Snow interview obviously also raises questions about Mao's views on *in bello* norms, policies, and practices. But before turning to these, I want to cite one other passage because it appears to tie together as well as deepen the *ad bellum* themes of just cause, right intention, legitimate authority, and (interestingly) last resort; again, I insert brief bracketed comments to identify these themes (SMW, p. 234):

> The way to oppose a war of this kind [aggressive war] is to do everything possible to prevent it before it breaks out [last resort] and, once it breaks out, to oppose war with war, to oppose unjust war with just war, whenever possible [just cause]...In our country the people and government, the Communist Party and the Kuomintang, have all raised the banner of righteousness in...the war against aggression [together constituting a legitimate authority]. Our war is sacred and just, it is progressive and its aim is peace. The aim is peace not just in one country but throughout the world, not just temporary but perpetual peace [right intention].

A passage such as this appears to put the lie to Mao's being solely concerned with the matter of just cause and nothing else. I believe that this passage sets the proper context for Mao's use of the phrase "War is the continuation of politics," for as he goes on to adumbrate (SMW, pp. 226–27):

> In this sense war is politics and war itself is a political action...since ancient times...The anti-Japanese war is a revolutionary war waged by the whole nation, and victory is inseparable from the political aim of the war – to drive out Japanese imperialism and build a new China of freedom and equality – inseparable from the general policy of persevering in the War of Resistance and in the united front, from the mobilization of the entire people, and from the political principles of unity between officers and men...between army and people and the disintegration of the enemy forces, and inseparable from the effective application of united front policy, from mobilization...from the efforts to win international support and the support of the people inside Japan. In a word, war cannot for a single moment be separated from politics.

Interestingly, another adage associated with Mao:

> "The object of war is specifically to preserve oneself and destroy the enemy' – is written in a context that implicitly raises *jus in bello* concerns, for Mao goes on to say, "to destroy the enemy means to disarm him or deprive him of the power to resist, and does not mean to destroy every member of his forces physically."
>
> *(SMW, p. 230)*

By pointing out this context, I do not mean to suggest that Mao is discounting the violence of military conflict and the quite proper use of force to defend oneself and one's forces, but this context does appear to strike a note of moderation similar perhaps to the idea of using proportionate means (SMW, p. 231):

> In the course of military operations...contrasting fundamental factors unfold themselves in the struggle by each side to preserve itself and destroy the enemy. In our war we strive in every engagement to win a victory, big or small, and to disarm a part of the enemy and destroy a part of his men and *materiel*. We must accumulate the results of these partial destructions of the enemy into major strategic victories and so achieve the final political aim of expelling the enemy, protecting the motherland and building a new China.

So what does Mao have to say specifically about *in bello* concerns? The most complete and integrated statement along these lines comes toward the end of "On Protracted War"; in a series of numbered paragraphs, Mao makes the following points in almost logical progression. I can do no better than to list these and use Mao's own words, though the thematic bracketed subheads are mine (SMW, pp. 260–61])

> [Internal military practices and policies:] A proper measure of democracy should be put into effect in the army, chiefly by abolishing the feudal practice of bullying and beating and by having officers and men share weal and woe. Once this is done, unity will be achieved between officers and men...and there will be no doubt of our ability to sustain the long, cruel war.
>
> [Civilian relations and recruitment:] The richest sources of power to wage war lies in the masses of the people...China's armies must have an uninterrupted flow of reinforcements, and the abuses of press-ganging and of buying substitutes...must immediately be banned and replaced by widespread and enthusiastic political mobilization...once the people are mobilized, finances too will cease to be a problem...The army must become one with the people so that they see it as their own army.[16]
>
> [Basic respect:] It is a question of basic attitude (or basic principle), of having respect for the soldiers and the people. It is from this attitude that the various policies, methods and forms ensue. If we depart from this attitude, then the policies, methods and forms will certainly be wrong, and the relations between officers and men and between the army and the people are bound to be unsatisfactory.
>
> [POW treatment:] we must start with this basic attitude of respect for the soldiers and the people, and of respect for the human dignity of prisoners of war once they have laid down their arms. Those who take all this as a technical matter and not one of basic attitude are indeed wrong.

Regarding POW treatment, Mao had earlier written that beyond attacking enemy troops physically, "the chief method of destroying them is to win them over politically," by "treating prisoners of war leniently," and leading them to see the "anti-popular character of the aggression committed by the Japanese rulers" [SMW, p. 251].[17] In this way, suggests Mao, the Chinese army will be undermining the prior indoctrination of the Japanese soldiers and also be psychologically redirecting their "pride" in a proper direction, while indirectly sapping the Japanese army's morale. In addition to the aforementioned policies, Mao, throughout the essay returns to the theme of command virtues, with a particular emphasis on practical wisdom, flexibility in decision, and maintaining the initiative. For example, he writes (SMW, p. 242):

> The ancients said: "Ingenuity in varying tactics depends on mother wit; this "ingenuity," which is what we mean by flexibility, is the contribution of the intelligent commander. Flexibility does not mean recklessness [which] must be rejected. Flexibility consists in the intelligent commander's ability to take timely and appropriate measures on the basis of objective conditions after "judging the hour and sizing up the situation."

In this connection, Mao also writes "it is often possible by various ruses to succeed in leading the enemy into a morass of wrong judgments and actions so that he loses his superiority and the initiative. The saying, 'There can never be too much deception in war' [a reference to Sun Tzu] means precisely this." And he later goes on to write that a big part of "our tactics of defeating the enemy [is] by misleading him and catching him unawares" and that "We are not Duke Hsiang of Sung and have no use for his asinine ethics" (SMW, p. 242).

The latter reference is to another episode in Chinese military history from the Spring and Autumn Era, when the state of Sung fought a war with the powerful state of Chu: when the Chu troops were crossing the river, a Sung officer suggested that this was the moment for attack, but Duke Hsiang said, "No, a gentleman should never attack one who is unprepared," and finally attacked only after the Chu troops were prepared. As a result of this chivalrous delay, the Sung troops were defeated and the Duke wounded (this brief account drawn from an extensive editorial footnote to "On Protracted War" in SMW, p. 267). I believe that Mao's reference to "asinine ethics" is sometimes misinterpreted as meaning he rejected all ethics *tout court*, but the context makes it clear that he is referring only to chivalric (and Confucian) military tactics that are especially inappropriate for modern warfare.

Despite their supposed united front in a just war of resistance against Japanese aggression, the CCP and the KMT continued with their mutual civil war struggle (including military operations) throughout the international war, and this matter is openly addressed in Mao's "Conclusions on the Repulse of the Second Anti-Communist Onslaught" (dated May 8, 1941) (SMW, pp. 287–93). But it also seems reasonably clear that neither Mao nor Chiang Kai-shek desired a final split prior to Japan's defeat. For Mao's part, he recommends a strategy for dealing with the KMT on the basis of a principle of reciprocity: "do unto them as they do unto us, stick for stick [i.e., attack and counter-attack] and carrot for carrot [i.e., more agreeable mutual cooperation]. Mao also characterizes this strategy against Chiang as a "tit-for-tat" struggle, adding that "but this struggle must be waged on just grounds, to our advantage, and with restraint" (SMW, pp. 288–289). In an "Interview with Mr. Liu [and other correspondents]" (dated September 16, 1936), Mao addressed the issue in these terms:[18]

> Our attitude is: "We will not attack unless we are attacked; if we are attacked, we will certainly counterattack." As the saying of an ancient Chinese sage goes, "What the rules of propriety value is...reciprocity. If I give a gift, and nothing comes in return, that is contrary to propriety if the thing comes to me, and I give nothing in return, that is also contrary to propriety." Our attitude is that we treat people in accordance with propriety. But our stand is strictly one of self-defense; no Communist is permitted to go beyond self-defense.

And a bit more colorfully (if that is the right word) and politically realistic, he later writes (SMW, pp. 290–91):

> [F]or the whole period of the anti-Japanese war the Party has a single integral policy – the national united front policy (a dual policy) which integrates the two aspects, unity and struggle – towards all those in the upper and middle strata who are still resisting Japan, whether they belong to the big landlord class and big bourgeoisie or the intermediate class. This dual policy should be applied even to the puppet troops, the traitors and the pro-Japanese elements, except for those who are absolutely unrepentant.

In some sense this policy appears to have a semblance of justice, but it is also clear that Mao continues to harbor another and different just cause continuing from the previous phase: namely, the elimination of the people's oppression and the institution of political reform according to his vision. This cause and vision appear to be reflected in his characterization of the base areas under Communist control (SMW, p. 291):

> [T]he social character of...the anti-Japanese base areas in northern and central China is already democratic. The main criterion in judging whether an area is new-democratic...is whether representatives of the broad masses...participate in the political power there...At present the political system in the base areas is a political system of the united front of all the people who are for resistance and democracy, the economy is one from which the elements of semi-colonialism and semi-feudalism have been basically eliminated.

About four years later in "On Coalition Government" (dated April 1945) in SMW, pp. 301–07 and shortly before the open resumption of hostilities between the CCP and the KMT, Mao reprises and expands the practices and policies of the Red Army during the war against Japan, particularly with respect to military discipline vis-à-vis civilians both in the battlefield and the base areas. Here he highlights the fact that in his view "the army is powerful because all its members have a discipline based on political consciousness" and are fighting "for the interests of the broad masses and on the whole nation." "Guided by this purpose [just cause], this army has achieved remarkable unity in its own ranks and those outside its ranks," and, further, claims Mao, "the army has a correct policy for winning over enemy officers and men and for dealing with prisoners of war" [just conduct]. He is quite emphatic about this latter point: "Without exception all members of the enemy forces who surrender, who come over to our side or who, after laying down their arms, wish to join in fighting the common foe, are welcomed and given proper education. It is forbidden to kill, maltreat or insult any prisoner of war." Moreover, Mao goes on to add that "the entire army is able to utilize the intervals between battles...and training periods to produce grain and other necessities...so that living conditions are improved and the burden on the people lightened" (again, expressing a policy of concern for ordinary civilians) (preceding quotations from SMW, p. 302). And this point is deepened when he writes a few days later ("On Production by the Army for Its Own Support" April 27, 1945) that the policy has a number of positive effects, for example: improved relations between officers and men, a better attitude to labor, strengthened discipline, and improved relations between the army and the people. In the latter regard especially, "once an armed force begins to 'keep house' for itself, encroachments upon the property of the people seldom or never occur...[and] friendship between them is strengthened" (SMW, p. 309–13). And most interestingly at the end of "On Coalition Government," Mao reframes the intention of the war in a way that appears to forecast some sort of reconciliation between the CCP and the KMT when he speaks of "thoroughly defeating the Japanese aggressors" and establishing "a new-democratic coalition government." Indeed, he goes so far as to write: "The moment a new-democratic coalition government comes into being in China, the Liberated Areas will hand their armed forces over it...[and] the Kuomintang armed forces will have to be handed over to it at the same time" (SMW, p. 306).[19] This appears to be an adjustment in intention that opens the door to internal political reconciliation in establishing a just peace.

Phase Three

About a year later (in "A Three Months' Summary" October 1, 1946) in SMW, pp. 321–26, however, the tone of Mao's prospective cooperation and reconciliation with the KMT changes to open hostility, since it becomes clear to him that after the defeat of Japan, civil war has broken out again. In fact, he charges the KMT with having violated a truce agreement, as being determined to wage civil war to destroy the people's democratic forces, and as being "the arch-criminal" of the civil war (SMW, p. 325). Here we see Mao casting the CCP as pursuing a self-defensive and just war against the KMT after having its implicit offer of negotiations rejected (with shades of an appeal to last resort). He also suggests early on that the KMT is pursuing oppressive unjust policies in its conduct of the civil war through its "resumption of conscription and grain levies" vis-à-vis the civilian population, later adding "press-ganging" to their list of crimes. By contrast, in a subsequent essay entitled "Strategy for the Second Year of the War of Liberation" (September 1, 1947), in addition to resuming self-defensive military operations against the KMT, Mao indicates the CCP's adoption of a policy of "replenishing our strength with all the arms and most of the soldiers captured from the enemy (80–90 percent of the men and a number of junior officers), who are won over by the Communist Party cause and its democratic military policies," in addition to resolutely carrying through land reform for the people (SMW, p. 333).

Possibly the most comprehensive statement of Mao's understanding of the just aspects of the resumed civil war is articulated in the extraordinary document entitled "Manifesto of the Chinese People's Liberation Army (also known as the "October 10th Manifesto; October 1947) in SMW, pp. 335–41. Here Mao makes clear the just cause of the civil war from the Communist perspective: the liberation of the Chinese people and nation from the structural and repressive violence of the KMT, with the intention of forming a democratic coalition government. He argues that the KMT has utterly foregone any legitimacy by not only failing to respond to the people's desire for structural reform but also in launching an all-out offensive against them involving "persecution and oppression by every possible means." Mao charges that under Chiang Kai-shek, corruption is rife, "taxes are crushing," and conscription and grain levies are so onerous that "the people are plunged into an abyss of suffering under the guise of serving the public." Moreover, Mao appears to claim something like last resort by the People's army, citing a series of instances over the years where Chiang was offered negotiated agreements which he signed and constantly broke, demonstrating "his lack of repentance and being devoid of all sense of gratitude and opportunity to turn over a new leaf." Moreover, Mao analogizes how Chiang and his army are conducting themselves to behaving "exactly like the Japanese bandits" – "wherever his troops go, they murder and burn, rape and loot, and carry out the policy of the three atrocities [burn all, kill all, loot all]." In short, Mao charges Chiang's supposed rule with "being traitorous, dictatorial, and against the people" (this and preceding quotations from SMW, pp. 336–37).

By contrast, Mao represents the "chief urgent demand of the Chinese people" as including the following (and I number these for convenience, drawing loosely from Mao's own language) SMW, p. 338): (1) uniting all classes, overthrowing the KMT, and establishing a democratic coalition government; (2) arresting, prosecuting, and punishing war criminals; (3) guaranteeing freedom of speech, freedom of the press, and freedom of assembly; (4) clearing out corrupt officials and establishing clean government; (5) confiscating the property of the chief war criminals and using the proceeds to improve the lives and livelihood of the people; (6) abolishing feudal exploitation and putting into effect land redistribution; (7) recognizing the right to equality and autonomy of the minority nationalities within China;[20] and (8) abrogating unjust treaties imposed on China and developing in their stead new treaties of trade and friendship with other countries on the basis of equality and reciprocity. Most of these "demands" appear to be a more precise articulation of Mao's intention to restore peace within the frame of more just society than China has heretofore had in much of its history, especially modern history.

And the Manifesto goes so far as to articulate three sorts of policies regarding the treatment of former enemies (the KMT) and military-civilian relations (see SMW, pp. 339). First, the chief war criminals (the architects of the civil war) shall be tried and punished, while those who were accomplices under duress shall be set free. Second, Mao calls upon those who have "being doing evil" to stop immediately, repent, and break with Chiang Kai-shek, and in turn they will be given a chance to make amends, along with the promise of not being killed or humiliated and being given the choice between joining CCP service or going home. Third, appended to the Manifesto is a list of "three main rules of discipline" and "eight points of attention" for the Red Army and its relations with civilians, including, for example: do not take a single needle or piece of thread from the masses; speak politely; pay fairly for what you buy; return everything you borrow; pay for anything you damage; do not hit or swear at people; do not damage crops; do not take liberties with women; do not ill-treat captives.[21] These injunctions appear quite consonant with Mao's earlier policies regarding relations with civilians (and POWs) in times of conflict, and they appear reasonably proximate to *jus in bello* norms regarding civilian immunity. It is reported that versions of these injunctions were issued repeatedly to the Red Army as early as 1928, thus the these rules are titled as being a "reissue."

Endgame

Mao's military writings end with an eight-point covenant with the people promulgated as a "Proclamation of the People's Liberation Army" (April 25, 1949) in SMW, pp. 407–10. In a nutshell, the eight points are these: (1) a promise to protect the lives and property of all the people conditional on their cooperative attitude toward the PLA; (2) a promise to protect the businesses of the national bourgeoisie and all privately owned enterprises without exception; (3) confiscation of the enterprises owned by the KMT with the exception of privately owned shares along with the offer of continued employment of their personnel if they so desire; (4) protection of all schools, hospitals, cultural institutions, athletic fields, and other public welfare establishments with the offer of continued employment to their personnel; (5) release without charge or arrest of all officials except those war criminals who committed heinous crimes; (6) protection of all stragglers and disbanded soldiers who report and surrender to the PLA or the people's governments in their localities; (7) gradual and phased abolition of the feudal system of land ownership, beginning with the reduction of rent and interest, with the redistribution of land coming later; and (8) protection of the lives and property of all foreign nations so long as they abide by the law. The covenant ends with the statement that:

> "the PLA is highly disciplined; it pays fairly for whatever it buys and is not allowed to take even a needle or piece of thread from the people. It is hoped that the people throughout the country will live and work in peace…This proclamation is hereby issued in all sincerity and earnestness,"
>
> *(SMW, pp. 409–10)*

It should not go unnoticed that this proclamation is cast as a "covenant," appears aimed at establishing a reasonably just peace, and concludes with a reference to the virtue of sincerity (highly regarded in Chinese society from its earliest history).

Further Reflections

Having sketched the outlines of what is arguably Mao's military ethics – or at least a semblance thereof – I want to add a few additional caveats and reflections. I will divide these, as I have done previously, into *ad bellum* norms, *in bello* concerns, and key military virtues.

With regard to *ad bellum* norms, I have already entered a caveat about Mao's apparent invocations of last resort, which at least for the civil war appear to be undermined by his political philosophy's commitment to achieve revolutionary change by violent means (but see footnote no. 7 above). By contrast, with regard to the Sino-Japanese war, the criterion of last resort may be a bit more plausible and applicable. By the same token, however, inasmuch as Mao links the defensive war aim against the Japanese to the civil war aim of achieving internal political reform, it appears that his political philosophy ineluctably predicts success in both aims against the backdrop of what has been called "a revolutionary sense of time": the sense of keeping the struggle going – however long it takes – until victory is achieved.[22] That is, the length of time envisaged by his philosophy of just war is unlimited, and, in the case of the Japanese invasion, space is virtually unlimited as well, since China is a huge territory that no smaller country, however technologically superior, could realistically hope to fully conquer or colonize. This philosophical vision and commitment combined with virtually unlimited time and space accounts for Mao's certitude of eventual victory in both cases. So in a significant sense Mao does not have much to calculate by way of weighing probabilities in order to gain a reasonable hope of success. Indeed, as a consequence of his philosophical commitment and vision, he has a firm expectation of – rather than just a reasonable hope of – success.

As an aside and possibly linking *ad bellum* to *in bello* concerns is the question of whether Mao construed the Japanese invasion as a Michael Walzer-like "supreme emergency" that might justify extreme measures (contravening war conventions) to be used against the enemy in order to avoid threatened enslavement and/or extermination of the Chinese people, their way of life, and their fundamental values (Walzer 2006, chap. 16 "Supreme Emergency"). As Walzer might put the question: does the Japanese invasion constitute such an imminent *and* extreme (horrifying, immeasurably awful) danger to the Chinese people to warrant such extreme measures? While in principle it might be possible to argue such a case – especially in light of, for example, the Nanjing massacre and other such atrocities – the closest that Mao comes to suggesting something like this is his contention that Japan's aim is make China into a Japanese colony and his occasional use of the rhetoric of the intended enslavement of the Chinese people.[23] But I do not see him as laying out a clear case for the Japanese invasion as being such an extreme emergency, especially since (as noted above) he so confidently predicts Japan's defeat.

With respect to *in bello* norms, Walzer raises a number of provocative points about Mao's proposed protections and treatment of civilians and prisoners of war. First, it should be noted that Walzer does regard Mao's "Eight Points" or rules of discipline as addressing the "moral quality" of the means Mao's army employs in pursuit of its just causes. But Walzer goes on to argue that these "rules of engagement...reflect only the utilitarian requirements of guerrilla war, and they cannot stand against the higher utility of winning, and he explicitly invokes as a case in point Mao's tactical advice that "guerrilla bands on the move cannot take prisoners." Suggests Walzer, clearly having Mao's tactical advice in mind, "Once soldiers are actually engaged...especially in a just war, a steady pressure builds up against the war convention and in favor of particular violations of its rules." Thus, says Walzer, "they are at most 'rules of thumb,' general precepts of honor (or utility) to be observed only until observing them comes into conflict with the requirements of victory" (Walzer, pp 181 and 227–28, respectively). Walzer subsequently suggests one of the four ways to deal with this tension between *jus in bello* and *jus ad bellum* is that "the convention yields slowly to the moral urgency of the case: the rights of the righteous are enhanced, and those of the enemies devalued" (Walzer, pp 231–32). Here I want to suggest that something like this strategy may account both for Mao's tactical advice for guerrillas on the move to disperse or execute captured soldiers and also his continuing to hold to the value of capture and humane treatment in all other cases (even in the same writing). This dual policy could be accounted for in strictly utilitarian terms, but the exception for guerrillas on the move (amid military action) appears to be tightly constrained by "military necessity" and confined to just that case. I am inclined to wonder whether Mao's norm of "respect" articulated in a virtually concurrent writing is doing a great deal of heavy lifting in framing such a tightly framed exception. Is such respect merely utilitarian

Another (or alternative) way to account for Mao's dual policy or apparent inconsistency involves a speculative suggestion on my part: namely, that Mao's *Basic Tactics* is derived from a somewhat flawed stenographic record of his lectures at a military academy, for in exactly the same year in which these lectures were published, Mao writes in an essay "On the Basic-Tactic of Anti-Japanese Guerrilla Warfare – The Surprise Attack" (dated January 11, 1938) in MRP, vol. VI, pp. 179–92; quotation from p. 184:

> [A]fter the surprise attack...[a]s for the prisoners, when you are close to our own base areas...escort all the prisoners to the rear areas or turn them over to the regular armies to handle. Otherwise they should be released immediately after questioning and given suitable propaganda. It is strictly forbidden to kill prisoners...Even if captives sometimes do not to go with the guerrilla troops, they may only be disarmed and let go...But one must take care to keep one's own actions secret while releasing prisoners. To achieve this objective, one must first release the prisoners who wish to go and lock in a room those who will not go, so that they will not see the direction in which the guerrilla unit withdraws.

These precise and rather detailed instructions contrast sharply with the rather crude advice of "either disperse or execute" POWs for guerrilla units on the move. The latter advice is, frankly, out of line with everything else Mao has written or said about the proper handling of prisoners of war (whether KMT or Japanese). So, on balance, I am inclined to think that the text of *Basic Tactics* is somewhat corrupted.[24]

There is a broader aspect of *jus in bello* that warrants greater attention, especially in light of the fact that Mao is sometimes criticized – perhaps unfairly – by Western commentators for employing techniques of psychological warfare aimed at deceiving and ambushing the enemy, demoralizing him, disintegrating his will to fight, and trying to win him over to Mao's cause (whether revolutionary, self-defensive, or both).[25] I do believe that we can see some of Sun Tzu's strategic thinking here, and, for Sun Tzu at least, they represent a less violent and less harmful way to achieve military objectives. For example, P.C. Lo suggests that Sun Tzu's preference for psychological strategies over the use of physical force implies "a certain sense of proportionality of means in warfare," and the question is whether the same might be said of Mao's apparently similar preference (Lo 2015, 75).[26] Given Mao's numerous references – both acknowledged and unacknowledged – to Sun Tzu and his own emphasis on psychological means as being particularly effective in fragmenting and undermining enemy units as well as a whole enemy army, I think it arguably plausible that in employing psychological warfare Mao is similarly using "proportionate means" in the manner of Sun Tzu and that this has a moral (even if only utilitarian) import.[27]

There are any number of commentators who have remarked on the positive leadership qualities of the CCP's Red Army commanders in contrast to the shortcomings of the KMT's Nationalist Army's commanders. Whereas Mao's commanders exhibit durability, flexibility, dedication, quick evaluation and decisive action, as well as knowledge of the locale and support from the people, the same is not said of Chiang's commanders who are characterized as inexperienced, mutually uncooperative, indecisive, as well as having uneven morale and stirring up popular discontent.[28] Mao's commanders display moral-military virtues on the battlefield – and here I include those other virtues identified earlier in this paper. It has often been remarked as well (by other commentators) that Mao himself displays a rather deep consistency over time, and this is no less true in the case of virtues. For example, in one of his first writings before he ascended to power – entitled "A Study of Physical Education" (1917) – Mao writes (and here I am selectively trying to identify a continuous theme throughout the essay by linking these selected quotations):[29]

> Morality, too, is valuable; it is the basis of the social order and equality between ourselves and others. But where does virtue reside? *It is the body that contains knowledge and houses virtue* [Mao's italics]. It contains knowledge like a chariot and houses morality like a chamber…When one's decision is made in his heart, then all parts of the body obey its order. Fortune and misfortune are of our own seeking. "I wish to be virtuous, and lo, virtue is at hand" [Analects]…The power of the sentiments is extremely great. The ancients endeavored to discipline them with reason. Hence they asked, "Is the master [reason] always alert?" They also said: "One should discipline the heart with reason" [Mencius]. But reason proceeds from the heart, and the heart resides in the body…Physical education not only harmonizes the emotions, it also strengthens the will. The great utility of physical education lies precisely in this. The principal aim of physical education is military heroism. Such objects of military heroism as courage, dauntlessness, audacity, and perseverance are all matters of will.

Thus at a relatively young age (he was twenty-four at the time), Mao was already beginning to identify the virtues that he later associates with command responsibility, and, intriguingly, he refers to the Confucian classics rather than

Sun Tzu, whose work appears to be a subsequent accretion and influence.[30] One might arguably add to the military virtues I have identified throughout this paper both "respect" (perhaps a stand-in for "benevolence") and "patience" as if (per my remarks above) one has "all the time in the world" to accomplish one's main objective of bringing about justice.

A final note seems in order here. I am fully aware that there are many scholars who will take this reconstruction of Mao's military ethics with a grain (or even tablespoon) of salt, particularly in light of Mao's actions *after* the establishment of the PRC in 1949 – not only internally (within China) but also externally (e.g., the Korean Conflict).[31] While they might concede that prior to the PRC's establishment Mao used the rhetoric of just war, including in bello norms regarding the discipline of soldiers and the treatment of civilians, these would, I think, say that Mao's moral rhetoric is insincere and that his subsequent actions showed his true colors as just another power-hungry, amoral, political realist without a serious moral compass. I candidly concede this possibility (even likelihood), but with Michael Walzer I would also say that moral hypocrisy in wartime discourse gives testimony to the staying-power, persuasiveness, and prevalence of just war categories (Walzer, ch. 1, especially pp. 19–20). Even if only for prudential reasons, it is important for political leaders in warfare to appear to be in the right both for internal audiences (e.g., the Chinese people) and for external ones (actual or potential allies abroad). I would also observe that Mao's military writings are presumably read by and taught to others, and it seems important that these others have an opportunity to reflect on the fact that Mao himself appears compelled to use just war categories in representing and justifying his policies to others. Putting aside Mao for the moment, if his followers were to appreciate and/or adopt his just war thinking – borrowing it as it were – then who is to say that such thinking might not come to be more seriously (and sincerely) their own? As Mencius wrote about the hegemon during the Warring States period, "if a man borrows a thing [humaneness] and keeps it long enough, how can one be sure that it will not become truly his" (Mencius 7A30). While this optimistic adage may not have been finally true of Mao himself, it is possible that others may come to instantiate it.[32]

Notes

1 On the first view, see, e.g., Fuller 1958, 139–45, especially 139; on the second, see, e.g., Boorman 1963, 1–55, especially 50.
2 See, e.g., Lo 2015, chap. 3, especially 78.
3 My principal sources are these: Mao 1967, (SMW); Mao 1965, (SW); Mao 1961, (GW); Mao 1966, (BT); and Schram 1992–2015, (MRP). Note parenthetical shorthand for intratextual references to these works followed by appropriate page numbers.

4 A useful discussion of Mao's intellectual background can be found in Meisner 2007, chap. 1 "Youth, 1893–1921".

5 Scheid 2015, esp. chap. 4 "The Just War Tradition and Revolution"; see also Morkevicius 2013, 401–11. In what follows, I simply list and follow Scheid's criteria in the sequence she uses to introduce and discuss them. In more standard renditions, for example, right or legitimate authority is usually discussed first (since it is the usual governing criterion for action in the public good), followed by the rest.

6 These quotations are from a piece entitled "Problems of War and Strategy" (November 6, 1938) in SMW, pp. 269–71. As a measure of how far Mao had traveled in the development of his philosophy, in his "Manifesto on the Founding of the *Xiang River Review*, July 14, 1919, Meisner claims that "neither the terms 'socialism' nor 'Bolshevism' appeared in Mao's 'Manifesto.' Rather, it was 'democracy' that Mao called 'the basic ideology' of resistance to oppression. Moreover, he insisted that the battle for democracy must take the form of a 'bloodless revolution,' eschewing bombs and chaos," and Meisner quotes Mao himself as writing, "We must accept the fact that the oppressors are people, are human beings like ourselves...[if] we use oppression to overthrow oppression...the result [will be] that we still will have oppression" (Meisner 2007, 18). The full text of this manifesto is available in MRP, vol. I ("The Pre-Marxist Period, 1912–1920"), pp. 318–20.

7 But, interestingly, Alexander V. Pantsov reports that in November 1920 although "[Mao] might still agree in principle with Xiao Yu [an old friend and correspondent] that it was a good thing 'to seek the welfare of all by peaceful means'...he now considered all these [other] positions as utopian fantasies," and Pantsov further quotes Mao as writing that "A Russian-style revolution...is a *last resort* when all other means have been exhausted. It is not that some other better means are rejected and we only want to use this terrorist tactic" (Pantsov 2012, 94, my italics). The full text of this letter addressed to Xiao Xundong, Cai Linbin, and others in France (dated December 1, 1920) is available in MRP, vol. II ("National Revolution and Social Revolution December 1920-June 1927"), pp. 5–14 (quotation can be found on p. 9).

8 There are many timelines for this period in Chinese history. One that I found useful and draw from here is "Chinese Civil War" available at http://new worldencyclopedia.org/enry/Chinese_Civil_War.

9 These are, respectively, "Analysis of the Classes in Chinese Society" (March 1926) and "Report on an Investigation of the Peasant Movement in Hunan (March 1927), both in SW, Vol. I, pp. 13–59. See also the translated texts in MRP, vol. II: "Analysis of All the Classes in Chinese Society" (December 1, 1925; note the corrected date), pp. 249–62; "Analysis of the Various Classes among the Chinese Peasantry and Their Attitudes toward the Revolution" (dated January 1926), pp. 303–09; and "Report on the Peasant Movement in Hunan" (dated February 1927; note corrected date), pp. 429–71.

10 All preceding quotations from SMW, pp. 29–34. As Mao also later wrote in "Eight Conditions for a Great Victory in the Second Campaign" (1931): "The White Army [KMT] soldiers who were invading the Red [CCP] areas...saw our slogans and leaflets encouraging them to struggle against local bullies and divide up the land, and telling them that the Red Army is the army of workers and peasants, within which everyone is equal and free. At the same time, they also witnessed the White army burning, killing, looting, and plundering...[T]wenty-three soldiers defected to the Red Army...Furthermore, the prisoners we took...spread...the word that we gave special treatment to them, and that we did not search their pockets but instead gave every one of them three *yuan*...As a result...their soldiers do not have the courage to fight the Red Army...[and]...are even inclined to come and hand over their guns to the Red Army..." (MRP, vol. IV ("The Rise and Fall of the Chinese Soviet Republic 1931–1934"), pp. 45–51; quotation from p. 49).

11 These policies are discussed in minute detail in MMW, pp. 17–52. It was during this period that Mao formulated and issued his Three Main Rules of Discipline and Six Points for Attention. Apparently, according to the most recent research, the entire set of principles was formally promulgated in April 1928: The three main rules were (in brief): (1) obey orders; (2) do not steal from workers and peasants; and (3) turn in whatever is confiscated from local bullies. The six points (again. in brief): (1) put back doors (used for bedding); (2) put back straw (used for bedding); (3) speak politely; (4) pay fairly for all purchases; (5) return everything borrowed; and (6) compensate for any damages. Subsequently, these were added: (1) do not bathe in sight of women; (2) use latrines; (3) do not search the pockets of captives. For more information, on these rules, see MRP, vol. III ("From the Jinggangshan to the Establishment of the Jiangzi Soviets July 1927–December 1930"), "Order on Rectifying Military Bearing and Discipline" (dated March 21, 1930), pp. 283–284 (especially footnote #2 dealing with historical matters). See also "Statutes of the Red Army Soldiers' Association" (dated September 25, 1930) in the same volume, pp. 536–41 (especially "Art. 22, Discipline," p. 539). See also below on the 1947 reissue of these rules.

12 Virtually all of the preceding points and others to follow are summed up by Mao in poetic form with ninety lines, with each line containing just four Chinese characters (for ease of memory) under the prosaic heading of "Notice Issued by the Fourth Army Headquarters of the Red Army" (dated January 1929), available in MRP, vol. III, pp. 136–38. The complete poem in translation is appended to this chapter.

13 Here it seems appropriate to mention that the language of "just and unjust wars" is significantly informed by the early Chinese communists' study of the Russian Soviet view of such a distinction. See, for example, "History of the Communists Party of the Soviet Union (Bolsheviks) Short Course" which was translated into Chinese and closely read by the CCP, including Mao. An excerpt from the English translation of this work (New York: International Publishers, 1939): "Just wars… are wars…of liberation, waged to defend the people from foreign attack and from attempt to enslave them, or to liberate the people from capitalist slavery….Unjust wars, wars of conquest, [are] waged to conquer and enslave foreign countries" (chap. 6, 167–68). My thanks to P.C. Lo for drawing my attention to this source.

14 As pointed out by Shu Guang Zhang in his *Mao's Military Romanticism: China and the Korean War, 1950–1953* (University Press of Kansas, 1995), p. 28: "[E]ven a cursory reading of [Mao's] works reveal that he heavily quoted from Sun Tzu and other ancient strategists [e.g., Wu Tzu]…Of these ancient writers Sun Tzu had the greatest impact on Mao and other CCP military leaders. Mao accepted some of Sun's principles, such as correct and flexible command, creation of favorable momentum [*zao shi*] in battle, 'ten against one,' and 'knowing your enemy and knowing yourself.'"

15 I will return to this point in "Further Reflections" below.

16 One piece of concrete evidence that Mao takes seriously army relations with the civilian population comes in the form of legal enforcement. Although I have not found any record of military court martials of soldiers who contravened the rules of discipline, there is one case of capital punishment of a decorated Red Army soldier for murder that is addressed by Mao in a letter to a judge, "To Lei Jingtian" (dated October 10, 1937). Mao writes in part about this case: "He receives capital punishment today and the comrades of the Party Central Committee and I feel sorry about it. But he committed an unpardonably serious crime. If a Communist Party member and Red Army cadre were pardoned for such a mean and brutal act, an act that deserted the stand of the Party, the revolution, and humanity, it would be impossible to educate the Party, the Red Army, and the revolutionaries, or to teach people how to be ordinary human beings…The Communist Party and the Red Army must impose stricter discipline on their members and soldiers than on ordinary civilians" (quoted from MRP, vol. VI, p. 89).

17 See also Mao's "Interview with Edgar Snow on Japanese Imperialism" (dated July 16, 1936), which he concludes by saying "Japanese officers and soldiers captured and disarmed by us will be welcomed and will be well treated. They will not be killed. They will be treated in a brotherly way. Every method will be adopted to make the Japanese proletarian soldiers, with whom we have no quarrel, stand up and oppose their own Fascist oppressors...Anti-Fascist Japanese troops are our friends, and there is no conflict in our aims" (MRP, vol. V ("Toward the Second United Front January 1935–July 1937", pp. 258–66; quotation from p. 266)) Also, in an "Interview with British Journalist James Bertram" (dated October 25, 1937), Mao responded to a question about his policy of giving special treatment to prisoners of war, saying "We will persevere in this policy of ours. For instance, the Japanese army has publicly announced that it will use poison gas against the Eighth Route Army [the CCP army of the second united front], but even if it does that, our policy of special (lenient) treatment will not change. We will go on looking upon Japanese soldiers...who have been forced to fight just as if they were our own brothers, showing sincere sympathy to them, and giving them special treatment immediately after they are disarmed. We will not humiliate or rebuke them, but will release them and allow them to go back after explaining to them the...interests of the two countries' people. Those who do not want to go back may serve in the Eighth Route Army" (again in MRP, vol. V, pp. 112–26; quotation from pp. 129–130).

18 The full title of this interview is "Mr. Mao Zedong's Interview with Mr. Liu, Correspondent for the Central News Agency, Mr. Geng, Correspondent for *Saodang Bao*, and Mr. Zhang, Correspondent for *Xinmin Bao*"; the text is available in MRP, vol. VII ("New Democracy 1939–1941"), pp. 201–06; quotation from pp. 204–05.

19 In the full version of "On Coalition Government" in SW, vol. III, pp. 205–70, Mao specifies what he means by "coalition government" in the following way: "a provisional central government composed of representatives of the Kuomingtang, the Communist Party, the Democratic League and people with no party affiliation...to promulgate a democratic programme of political action...[and] in the next stage...convene a national assembly after free and unrestricted elections and form a regular national coalition government in which the representatives of all classes and political parties...take part" (quotations from pp. 240–42).

20 This particular demand stems from Mao's long-standing charge of Chiang Kai-shek's Han chauvinism against minority groups in China. For example, in the "Declaration to the People of Muslim Nationalities by the Central Soviet Government" (dated May 25, 1936), Mao writes: "For more than ten years, the Muslim peoples have repeatedly suffered the oppression and exploitation of the Chinese ruling class. In particular, the exorbitant taxes and levies under the rule of Chiang Kai-shek...plunge the lives of the Muslim people into the deepest misery...Moreover, they often stir up bad feelings between the Muslim and Han peoples, even to the point where they murder one another, leading ultimately to suppression by large military forces...The Nanjing [KMT] army's entry into Gansu is a major step toward destroying the Northwest and suppressing the independence movements of the Muslim, Mongolian, and Tibetan nationalities...Therefore the westward movement by the Anti-Japanese People's Red Army...provides an excellent opportunity for the independence and liberation of the Mongolian, Muslim, and other small and weak nationalities, especially the Muslim nationalities" (MRP, vol. V, pp. 201–02). And Mao later writes in "On the New Stage in the Development of the National War of Resistance against Japan and the Anti-Japanese National United Front" (dated October 12–24, 1938): "the...task at the moment is to unite all the nationalities as one and resist the Japanese bandits together. For this purpose, we must...give the Meng [Mongolian], Hui [Muslim], Zang [Tibetan], miao, Yao, Yi, Fan, and all the other nationalities equal rights with the Han...to manage their own affairs" (MRP, vol. VI, pp. 458–541; quotation from p. 506).

21 "On the Reissue of the Three Main Rules of Discipline and the Eight Points for Attention – Instruction of the General Headquarters of the Chinese People's Liberation Army" October 10, 1947) in SMW, pp. 343–44.

22 See, for example, Katzenbach and Hanrahan 1955, 321–40, especially 324–30.

23 See, for example, Mao 1935, 153–78.

24 Schram himself reports in the introduction to his translation of Mao's BT, p. 15: "the book is probably based on the stenographic report of what he said, rather than on a copy completely written out by Mao himself...[p]articularly noticeable because a character is sometimes wrongly replaced by another character pronounced the same way but with a totally different meaning. There are also one or two sentences that appear out of place."

25 See, for example, Fuller 1958, 139–45, especially 142 and 145.

26 I should add that P. C. Lo, an expert regarding Sun Tzu's work, disagrees with my interpretation here. At least at this point I stand behind my claim, but am open to persuasion otherwise.

27 Interestingly, there are contemporary Chinese scholars of military who agree that Sun Tzu's thought significantly influenced Mao's thinking (or at least rhetoric) about military ethics, including the prioritization of "a strategy of preservation over one of destruction" and "the role that psychology plays in fighting a war." As Li Ke further elaborates: "Mao's military ethics are based on Sunzi's ideas concerning the recognition of the people as the foundation of military affairs... righteous war involves 'keeping a nation at peace, and an army intact'...the strategy of 'commanding the army with civility and rallying them with martial discipline; and preparing the army and waging the war with a humanitarian spirit" (all quotations from Li 2012, 59–62).

28 See, for example, Boorman and Boorman 1966, 171–95, especially 188–94.

29 In Stuart R. Schram, *The Political Thought of Mao Tse-tung*, Rev. Edition (Frederick A. Praeger, 1969 [1963]), essay translated from the Chinese by Stuart Schram, pp. 152–162; quotations from pp. 154, 155, and 157, respectively. See also Mao's "Marginal Notes to: Friedrich Paulsen, *A System of Ethics*" (dated 1917–1918), in MRP, vol. I, pp. 175–313, and especially pp. 263–264, where Mao writes: "The truly great person develops the original nature with which Nature has endowed him, and expands upon the best...the great motive power that is contained in his original nature...the spring that fulfills his character...He then judges whether. .it is right and proper to exercise this motive power...This comes solely from his own judgment...The great actions of the hero are his own, are the expressions of his motive power, lofty and cleansing...His force is like that of a powerful wind arising from a deep gorge...All obstacles dissolve before him. I have observed from ancient times the fierce power of courageous generals on the battle line undaunted...This is because...his motive force presses forward in a straight line...This is true also of the great man and of the spirit of the sage."

30 Stuart Schram claims in his *Thought of Mao Tse-tung* (1989) that "Mao said in 1968 that...he had taken a look at Sun Tzu before writing his own works on military tactics in 1936–8" (p. 54). Lest it be thought that Mao was unacquainted with Sun Tzu prior to that period, it should be observed that in his "Classroom Notes" (dated October-December 1913), Mao summarized in his own words some of Sun Tzu's major points; see MRP, vol I, pp. 9–56; see especially pp. 30–31. Interestingly, on October 22, 1936, in a letter "To Ye Jianying and Ling Ding," Mao requested that they "Buy a copy of Sunzi's Art of War" (MRP, vol. V, p. 414).

31 My gratitude to P. C. Lo for his own expression of skepticism about Mao's sincerity in his military ethics, which prompted me to add this explicit reflection.

32 I take this opportunity to thank Mark Mecalf, John Kelsay, P. C. Lo, and Ximi and Richard Harry for their comments on various drafts of this paper.

Appendix: "Notice Issued by the Fourth Army Headquarters of the Red Army" (January 1929)

The aim of the Red Army
is democratic revolution,

Our western Jiangxi First
Army reputation has
spread far

and wide.

The present plan is to
move forward by
divisions.

Be they officers or foot
soldiers,

All must obey commands.

Be fair in dealings with
the people,

Thus proving ourselves
trustworthy.

Wanton burning and
killing

Must be strictly
forbidden.

All over the nation,

Oppression is unbearable.

The workers and the
peasants

Endure bitter sufferings.

Local bullies and gentry

Are tyrannizing over vil-
lages and towns.

High interest and heavy
taxes

Rouse everyone's anger.

White Army soldiers

Go hungry and cold.

The petty bourgeoisie

Pays extremely heavy
taxes.

Conflicts have arisen
among them,

The warlords will meet
their fate.

Food is what alleviates
hunger,

Medicine is what cures
disease.

What the Communist
Party advocates

Is exceedingly just.

The fields of the
landlords

Should be given to the
peasants to till.

Debts need not be paid
back,

Rents need not be paid.

An increase in workers'
wages

Must be borne by the
bosses.

Eight hours of work a day

Is just the right amount
of time.

The way the troops are
treated

Urgently requires
improvement.

In distributing the land,

Soldiers are entitled to a
share.

Enemy officers and their
soldiers

Must be allowed to
switch sides.

What they've done in the
past

Their factories and their
banks

Must be confiscated and
taken over.

Foreign investments
and foreign debts

All are declared null and
void.

Foreign troops and for-
eign ships

Are not allowed to
enter our borders.

Overthrowing the big
powers

Will bring joy to every-
one's heart.

Overthrowing the
warlords

Means a thorough pur-
ging of evil.

Unifying the whole of
China

Is reason for the nation
to rejoice.

As for the Manchus,
Mongols, Hui, and
Tibetans,

They will determine
their own statutes.

The Guomindang
government

Is nothing but a pack of
scoundrels.

Uniting to get rid of
them,

We thoroughly purge
the corrupt regime.

The more imported goods there are,

The harder it is to sell domestic ones.

As for imperialism,

Who is there that doesn't hate it?

The Guomindang bandit party

Is completely reactionary.

It says one thing and means another,

They can't be too strong.

Chiang, Gui, Feng, and Yan

Share a bed but not their dreams.

Will not be held against them.

The method of progressive taxation

Is by far the most suitable.

All exorbitant taxes and levies

Must be thoroughly swept away.

As for merchants in the cities,They have hoarded bit by bit.

As long as they are obedient,

The rest does not matter.

The treatment meted out to foreigners

Must be exceedingly strict.

The workers and peasants of the entire nation

Are swift as the wind and powerful as thunder.

The day when we will seize political power

Is not far away.

The success of revolution

Depends on the popular masses alone.

Let this be proclaimed on every hand,

And everyone be roused to action.

Source: MRP, vol. III, pp. 136–38

Bibliography

Mao, Tse-tung. 1966. *Basic Tactics.* Translated and with Introduction by Stuart R. Schram. New York: Frederick A. Praeger. [Textual citation: BT]

Mao, Tse-tung. 1961. *On Guerrilla Warfare.* Translated by Samuel B. Griffith. New York: Praeger Publishing. [Textual citation: GW]

Mao, Tse-tung. 1967. *Selected Military Writings of Mao Tse-tung.* Second Edition. Peking: Foreign Languages Press. [Textual citation: SMW]

Mao, Tse-tung. 1965. *Selected Works of Mao Tse-tung.* 3 Vols. Peking: Foreign Languages Press. [Textual citation: SW]

Mao, Tse-tung. 1935. "On Tactics Against Japanese Imperialism." In *Mao's Selected Works.* Vol. I, 153–178, December 27, 1935.

Schram, Stuart R., ed. 1992–2015. *Mao's Road to Power: Revolutionary Writings*1912–1949. 8 Vols. Armonk, NY and London. [Textual citation: MRP]

Sun, Tzu. 2011. *Master Sun's Art of War.* Translated with Introduction by Philip J. Ivanhoe. Indianapolis, IN: Hackett Publishing Co.

Boorman, Howard L. 1963. "Mao Tse-tung: The Lacquered Image." *The China Quarterly* 16 (Oct.-Dec.): 1–55.

Boorman, Howard L. and Scott A.Boorman. 1966. "Chinese Communist Insurgent Warfare, 1935–49." *Political Science Quarterly* 81, no. 2(June): 171–195.

Fuller, Francis F. 1958. "Mao Tse-tung: Military Thinker." *Military Affairs* 22, no. 3 (Autumn): 139–145.

Katzenbach, Jr., Edward J. and Hanrahan, Gene Z. 1955. "The Revolutionary Strategy of Mao Tse-tung." *Political Science Quarterly* 7, no. 3(September): 321–340.

Li, Ke. 2012. "Mao Zedong's Military Ethics and Sunzi's *Art of War*." *Journal of Wuhan University of Science and Technology (Social Science Edition)* 14, no. 1: 59–62 (translated from the Chinese by Ximi Zuo Harry with the assistance of Richard Harry).

Lo, Ping-cheung 2015. "Warfare Ethics in Sunzi's Art of War? Historical Controversies and Contemporary Perspectives." In *Chinese Just War Ethics: Origin, Development, and Dissent*, edited by Ping-cheung Lo and Sumner B. Twiss, chap. 3. Abingdon: Routledge.

Meisner, Maurice. 2007. *Mao Zedong: A Political and Intellectual Portrait.* Cambridge: Polity Press.

Morkevicius, Valerie. 2013. "Why We Need a Just Rebellion Theory." *Ethics & International Affairs* 27, no. 4(Winter): 401–411.

Pantsov, Alexander V. 2012. *Mao: The Real Story.* New York: Simon & Schuster.

Scheid, Anna Floerke. 2015. *Just Revolution: A Christian Ethic of Political Resistance and Social Transformation.* Lanham, MD: Lexington Books.

Schram, Stuart R. 1969. *The Political Thought of Mao Tse-tung.* Revised ed. New York: Frederick A. Praeger.

Schram, Stuart R. 1989. *The Thought of Mao Tse-tung.* Cambridge: Cambridge University Press.

Walzer, Michael. 2006. *Just and Unjust Wars: A Moral Argument with Historical Illustrations.* 4th ed. New York: Basic Books.

Zhang, Shu Guang. 1995. *Mao's Military Romanticism: China and the Korean War, 1950–1953.* Lawrence, KS: University Press of Kansas.

11

CHIANG KAI-SHEK'S MILITARY ETHICS

An Analysis of His Wartime Rhetoric

Sumner B. Twiss

Introduction

The preceding chapter explored Mao Tse-tung's military ethics (1927–49). It examined the rhetoric surrounding Mao's use of just war language in connection with the two phases of the civil war in China (1926–37 and 1945–49) and the interregnum period of the Second Sino-Japanese War (1937–45). It argued that despite common misperceptions to the contrary, Mao's military writings displayed remarkable consistency in the use of *jus ad bellum* criteria, the articulation of *jus in bello* norms, and the identification of "command virtues" of a moral nature. Mao's principal rival and counterpart – and occasional ally – during these periods was Chiang Kai-shek, the commander of the Kuomintang's Nationalist Army and, later, the President of the Republic of China. Like Mao, Chiang wrote a lot about his military strategies and actions, exhibiting distinctive cultural influences in his rhetoric. What is truly remarkable about Chiang's writing in this regard is the way that he combines references to classical and neo-Confucian texts and figures, Sun Tzu's *Art of War*, Sun Yat-sen's political philosophy, the Japanese Bushido code, Christianity, and the law of war.

After sketching a brief and very selective biography of Chiang and his background, I will focus on ferreting out the ethical themes of his wartime messages and addresses during the Second Sino-Japanese War, which, of course, overlaps the Pacific phase of the Second World War more generally. In so doing, I will employ as a heuristic organizing device the just war typology of *jus ad bellum, in bello*, and *post-bellum* in a rough and ready way so as to reconstruct a reasonably coherent and accessible account of Chiang's military ethics. It should be noted that as, in the case of other Chinese military and political thinkers, Chiang uses the language of virtue and character as well as

DOI: 10.4324/9781003336372-13

invoking moral and military exemplars, including here both historical figures and episodes. Here then is a brief biographical overview of Chiang's background and career.

Chiang Kai-shek was born in 1887 during the late Qing Dynasty, received a traditional schooling in the Confucian classics, classical poetry, and Chinese history, attended the Baoding Military Academy before moving on to the Tokyo Shinbu Gakko (an imperial Japanese Army academy for Chinese students), served a stint in the Japanese Army (field artillery unit), and returned to China to join in the revolutionary struggle to overthrow the Qing Dynasty.[1] He joined the Chinese Nationalist Party of the Kuomintang (KMT) and became a close supporter and associate of Sun Yat-sen, who appointed him as Commandant of a new military academy at Whampoa to train officer cadets for a new revolutionary army. After Sun died in 1925, the Nationalist government named Chiang as commander of all its forces, charging him to lead a Northern Expedition against the warlords in that area with the objective of unifying China under the KMT. By the fall of 1928, Chiang largely succeeded in this endeavor and then became head of a newly organized government of the Republic of China, based in Nanking and dedicated to reconstructing China according to Sun's "Three Principles of the People" (nationalism, democracy, and the people's livelihood). Although Chiang was acknowledged by the international community as China's President, his rule was seen as corrupt by the Chinese Communist Party (CCP), and thus began the first phase of civil war between the communists and the nationalists. After marrying Mayling Soong (the sister of Sun Yat-sen's widow) and then studying the Christian bible and related works, Chiang officially converted to Christianity (Methodist) in 1930. In 1934, after reinterpreting Sun's political thought as fundamentally Confucian at heart, Chiang launched a New Life Movement to help infuse a moral spirit into the Chinese populace based on the Confucian virtues of propriety, justice, integrity, conscientiousness, and benevolence. In 1937, after a number of military provocations by the Japanese (including their earlier invasion and annexation of Manchuria), the Second Sino-Japanese War broke out, which resulted in a temporary uneasy "alliance" of sorts between the KMT and the CCP, until the Japanese were defeated by the allied powers in 1945. The internal civil war between the KMT and CCP resumed in 1946, and after three years the communists established the People's Republic of China (PRC) in 1949, with the result of Chiang fleeing mainland China to Taiwan with remaining KMT leaders and military forces.

It cannot go unnoticed that I have intentionally crafted this brief "bio" to highlight the influential sources of Chiang's thought. He had a classical Chinese upbringing that steeped him in the Confucian classics (e.g., *Analects, Mencius, The Great Learning*, and *Doctrine of the Mean*) and exposed him to ancient Chinese military-philosophical texts (e.g., Sun Tzu) and neo-Confucian scholar-generals (e.g., Wang Yang-ming, Zeng Goufan). He had most of his original military training in Japan, exposing him to the Japanese

Bushido code, which, in turn, was reportedly influenced by neo-Confucian thought (e.g., Wang Yang-ming). His relationship with Sun Yat-sen, while principally action-oriented, nonetheless instilled in Chiang Sun's Three Principles. In serving as the first Commandant of the Whampoa Military Academy, Chiang controlled the curriculum, and it is reported that he specifically instructed the military cadets "in the principles of the *Great Learning*, constitut[ing] the most important political philosophy of our nation" [Chiang's own characterization] and that he published an enlarged version of selections of the maxims of Tseng Kuo-fan (Zeng Goufan) and Hu Lin-I, using it as "a required text for the ideological edification of the Whampoa cadets." With respect to Chiang's reinterpretation of Sun Yat-sen's Three Principles and the launching of the New Life Movement, it is reported that Chiang relied on Tai Chi-t'ao's published exegesis, *The Philosophical Foundation of the Teachings of Sun Yat-sen*, in which Tai related a conversation in which Sun claimed "There is an orthodox system of moral philosophy in China that began with Yao, Shun...and ended with Confucians. My thought is in the tradition of this orthodox system...which I seek to develop and illumine." And, with respect to how his conversion to Christianity could be reconciled with his Confucian leanings, Chiang, in a speech to a group of students in 1934, said "These principles [of brotherhood and public service] were taught by Confucius to his disciples and by all saintly and worthy men through the ages. Even Jesus' giving of himself to the world was based on the spirit of those principles. We must emulate the self-sacrifice of Confucius and Jesus in our own service to the community, the nation, and even...entire humankind."[2]

Chiang's military thinking is available in English translation in a number of out-of-print collections, such as *The Collected Wartime Messages of Generalissimo Chiang Kai-shek, 1937–1945*, compiled by the Chinese Ministry of Information in 1946; Chiang Kai-shek, *Resistance and Reconstruction: Messages During China's Six Years of War 1937–1943*, published by Harper & Brothers in 1943; *President Chiang Kai-shek's Selected Speeches and Messages 1937–1945*, published by the China Cultural Service in Taipei (no date); Chiang Kai-shek, *All We Are and All We Have: Speeches and Messages Since Pearl Harbor, December 9, 1941–November, 1942*, published by John Day Co. (no date); *The Voice of China: Speeches of Generalissimo and Madame Chiang Kai-shek Between December 7, 1941 and October 10, 1943*, issued by the Chinese Ministry of Information (no date); and *Selected Speeches on Religion by President and Madame Chiang Kai-shek*, compiled and published by the Office of the Government Spokesman in Taipei, July 1952. These messages and speeches have varied audiences – e.g., government officials, the armed forces, schools, the public (both national and international), and Japan itself (including its government, military, and public). In what follows, I will attempt to highlight and illustrate the main military ethical themes according to the usual template of just war categories, beginning with *jus ad bellum* concerns.[3]

Jus ad bellum

With respect to *ad bellum* criteria, I will focus primarily on just cause, right intent, last resort, and probability of success, since it is patently obvious that Chiang himself was regarded as China's legitimate authority during the Sino-Japanese conflict. I will also under this general rubric say something about Chiang's overall military strategy and his critique of Japan's strategy, since in both cases he is at pains to invoke what is arguably the "ethical" side of Sun Tzu's *Art of War*.[4] What I mean by Sun Tzu's "ethical side" is his emphasis on "command" moral virtues in strategic thinking, with particular emphasis on sagacity (practical wisdom), including self-examination in comparison with the other's position, prudence, flexibility in decision-making, and maintaining the initiative in the field, all of which are also relevant to Chiang's *in bello* thought, as we shall see later. Although the Japanese had earlier invaded Manchuria (the Northeast Provinces of China) in 1931 and then continuing to occupy it until 1945, the war formally began with the Lukouchiao (Marco Polo Bridge) Incident in July 1937, when, unprovoked, Japanese soldiers attacked and killed Chinese soldiers stationed there. Up until this incident, the Chinese National Government had been dealing diplomatically with Japan's belligerent activity. From China's point of view, this incident was the last straw. As Chiang says:

"For the last two years the National Government...has consistently followed these principles [of diplomacy] in the hope that the confusion caused by Japan's arbitrary actions might be overcome and all problems might be dealt with through recognized diplomatic channels...[T]here is only one thing to do when we reach the limit of endurance...throw every ounce of energy into the struggle for our national existence and independence....Let our people realize to the full the meaning of 'the limit of endurance' and the extent of sacrifice implied...There is an old saying, 'He is the sacrificial knife and bowl, and I am the sacrificial meat and fish.' We are about to reach this most terrible condition."
("The Limit of China's Endurance" (1937), SSM, pp. 3–5)[5]

A few months later, Chiang becomes even more pointed about the nature of Chinese resistance:

"We are...suffering from the invasion of a cruel foe; an invasion carried out with unparalleled ferocity...[W]e are fighting this war of self-defense in order to save ourselves from annihilation...This war is not simply for the survival of our own race, it is a struggle for justice...and for international faith and righteousness. The Japanese started this war to satisfy their lust for aggression...[T]hey...have become the enemy of all mankind. Such an inhuman and unjust war of aggression, such an unwarranted attack upon another country cannot but end in [Japanese] defeat and ruin."

("Fight to Win" (1937), RR, pp. 22–23)

It is patently clear from these brief excerpts that the just cause for this war is self-defense against unjust aggression and that it was preceded by efforts at peaceful accommodation. In effect, China's self-defensive war is taken only reluctantly as a last resort. With respect to right intention and probability of success, Chiang goes on to write:

"There is an old saying: 'In war a righteous cause is strength, but an unjust cause is weakness,' and again, 'Many come to the aid of the man who has right on his side, but none helps the man who flouts all moral principles.' This war has already shown that the Japanese, in spirit and in reality, are defeated, and that their end is at hand. If we but sacrifice to the end, our cause will certainly triumph...[W]e have behind us five thousand years of history and culture; we have the resources of four million square miles of territory; we have a population of 450,000,000 – the greatest in the world. We may be sure that so great a nation cannot be destroyed...We must not only save ourselves; we must save the world. This is the spirit of Christ – His spirit of self-sacrifice, of love, and of peace...We must realize that our struggle today has as its basis our determination to establish permanent peace."

("Fight to Win," RR, p. 24)

The intention to re-establish peaceful and just conditions and maintain them seems clear enough, but, with respect to probability of success, Chiang's comments might be seen as pious platitudes, but in another couple of addresses, Chiang shows the pragmatic underpinnings of his claim of China's probable success in its defensive war. This pragmatic line of thinking is notable for the way in which Chiang's uses Sun Tzu's *Art of War* to criticize Japan's lack of a coherent and effective strategy and for his own positive deployment of Sun Tzu in China's strategy.

Before I turn to his explicit invocations of Sun Tzu, let me briefly introduce the way that Chiang characterized China's broad self-defensive strategy in an address to the Chinese people. In his 1938 address, "A Turning Point in Our Struggle" (RR, pp. 49–53; the following quotes are from these pages), Chiang writes: "From the beginning, our plan has been to establish the bases of our resistance, not along the coast or rivers, or at the centers of communication, but in the vast interior," later adding, "[W]e must keep in mind the consistent program and policy...resolved upon at the outset...[which]...consists of three essentials: (1) the war must be prolonged; (2) the war must be fought on many fronts; (3) we must keep the initiative." And, he then goes on to clarify what he means by keeping the initiative by prolonging the war:

"Our way of dealing with the enemy is to prolong the war without surrender; for the deeper the enemy penetrates into our territory, the more he will himself be in a defensive position…Only by securing and maintaining the initiative shall we be able to defeat the enemy's attempt to win a short decisive War… Ours is a nation of vast size, huge population, and immense resources. The wider the sphere of hostilities extends, the stronger will our initiative become." That is, Chiang's strategy is to use the advantages of time and size to bog down the Japanese military down in such a way that it will be forced to scatter and dissipate its forces in a vain effort to control China's vast territory and population. As reported by an early commentator on the war, P. C. Chang (of later UDHR fame) described Chiang's strategy as:

> "It is not like one fist striking another. It is like a fist striking soft glue. As the fist advances the glue engulfs it. The fist may feel very proud that it going ahead. But the farther it goes in, the harder to get it out."
>
> *(Mallory 1939, 699–711)*[6]

This approach involves using the natural resources and features of China itself, as well as time, to undermine the enemy's strategic potential and undercut its ability to take the initiative – an insight worthy of Sun Tzu's counsel in the *Art of War*.

In a 1939 address to the Executive Committee of the KMT, Chiang becomes more precise about his reasons for believing that China cannot be conquered ("China Cannot Be Conquered," RR, pp. 70–83; subsequent quotes are from these pages). He first re-characterizes China's military strategy in relatively simple terms: "[O]ur war may be divided into two periods. During the first period…we tried to wear down the enemy's strength and, at the same time, to protect our rear so that solid foundations might be laid for the second period of protracted resistance…Our present task is to build upon the accomplishments of the first stage, to carry out the plans…for the second stage, and to concentrate our efforts upon victory and reconstruction. We are now turning defense into attack." Then he goes on to adumbrate the "reasons why our enemy will surely be defeated" (pp. 72–73):

> In the first place…his present campaign of aggression is an attempt to carry out his long-cherished continental policy [of dominating the Orient]…But his precipitate and premature invasion of China…ran directly counter to the prepared strategy of the Continental Plan. It not only placed him in a dilemma regarding future moves…He came to a point where he could not advance further without arousing the enmity of other Powers…Secondly, his scheme for defeating us depended primarily upon our yielding without resisting…As he pushed his way step by step inland his hopes of a short war and quick victory vanished…In the third place…Japan has not only gone contrary to three fundamental military

principles but has also made three serious military blunders...(1) [No] accurate appraisal of their own strength or of their opponent's strength... (2) They have relied too much on good fortune to win, and have dis- regarded the fundamental principle that victory is achieved by swift and agile movement of troops...prov[ing] the truth of Sun Tzu's famous saying, "A prolonged military campaign is disadvantageous to the state... (3) The old caution against "massing a big force in the enemy's inter- ior"...and yet the Japanese...have advanced recklessly...hundreds of miles into China's hinterland. The occupied areas have become fields of activity for our regular and guerrilla forces...the Japanese forces [now] find themselves in a position where advance and retreat are equally dangerous. They have become, in the words of Sun Tzu, "an army with fighting edge dulled by failing strength and exhaustion of supplies." They are face to face with the critical moment when, as the same writer puts it, "oppo- nents are awaiting their collapse to attack them." The enemy is now "suspended in mid air," out of touch with both earth and sky...the point where "having failed to win he can hardly return."

A bit later, Chiang also writes (p. 74):

Sun Tzu, in his treatise on topography, outlines the three military princi- ples of knowing "where the enemy cannot be attacked, where one's own troops cannot attack, and upon what ground a battle cannot be fought." I would state Sun Tzu's principles thus: "Know where the enemy can be attacked, know where one's forces can attack, and know upon what ground a battle can be fought," and would add, "Unless one knows the right time to attack it is useless to expect victory." Our enemy does not know whom he is attacking and he also does not know when to attack. For all these reasons Japan will unquestionably fail to conquer China.

Q.E.D.: China has a reasonable probability of success from Chiang's perspective.

And he drives the point home over and over again: "Sun Tzu says, 'A vir- tuous government has the support of the people; it can command life or death from the people without exciting fear or complaint.' Concord between Gov- ernment and people is the first essential to victory." Furthermore, says Chiang, "From the geographical point of view, our country possesses natural advantages for defense...we have always considered typography and climate of great importance...Economically we have the advantage...Our enemy has not reckoned with our endurance as an agricultural nation." And he adds to Sun Tzu another more contemporary insight or principle:

"In prosecuting a modern war, it is essential not only to have a thorough knowledge of the enemy and of oneself, but also to understand the trend of international developments...the main undercurrent [of which] has been in the direction of maintaining international obligations and world peace...In the course of time [China] is bound to earn the sympathy and cooperation of other just nations."

(pp. 77–79)

A Special Word about Just Cause

Before I turn to Chiang's reflections in the vein of *jus in bello* concerns, I want to say a few more words about the nature of China's just cause for two reasons. First, Chiang's assessment of just cause arguably employs the thought of neo-Confucian Zeng Goufan. Second, his assessment also bears some comparison with Michael Walzer's notion of a supreme emergency, which might, in turn, lay some groundwork for explaining (perhaps excusing) some of Chiang's deviation from the proper *in bello* treatment of Chinese citizens, which will be discussed at the end of this chapter. I have already mentioned that Chiang uses the phrase "annihilation of the Chinese people" in connection with the theme of self-defense, but this phrase pales in comparison with others that he uses. For example, in his 1938 address to the people entitled "Our Own Soil, Our Own people" (RR, pp. 41–42), Chiang writes:

> With their endless atrocities of indiscriminate bombing, killing, slaughtering, pillaging and raping, the Japanese invaders have broken the black record in the history of mankind. But these are not all. For, in addition... the Japanese have been adopting wholesale drugging and enslaving policies. They have driven our compatriots to fight for them at the battlefronts in pursuance of their pernicious principle of making Chinese kill Chinese. They have also kidnapped large numbers of Chinese children... and sent them to Japan to teach them to fight against their own kin. Evidently, the Japanese have determined not only to annex Chinese territory but also to exterminate the Chinese race...In [so] attempting, the Japanese have resorted to every pernicious means they can think of. Witness the rigid control of educational policies, the suppression of our language...the compulsion of adopting enslaving textbooks...the forcible allotment of bonds...their mean and malicious policy of drugging...[O] pium addicts in the four Northeastern Provinces numbered 13,000,000, representing more than one-third of the total population there.

For another example, in his 1937 broadcast, "After the Fall of Nanking" (RR, pp. 27–28), Chiang says, "The present Japanese invasion...has for its chief objectives not merely the occupation of our territory, the massacre of our people, and the destruction of our culture and civilization, but also the eradication of our Three Principles...[and imposing]...a life of slavery, a life no better than that of a beast of burden..."

These comments are deepened considerably in Chiang's 1940 message to the people of Manchuria, aptly entitled "Manchuria: Hell on Earth" (RR, pp. 198–201; subsequent brief quotes from these pages), in which the list of abuses includes, for example: "no security for life and property and no freedom of movement"; "[arbitrary] Japanese desires defining the scope of laws"; "special permits for marriage and giving in marriage"; "compulsory labor and military service for all men over 19 years of age"; "[arbitrary] police imprisonment and torture"; "confiscation of businesses and land and their reassignment to Japanese colonists"; "oppressive and extortionate taxation"; "deprivation of the right to purchase food staples (rice and flour)"; "dehumanization by hunger"; "arrest, detention, and killing of Chinese intellectuals"; "the closing of missionary schools"; "replacement of Chinese teachers with Japanese teachers in post-middle-school educational institutions"; "compulsory teaching of the Japanese language"; "replacement of Chinese language by a system of phonetic symbols similar to Japanese"; "film censorship"; "censorship of library holdings"; "denial of freedom of residence, livelihood, speech, education, marrying and burying"; "the establishment of opium and gambling dens, brothels, and wine shops"; monopolization of trade and the professions"; and the list could continue.

The first thing I want to say about these abuses is that they eerily parallel Raphael Lemkin's list of the techniques of genocide in his classic ninth chapter defining "genocide" in *Axis Rule in Occupied Europe* (Lemkin 2008, chap. 9). Consider those that Lemkin identifies with respect to Nazi impositions on occupied European countries: political techniques involving the replacement of local rule and self-governance; social techniques entailing attacks on intellectuals; cultural techniques banning or suppressing the use of native language in education; economic techniques shifting resources from the occupied to the occupier; biological techniques limiting marriage among the occupied and thus decreasing their birth rate; physical techniques involving the rationing of food, endangering of health, and killing; religious techniques disrupting the national and religious (or civilizational) influences of the occupied people; moral techniques or policies designed to corrupt the national group including encouragement of pornography, debasement of sexual mores, and alcoholism or drug abuse. I think it obvious that these parallels strongly suggest that, from Chiang's perspective, the Japanese occupation was "genocidal" in character (though that term would not been available to him at that time) and that Manchuria's "hell on earth" was being envisioned by Japan as its plan for the entire Chinese people. This observation seems to suggest that Chiang's China was faced with a supreme emergency of a sort not terribly different in scale from what was facing European civilization in the prospect of a Nazi victory. Again, I will return to this point in my later discussion at the end of the chapter.

The second thing I want to say is that, as laid out by Jonathan Chan in his chapter on Zeng's military ethics (chap. 9 of the present volume), Zeng himself had envisaged his counterinsurgency campaign against the Taiping rebels as partly justified by their threat of tyranny and destruction of the Confucian socio-political order (or Chinese civilization as it existed at the time) (Tien 1992, chap. 3; Guo and He 1999, 142–70, especially 154). Given that Chiang was clearly enamored of Zeng's military and political thought (as mentioned earlier), I think that it is entirely possible that Chiang was thinking of Zeng's just causes when he characterized the "hell" of occupied Manchuria as paradigmatic and indicative of Japan's broader intentions for the whole of China. To repeat Chiang: "The Japanese are clearly out to destroy the Chinese language and culture in their extinction of the Northeast as a part of China" ("Manchuria: Hell on Earth," RR, p. 201), and, more broadly, "The present Japanese invasion of China has for its chief objectives...the occupation of our territory, the massacre of our people, and the destruction of our culture and civilization" ("After the Fall of Nanking," RR, p. 27).

Jus in bello

Chiang's characterization of occupied Manchuria as a hell on earth segues nicely into his discussion of *jus in bello* norms and their violation. His discussion has two parts: firstly, his charges against the Japanese military forces for their contraventions of the laws of war and humanitarian law, followed by, secondly, his positive description of Chinese military training, discipline, and virtues of leadership (in contrast with the Japanese military). I begin by reminding us that the main principles of war law – then embodied in the Hague Treaty of 1907 and the 1929 Geneva Convention – include military necessity, distinction, and proportionality. The Hague Treaty in particular prohibits the attack or bombardment of undefended towns, villages, dwellings, or buildings, as well as mandating, so far as possible, the sparing of buildings dedicated to religion, art, science, or charitable purposes, historical monuments, hospitals and places where the sick or wounded are collected. Interestingly, in 1937, the League of Nations also adopted a resolution calling for the protection of civilian populations against aerial bombardment. In addition, the rules of land warfare make it clear that military necessity does not admit of cruelty, maiming or wounding except in combat, the use of torture, the use of poison in any way, or the wanton devastation of a district. Further, captured subjects are not to be compelled to take part in operations against their own country. Moreover, pillage is forbidden; prisoners of war are to be treated humanely; women are to be treated with due regard; and military occupation is to be temporary and its administration is hedged about by humane provisions too numerous to be listed here.[7] Both Japan and China were signatories to these conventions as well as to the 1928 Kellogg-Briand treaty "outlawing" wars of aggression and mandating peaceful means of dispute-settlement between states.

Japanese Conduct

In the face of all of these provisions, Chiang goes to great pains in various addresses – with such suggestive titles as "Bombing and Open Towns" (1939) (RR, pp. 101–105), "Japan: Enemy of Humanity" (1938), RR, pp. 34–36) and "To the People of Japan" (1938),RR, pp. 39–40) – to charge the Japanese military with massive war crimes and civilian abuses from which one can fairly infer norms that the Chinese military regarded itself as being committed to in its own behavior. I can no better than to quote Chiang about his charges against the Japanese military, and then to follow these up with Chiang's own reflections on the implications (selected from the pieces and pages mentioned above and numbered here for convenience):

1. "The following unfortified towns containing no military establishments whatsoever have been bombed, their streets covered with blood, their skies lit up with conflagrations [thence followed by a long list of such towns and cities; (RR, p. 101)."
2. "By incessant air raids, [the enemy] hopes to strike terror into the hearts of people...to hinder or destroy production enterprises, destroying the livelihoods of the masses...[H]e seeks to destroy our people, raining death on peaceful non-combatants in crowded centers...[P]laces with no military objectives and little air defense are...being bombed, incessantly and indiscriminately" (RR, p. 101).
3. "The enemy has not confined his atrocities to aerial massacres. Our fellow countrymen in the occupied areas are suffering more...Men are being enslaved, women maltreated. Many are being forced to take poisonous injections and narcotic drugs" (RR, p. 105).
4. "Countless [civilian] industries and vast quantities of raw material, at the front and in the occupied areas have been totally destroyed, and young men and girls, women and children, the old and the weak, have been subjected to unspeakable horrors, to rape and plunder and burning and death...The Japanese air force...[is]...taking as its special targets our cultural, educational, and philanthropic institutions and our residential areas" (RR, p. 36).
5. "Poisonous gases are relentlessly used and principles of justice are trampled...There was looting and burning of property. They massacred innocent civilians and wounded soldiers. They slung these poor creatures together by the hundreds and mowed them down with machine-gun fire...drove scores of people into a room and set fire to it...made competitions...to see who killed most simply for fun" (RR, p. 39).

I could continue in this vein, but you get the drift – and Chiang also makes the point that there is clear and impartial evidence for these atrocities and their being made known to the world by virtue of "officials and nationals of the friendly Powers conduct[ing] investigations on the spot...describing in detail what they have witnessed." His summary judgment:

> "If the savage cruelty of these Japanese...perpetrated in the name of civilization is allowed to continue unchecked and unpunished, then the world will never know permanent peace or justice, and we shall be left with an indelible stain on our consciences."
>
> *(RR, p. 36)*

Perhaps the most poignant way that he couches this judgment is in terms of both Confucian and Bushido traditions. In the case of the first: "What I do want to emphasize is the barbarism of the Japanese militarists, a barbarism which could wipe out those basic human qualities that Heaven has implanted in man" (RR, p. 36]) This reference to "qualities implanted by Heaven" is clearly reminiscent of Mencius's understanding of human nature as being imbued with the seeds of virtue that guide the moral development of human nature if they are nurtured and not interfered with. Note that Chiang is speaking of the Japanese militarists as fellow human beings who are in manifest danger of thwarting and losing their implicit moral destiny.

The second way that Chiang expresses his summary judgment is in his 1938 address to the Japanese people where he plaintively suggests: "nothing has [been] left of the 'Soul of Yamoto' and 'Bushido', of both of which your country had been so proud" (RR, p. 39). "The Soul of Yamoto" invokes the Japanese version of the heart-mind (spirit, conscience), and "Bushido," the Japanese concept of honor, valor, loyalty, and self-sacrifice to which Chiang had been exposed during his military training in Japan and which he had wished to instill in his cadets at the Whampoa Military Academy. A secondary source reports that Chiang believed that Wang Yang-ming's notion of the unity of thought and action – in the sense of true moral knowing as experiential action – lay behind the Japanese concept of Bushido, though perhaps lacking Wang's Mencian-like emphasis on the virtue of *ren* (benevolence, compassion, humaneness) (Benesch 2014, 129–68; Yamada 2017, 13–35, especially 28). If this representation is accurate, then Chiang appears to be reiterating his claim that the Japanese military is in grave danger – if it had not already done so – of losing its human moral nature, and he is rephrasing this claim in terms that the Japanese people in general could easily understand and appreciate.

Chiang links this claim or warning to a brief reprise in that 1938 address to the Japanese people of what he regards as Japanese war crimes, including an appeal to the notion of moral reciprocity ("To the People of Japan," RR, pp. 39–40): "The objects of their bombing were defenseless civilians, and cultural, educational, and charity institutions. My friends, you must have known that our military planes have flown around your principal cities. But we only conveyed to you our sympathy [in the form of propaganda leaflets dropped], not destructive bombs." This is clearly a moral appeal that has the additional

virtue of expressing the Chinese military's own recognition of and commitment to the principle of non-combatant immunity (or discrimination) in at
least aerial warfare. Let me now turn to Chiang's discussion of Chinese military training, discipline, and virtue, which I believe is steeped in distinctive
Confucian and neo-Confucian values.

Chinese Conduct

More accurately put, Chiang's own constructive understanding of the funda-
mentals for a well-trained, disciplined, and appropriately motivated military
appears to involve a combination of influences: Sun Tzu's *Art of War* and the
military ethics of Zeng Guofan, with "pinches" of Japanese Bushido and
Wang-Yang's thought possibly added to the mix. The influence of the Bushido code appears to be sounded in Chiang's emphasis on the spirit of self-
sacrifice, loyalty, and courage; for example:

> "Although our military equipment is not equal to that of the enemy, yet if
> we retain the...spirit that is ready for any sacrifice and keeps loyal and
> brave to the end, and in that spirit go forward against the foe, there will
> be no question of Japan's defeat."
>
> *("Drive Out the Invader" (1937)*

, RR, pp. 16–19; subsequent quotes are from these pages). Although admit-
tedly these traits are highlighted by most forms of military ethics, particularly
in exigent circumstances, this formulation is repeatedly used by Chiang in
various of his addresses and admonitions to the Chinese armed forces and is
strikingly redolent of the essentials of the Japanese Bushido code in which he
was trained.

More definitively, what I am calling the "ethical side" of Sun Tzu (see
above) is sounded in such injunctions as:

> "We must make full use of our mental powers and take the initiative. In
> the history of war, general strategy and tactics have naturally been the
> responsibility of the highest authority, the Commander-in-Chief...But the
> officers of each unit must on their own initiative study the situation
> before them and implement the orders of headquarters, for example in
> matters that concern local topography, the details of the enemy's condi-
> tion and of our own, the organization of guerrillas, and the use of spies.
> Ways and means of meeting emergencies...must be devised by the officers
> of each unit, using their own mental powers to secure the victory. All
> officers from army, division, and brigade commanders down to subalterns
> must learn to take the initiative."
>
> *(RR, p. 17)*

This formulation of military traits and virtues appears to reflect classical Sun Tzu thinking, now being extended by Chiang to include subordinate officers.

In examining the military ethics of Zeng Guofan, Jonathan Chan identifies a distinctive Confucian virtue-oriented emphasis on "loving the people," even to the point of Zeng's writing a lyric poem on the subject [chapter 9, this volume]. Arguably, this aspect of Zeng's ethics is encapsulated, though less poetically, in Chiang's admonition and appeal to the military:

> "Soldiers and civilians must be united in a common bond of love and sincerity...[I]f we expect the ordinary people and the soldiers whole-heartedly to unite their efforts...to work in perfect harmony and to help each other, then...you soldiers must...show your genuine love for the people and win their trust and confidence...treat them as though they were members of your family...[Y]ou must, wherever you are, help the civilian population, instruct them and guide them, protect them and save them, and so give evidence of your love and sincerity. If you sympathize with them in their sufferings, and share in their joys and sorrows, then soldiers and civilians will form a closely knit body...and the enemy will meet with defeat everywhere."
>
> *(RR, pp. 17–18)*

Another distinctive aspect of Zeng's military ethics that Chan discusses is his conceptualization of the military as forming a family whereby the superior officers act like fathers toward their soldiers and subordinates are expected to act like sons [chapter 9, this volume]. Again arguably, this particular aspect of Zeng's thought is discernable in Chiang's reasoning behind having to discipline and disband the New Fourth Army. In this episode, superior officers had repeatedly disobeyed orders issued by headquarters to the detriment of subordinate soldiers, military position, and even the civilian population. I need not enter into the details, but I do want to draw attention to how Chiang represents his disciplinary role ("The Function of Revolutionary Discipline" (1941), RR, pp. 218–24):

> "I have often compared the army to a family wherein I look upon the soldiers under me as a father regards his children. If his children behave well the father feels they reflect honor upon him; if badly, they disgrace him. I attempted to discharge my responsibility towards the New Fourth Army in the past by repeatedly warning it and imploring it to make a fresh start in the genuine service of the nation. I feared a premature revelation of its misdeeds might cut off its way to reform. My solicitude failed, however, to move them; they interpreted it as weakness and even timidity...It was certainly not, therefore, for fear of letting them or the world know that we abstained from publishing the state of affairs for so long. All along the motive lay in the moral precept, held so important in Chinese

society, of 'keeping evil out of sight and bringing good to the fore.' I have always observed this principle…only the more studiously in dealing with soldiers under my command, to whom I feel bound in an intimacy equal to that of family relationships…[T]here was no further room for pardon, if I myself were not to become criminally negligent of my country's welfare."

So, in sum, the themes of the army being a family, and, earlier, love of the people are Zeng's themes being recapitulated by Chiang. Moreover, Chiang's patient treatment of the recalcitrant Fourth Army – issuing warnings and giving it opportunities for reform – bears some similarity to Wang Yang-ming's military ethics in Wang's "discriminate pacification" of bandits in the borderlands during the late Ming Dynasty when he offered the bandits opportunities for reform as a father would to his beloved children (Twiss and Chan 2015, chap. 7).[8] Chiang does not, so far as I can determine, lay out a specific set of rules for the military's interaction with civilians – as did Wang Yang-ming and even Mao Tse-tung – but these are surely implied in what he does say about such interaction. Moreover, again as far as I know, although he does not explicitly discuss the topics of non-combatant immunity and proportionality in the conduct of war or norms governing the treatment of POWs, these were evidently embraced in his forceful critiques of Japanese military behavior. I think that one can safely assume that Chiang does not need to do so, for his military appears committed to the rules of war and humanitarian law as laid out in the conventions which China had earlier signed.

It appears, then, that Chiang's *in bello* ethics involves the following virtues and principles, indicating parenthetically the principal influences for each cluster:

1. Being of one mind, unified, and resolved to be loyal, self-sacrificial, and courageous to the end (Bushido code).
2. Making full use of one's mental powers and taking the initiative based on careful study and comparative assessment of the enemy and oneself (Sun Tzu).
3. Soldiers and citizens being united in a common bond of love and sincerity, with soldiers' treating ordinary citizens humanely and benevolently as if they all are family members (Zeng).
4. Officers and soldiers construed as a family, with the commanders treating subordinates with both solicitude and firmness, disciplining them severely only after giving them the opportunity to reform (Zeng and Wang Yang-ming).
5. Conforming to extant rules of warfare and humanitarian law operative at the time in lieu of formulating explicit lists of regulations (Hague and Geneva treaties).

Jus post bellum

Simultaneous with his military planning, Chiang also addresses the issue of his vision for *post-bellum* national reconstruction, or phrased more precisely, his on-going vision of such reconstruction both during and continuing after the war. As he put it early on in 1938: "These phases of our evolution, from unity to resistance and finally to reconstruction, have long been foreseen" ("A Turning Point in Our Struggle," RR, p. 52). Much of this vision is framed in terms of Sun Yat-sen's Three Principles of the People, namely: the achievement of national independence, the attainment of a true democracy after a period of political tutelage of the people, and economic revitalization able to meet the basic needs of all the people in a equitable manner. Chiang's plans for meeting these goals are quite detailed and need not be developed here, but it is important to emphasize that he thinks it essential for the people themselves to be morally and socially reconstructed – or to use a more Confucian term, "rectified" – and so he launched the previously mentioned New Life Movement, "the chief aims of which are to change the moral atmosphere, to eliminate bad social customs, and to develop a vital and healthy social organism involving the principles of propriety, justice, integrity and conscientiousness" ("On National Reconstruction", 1937, pp. 7–14).

As he puts it in the conclusion to "National Reconstruction": "the driving force we need is none other than the ancient moral principles of our race: loyalty, filial devotion, kindness, love, faithfulness, justice, harmony, and peace," supplemented by "sincerity and strict self-discipline" (RR, p. 14]) Remarkably in yet another place ("Mobilizing Our Spiritual Resources" (1939), SSM, pp. 112–15; the following quotes are from these pages), Chiang links Sun's Three Principles to Confucian (and later neo-Confucian) socio-moral and political thought: "In order to mobilize the national spirit it is necessary for the people to have a common standard of ethics, to share a common faith in the future of the country, and to labor, struggle, and sacrifice for this common standard…What is this common standard of ethics? Nothing but the abandonment of the small self in favor of the large self [thus invoking the language of Mencius VI.A.14]. What is the common faith? Nothing but the *San Min Chu I* (Three Principles of the People), the ultimate goal of which is China's highest political ideal – universal brotherhood."

This linkage is followed by a well-known passage from the *Book of Rites* about the Great Way: "When the Great Way is followed all under heaven work for the common good. The wise and capable are chosen for office. They speak the truth and cultivate harmony. Men do not limit their filial piety to their own parents nor their parental love to their own children. The aged are provided for…the able-bodied are usefully employed…Kindness is shown to the…afflicted. Every man has his work…People do not waste…or hoard… [W]hen they toil it is not simply for their own advantage…" Chiang concludes this quotation by saying, "These are the objectives of our program of national regeneration. In their attainment the *San Min Chu I* will find realization and our own people will find happiness. Today we are giving our sweat and blood upon the battlefield, and at the same time we are marshaling all our intellectual forces for reconstruction." And Chiang's subsequent admonition to People's Political Council expresses "the hope that not only members of KMT but also the entire nation, especially the leaders in various fields, would take up together these plans for reconstruction and turn them quickly into reality." And, he is not beyond invoking – at least implicitly – Huang Tsung-hsi: "the fall of the Sung and Ming Dynasties was due largely to selfish[ness] …The leaders of the state sacrificed national freedom for personal interests .. Let us not repeat their mistake but rather be a noble example to our descendants" (p. 116] see Huang 1993, 91–92).

In another address to the People's Political Council ("China's March Toward Democracy" (1939), RR, pp. 89–92; following quotes from these pages]) Chiang links Sun Yat-sen's Three Principles to not only Huang – again implicitly – but also past classical examples of political tutelage:

> "The entire aim's of Dr. Sun's life and struggle was to carry out the Three Principles and to restore the government to the people…call[ing] upon the intellectuals and his comrades…to become guardians of the people, and as public servants to teach and guide [them]. In Chinese history there are two good examples of political tutelage: Premier Yi Yin and King Tai Chia of the Shang Dynasty, and Duke Chou and King Cheng of the Chou Dynasty [who taught and guided their monarchs before turning over to them the reins of power]…Dr. Sun…often likened himself to Premier Yi Yin and Duke Chou…We [of the Political Council] are the Premier Yi Yins and the Duke Chous of today and the people are the King Tai Chias and the King Chengs. We must look upon the people as our masters [reminiscent of Huang's famous image of the people as masters and the princes their servants], and teach, train, and protect them until they are mature, and governing powers can be returned to them… We…must bear ourselves in an exemplary manner."
>
> *(Huang 1993, 92–93)*

It should not go unremarked that the political thought of Zeng Guofan also informs Chiang's envisioned reconstruction, for, as Mary Wright has shown in her extensive scholarship on the Taiping Rebellion period, Zeng was not only an exemplary scholar-general with military prowess but also an exemplary political leader who, after successfully quelling the rebellion, was important in the T'ung-chih Restoration (Wright 1955, 515–32; Wright 1966[1957]). That Restoration precisely involved re-establishing the Confucian socio-political and moral order not unlike Chiang's own reconstructive vision. Finally, I want to observe that Chiang is able to reframe his reconstructive vision in terms of his religious beliefs ["The Truth of Life" (1943), SSM, pp. 227–29].

He does so by speaking of Sun's Three Principles as part of "the object of faith" and having the character of "ever-living truth." Similarly, he argues that the three theological virtues of faith, hope, and love are instrumental in gaining self-control, self-respect, and strength of character. Indeed, in one place ["My Religious Faith" (1938), SSM, pp. 20–23; following quotes from these pages], he writes, "if we believe with all our hearts that the Three Principles of the People are essentially true and just principles, then we shall have the power to put them into effect." Moreover, he goes so far as to say that "Our late leader Sun Yat-sen, with his universal sympathy for all oppressed and his profound understanding of Jesus' revolutionary spirit of love and self-sacrifice, carried on his work for forty years and brought about at last the liberation of the Chinese people." And, he concludes by mentioning the New Life Movement and relating it to the promotion of ways of living with "a new spirit... [and] the quality of life that is inspired by the love and sacrificial purpose of Jesus." Thus, in addition to Confucian and neo-Confucian values, Chiang also frames (or reframes) his reconstructive vision in Christian terms and concepts.

Further Discussion

This completes my brief reconstruction of Chiang's military ethics, but there are a few additional observations and two issues to discuss, however tentatively, regarding Chiang's military ethics. The observations are simply put. First, Chiang's military ethics explicitly spans *ad bellum, in bello*, and *post bellum* concerns in a sweeping and integrated way that is fairly distinctive of Confucian ethics, traceable back to at least Wang Yang-ming's military thought and arguably Mencius and Xunzi on punitive/rescue missions. Second Chiang consistently uses the language and virtue and character, rather than codified principles and rules, which is again distinctive of Confucian ethics. Third he invokes and interlaces different ethical traditions and their terms, seeming to move with ease from one to another or even blending them within the same sentence or paragraph, which, in turn, raises the issue of consistency in his thought. That is to say, how are we to understand and characterize Chiang's framing and reframing of ethical themes in alternative languages (so to speak) – e.g., classical military thought, Confucianism, the Bushido code, Sun Yat-sen, international war law, and Christian belief? Is he simply using these traditions in a random and *ad hoc* manner, drawing on a collection of assorted odds and ends, making a loose assemblage that appeals to different audiences with differing backgrounds and commitments? Is there some overall thread of unity in this patchwork

This issue may be addressed by the possible fact, as argued by Jeffrey Stout, that "we are all *bricoleurs*, insofar as we are capable of creative moral thought at all" (Stout 2001, 74; also LaFleur 1992, 12). That is, in dealing with moral dilemmas and crises, we tend to retrieve, adapt, and reconfigure available moral concepts and categories to make sense of and respond to crises from our cultural toolboxes, which, in turn, are comprised of differing traditions, ranging from those in which we were raised to those subsequently learned. Chiang himself was originally raised in the Confucian tradition but then subsequently learned in his military training Sun Tzu, Bushido, and internationally accepted rules of war and, yet again even later, aspects of the Christian tradition. So, perhaps we should expect him to draw upon all of these sources as a practical and creative thinker and person of action. Chiang is not striving to be any sort of consistent philosopher, but rather attempting to save his nation from catastrophe within pragmatic, moral, and legal constraints. That said, it appears that there is a consistent, unified thread (even a rope) woven throughout his military ethics, and it is the constant touchstone in his *ad bellum, in bello,* and *post bellum* thinking – this thread is, of course, Confucianism, and other resources are interpreted and utilized insofar as they are consistent with this primary traditional commitment, Chiang's "first language" so to speak.

A second issue is more substantive and difficult to resolve, and I need to be careful in expounding it properly, as it is one that Chiang does not discuss in his writings (as far as I can determine), and it is claimed by other scholars that he never discussed it in his diaries. Hints of this problem were known and referred to in the West early on in the Second Sino-Japanese War, though not characterized as a moral or legal problem. For example, in his 1939 article on "The Strategy of Chiang Kai-shek," Walter Mallory wrote that:

> "Chiang has followed a plan of deliberately destroying everything of even potential economic value to the Japanese before he has retreated from a given area. This has come to be called the 'scorched earth policy.' Factories have been dismantled and what machinery could not be moved wrecked; railway rolling stock…and the rails have been taken by the Chinese as they have fallen back. Even the peasants in many regions have deserted their farms and villages, so that when the Japanese moved in they found a barren land."
>
> *(Mallory 1939, 711)*

So far as this report goes, this policy appears to be justified by military necessity in denying the Japanese the opportunity to appropriate and use Chinese equipment and to live off China's land and agricultural products. But there is a danger inherent in such a policy – namely, it risks imposing on ordinary Chinese citizens considerable harm (even disproportionate harm) if the policy is taken too far or events get out of hand.

One case in point seems especially relevant here. It is known as the "1938 Yellow River Flood," and it occurred at the beginning of the war, shortly after the "Nanking Massacre" (which involved Japanese soldiers indiscriminately killing over 300,000 Chinese civilians and POWs in that city). In the case at issue, after a number of military defeats by a rapidly advancing and better equipped Japanese army, Chiang explicitly authorized the Chinese military to destroy dams around the city of Zhengzhou in order to delay the Japanese advance, resulting in a vast flood that did in fact work for a limited period, making it impossible for the Japanese to reach areas west of the flood zone for five months, thus having the tactical effect of giving Chiang's government sufficient time to move from Wuhan, which the Chinese later moved to Chongqing. But this action came at the cost of 500,000 Chinese civilian lives (some estimates count up to 800,000 lives). As reported by Jay Taylor in his recent biography of Chiang, the flooding covered thousands of square kilometers of farm land, destroyed several thousand Chinese villages and made homeless a few million rural dwellers, in addition to the death toll (Taylor 2009, 155).[9] The question in my mind is whether Chiang's authorization could be justified as a supreme emergency in this early phase of the war, when the Chinese alone were fighting a desperate war against a technologically superior invader intent on decimating the Chinese and their civilization as well as envisioning turning the whole of China into a vast "Manchuria." Is the Chinese situation at this point in any way comparable to the British situation in the early years of the European war where, using Michael Walzer's words, "the decision to bomb [German] cities was made at a time when victory was not in sight and the specter of defeat was present" (Walzer 2006, 258).[10]

Obviously, the situations are different in this respect: the British were contemplating the violation of a war convention in directly attacking enemy civilians, whereas Chiang must have been contemplating the likelihood of violating his own *in bello* norms regarding humane treatment of Chinese citizens, which presumably includes not imposing foreseeable and disproportionate harm on one's own people. Indeed, it is reported by Rana Mitter that "dike-cutting is the blackest of Chinese crimes" (Mitter 2013, chap. 9, quotation from p. 163). But consider the parallel between Chiang's awareness of the "genocide" facing China and Walzer's characterization of Nazism as "an ultimate threat to everything decent in our lives, an ideology and a practice of domination so murderous, so degrading even to those who might survive, that the consequences of its final victory were literally beyond calculation, immeasurably awful...as evil objectified in the world...a threat to human values so radical that its imminence would surely constitute a supreme emergency." Or again consider whether or not the following could not be equally applicable to China: "to save a nation we can violate the rights of a determinate but smaller number of people...[and] .. a world where entire peoples are enslaved or massacred is literally unbearable. For the survival and freedom of political communities – who share a way of life, developed by their ancestors, to be passed on to their children – are the highest values of international society."

Of course, as Walzer frames it, "danger makes only half the argument; imminence makes the other half." So, the question for Chiang is: were the early years of the war with Japan "a time when victory was not in sight and the specter of defeat ever present?" Chiang's own strategic plans regarding the two phases of the war and all that I earlier recounted in his use of Sun Tzu may appear to undercut a claim of imminent defeat, but what if this representation at the time was simply the expression of bravado on Chiang's part, designed to shore up the morale of his soldiers, the population, and even himself? In the war's early years, the Japanese advance surely must have been daunting – and the Japanese had succeeded in invading, occupying, and imposing tyrannical rule in Manchuria. Many historians of the war appear to confirm this prospect of China's imminent defeat at that time (see again Mitter 2013, chap. 9) [11] Whether this prospect justifies, excuses, or simply explains Chiang's decision to unleash the flood, I will leave for others to judge.

Notes

1 For useful biographies of Chiang Kai-shek, see, e.g., Hahn 1955; Fenby 2004; Taylor 2009.
2 Quotations in this paragraph are from Loh 1970, 211–38, specifically 212, 215, 222, 235. See also Loh 1971.
3 For convenience, the intratextual references in this chapter are to only two of these published sources (with message/address title and date, followed by page numbers in the respective volumes): *Resistance and Reconstruction* [RR] and *President Chiang Kai-shek's Selected Speeches and Messages* [SSM]. Additional primary sources consulted include: Chiang 1947a; Chiang 1947b as well as the other volumes mentioned above.
4 See Ivanhoe 2011, 95; Lo and Twiss 2015, chap. 3.
5 I have not yet been able to identify the precise source for this saying.
6 Quote from p. 711.
7 A convenient way is to access these rules is to consult U.S. War Department (1940).
8 One must not, however, overly romanticize Chiang's use of familial imagery with respect to the military, for as indicated by Robert Bedeski, "Chiang relied on concepts such as law and order, and noted the parallel between family relationships and the army chain of command. He didn't advocate the family system as the basis of military organization. Rather, he was proposing that affective orientations were needed to encourage a collective spirit" (Bedeski 1980, 149–70, quotation from p. 154).
9 This toll was entirely foreseeable as the Yellow River had periodically flooded over the centuries, with one flood occurring in 1887 that resulted in over a million deaths.
10 Following references to Walzer are from pp. 253–55 of this book.
11 It is also interesting to note that Mitter claims that "Chiang's government had committed one of the grossest acts of violence against its own people, and [Chiang] knew that the publicity could be a damaging blow to his reputation. He decided to divert blame [from himself] by…blaming the breach [of the dikes] on Japanese aerial bombing" (p. 162).

Bibliography

Primary Sources

Chiang, Kai-shek. 1947a [1943]. *China's Destiny*. 1st English ed. Translation by Wang Chung-hui with Introduction by Lin Yutang. New York: Macmillan Company.
Chiang, Kai-shek. 1947b. *China's Destiny and Chinese Economic Theory*. Combined in one volume. With Notes and Commentary by Philip Jaffee. New York: Roy Publishers.

Chiang, Kai-shek. 1946. *The Collected Wartime Messages of Generalissimo Chiang Kai-shek 1937–1945.* Two Vols. Compiled by Chinese Ministry of Information. New York: John Day Company.

Chiang, Kai-shek. n.d. [1946?]. *President Chiang Kai-shek's Selected Speeches and Messages 1937–1945.* Taipei: Chinese Cultural Service. [Textual citation: SSM]

Chiang, Kai-shek. 1943. *Resistance and Reconstruction: Messages During China's Six Years of War 1937–1943.* 3rd Ed. New York: Harper & Brothers. [Textual citation: RR]

Huang, Tsung-hsi. 1993. *Waiting for the Dawn: A Plan for the Prince.* Translated with Introduction by Wm. Theodore de Bary. New York: Columbia University Press.

Sun, Tsu. 2011. *Master Sun's Art of War.* Translated with Introduction by Philip J. Ivanhoe. Indianapolis, IN: Hackett Publishing Co.

Secondary Sources

Bedeski, Robert E. 1980. "Pre-Communist State-Building in Modern China: The Political Thought of Chiang Kai-shek." *Asian Perspective* 4, no. 2(Fall-Winter): 149–170.

Benesch, Oleg. 2014. "The Samurai Next Door: Chinese Examination of the Japanese Martial Spirit." *Extreme-Orient Extreme-Occident, War in Perspective: History and Military Culture in China* 38: 129–168.

Fenby, Jonathon. 2004. *Chiang Kai-shek: China's Generalissimo and the Nation He Lost.* New York: Carroll & Graf Publishers.

Guo, Yingjie, and Baogang He. 1999. "Reimagining the Chinese Nation: The 'Zeng Guofan Phenomenon'." *Modern China* 25, no. 2(April): 142–170.

Hahn, Emily. 1955. *Chiang Kai-shek: An Unauthorized Biography.* Garden City, NY: Doubleday & Co.

LaFleur, William R. 1992. *Liquid Life: Abortion and Buddhism in Japan.* Princeton, NJ: Princeton University Press.

Lemkin, Raphael. 2008 [1944]. *Axis Rule in Occupied Europe.* 2nd ed. Lawbook Exchange [chapter 9 "Genocide" also available on internet].

Lo, Ping-cheung. 2015. "Warfare Ethics in Sunzi's Art of War? Historical Controversies and Contemporary Perspectives." In *Chinese Just War Ethics: Origins, Development, and Dissent,* edited by Ping-cheung Lo and Sumner B. Twiss, chap. 3. London: Routledge.

Loh, Pichon P. Y. 1971. *The Early Chiang Kai-shek: A Study of His Personality and Politics, 1887–1924.* New York: Columbia University Press.

Loh, Pichon P. Y. 1970. "The Ideological Persuasion of Chiang Kai-shek." *Modern Asian Studies* 4, no. 3: 211–238.

Mallory, Walter H. 1939. "The Strategy of Chiang Kai-shek." *Foreign Affairs* 17, no. 4(July): 699–711.

Mitter, Rana. 2013. *Forgotten Ally: China's World War II, 1937–1945.* New York: Houghton Mifflin Harcourt.

Stout, Jeffrey. 2001. *Ethics After Babel: The Languages of Morals and Their Discontents.* With new postscript. Princeton, NJ: Princeton University Press.

Taylor, Jay. 2009. *The Generalissimo: Chiang Kai-shek and the Struggle for Modern China.* Cambridge, MA: Harvard University Press.

Tien, Chen-Ya. 1992. *Chinese Military Theory: Ancient and Modern.* Oakville, ON: Mosaic Press.

Twiss, Sumner B., and Chan, Jonathan K. L. 2015. "Wang Yang-ming's Ethics of War." In *Chinese Just War Ethics: Origins, Development, and Dissent*, edited by Ping-cheung Lo and Sumner B. Twiss, chap. 7. London: Routledge.

U.S. War Department. 1940. *Field Manual FM 27–10, Rules of Land Warfare*. Issued October 1, 1940, by Order of Generals G. C. Marshall and E. S. Adams.

Walzer, Michael. 2006. *Just and Unjust Wars: A Moral Argument with Historical Illustrations*. 4th ed. New York: Basic Books.

Wright, Mary C. 1955. "From Revolution to Restoration: The Transformation of Kuomintang Ideology." *Far Eastern Quarterly* 14, no. 4(Special Number on Chinese History and Society, August): 515–532.

Wright, Mary C. 1966 [1957]. *The Last Stand of Chinese Conservatism: The Tung-Chih Restoration, 1862–1874*. With new Preface and additional notes. New York: Atheneum.

Yamada, Tatsuo. 2017. "Chiang Kai-shek's Study in Japan in His Memories." *Sinica Venetiana* 4: 13–35.

12

A SURVEY OF 21ST-CENTURY PLA SCHOLARSHIP ON THE ROLE OF MILITARY ETHICS IN WARFARE

Mark Metcalf

Introduction

When comparing concepts between cultures, a fundamental requirement is ensuring that one has a clear understanding of the terminology that is in play. Without an awareness of subtle differences between the ways in which seemingly similar terms are construed, it is likely that cultural projection or other misconceptions will result. This is a particularly important factor when considering how the military of the People's Republic of China (PRC) – the People's Liberation Army (PLA) – understands military ethics. While PLA authors frequently address military ethics in ways that seem consistent with Western concepts, terminology, and practices, upon closer examination it becomes apparent that there are significant differences in both their motivation for developing military ethics and how it is implemented.

Since the turn of the millennium, a growing number of Chinese language books and journal articles have been written by PLA-affiliated academics over a broad range of military ethics topics.[1] The scope of these writings ranges from comparative ethics to the ethics of military technology to the ethics of the use of nuclear weapons. While the large number of such publications attests to the PLA's prioritization of military ethics research, an evaluation of the writings reveals that the PLA perspective regarding the topic is certainly (and understandably) different than those in the West – "military ethics with Chinese characteristics". Their writings are overwhelmingly infused by an unwavering insistence that, as understood and implemented by the PLA, Marxist ideology is the principal source of military ethics. Many writings also explain that an important contribution of military

DOI: 10.4324/9781003336372-14

ethics is its unarguable role in improving the military effectiveness and political dependability of PLA forces.

This chapter presents a brief survey of PLA perspectives regarding military ethics as reflected in publicly available PLA writings published since the late 1990s. The decision to limit sources to teaching or research faculty at PLA institutions was made because military ethics is considered to be political (i.e., CCP) concern. So, although theoretical military ethics research is performed at non-military institutions in the PRC, the PLA assigns responsibility for developing practical military ethics policies to its political officers.

Such an opinion is consistent with the PLA's general views regarding military scholarship –research performed by the PLA produces practical results. For example, when discussing why the PLA assigns many of its own academics to study Sunzi's *Art of War*, PLA historian Fu Chao explains:

> Different scholars and military experts all learn to research the *Sunzi*, but academic conclusions and practical conclusions are entirely different. Apart from differences in educational accomplishments, this is because the relative superiority of theoretical thought is different.
>
> *(Fu 2010, 211)*

As the PLA sees things, its academic personnel provide unique research that is of practical value to the PLA. Consistent with this perspective, this chapter focuses on evaluating military ethics writings of PLA-affiliated authors.

Sources of Military Ethics

A nation's military ethics culture is shaped by a variety of factors, including history, tradition, culture, philosophy, religion, politics, etc. As a result, and at risk of stating the all-too-obvious, military ethical considerations and practices may greatly vary among different militaries. In an attempt to highlight unique Chinese perspectives, this section discusses the sources of military ethics that are discussed in PLA writings.

Political Sources

While certain aspects of contemporary Chinese military ethical thought can undoubtedly be traced to philosophical ideologies and cultural norms which have developed over the several millennia of recorded Chinese history, the preponderance of recent PLA writings on Chinese military ethics affirm their "Marxist military ethical" roots. As the PLA was established by the Chinese Communist Party ("the Party") and has always been subordinated to the Party, adherence to Marxist precepts for such a crucial issue as military ethics is certainly understandable.

In her explanation about the sources used to establish PLA's military ethics culture, Tang Fang explains why its roots are resolutely political:

> Mao Zedong once remarked "The core force that leads our endeavours is the Chinese Communist Party, the theoretical basis that guides our ideology is Marxism-Leninism." When the authors of Marxism and proletariats examined the military question, all permeated the distinct proletariat ethical values orientation and integrated the modern ethical requirements regarding military culture. These ideologies are the ethical foundation of the military ethics culture building that we carry out.[2]
>
> *(Tang 2016, 1)*

According to Tang, PLA military ethical thought was originally based on the "military ethics culture ideologies" of Marx, Engels, and Lenin (Tang 2016, 18–20). The following observations regarding science and technology (hereafter, 'S&T') ethics research principles, citing Engels, provides an example of this political orientation:

> Military S&T ethics must be guided on the basis of historical materialism, firmly proceeding from the actual benefits of defending national security and not research questions put forth on the basis of feelings divorced from reality, abstract feelings, or humanitarian positions, otherwise research can end up in a black hole. This is because good and evil are moral concepts and "all previous moral debates are all, in the final analysis, the result of the social economy of their time. And society, up to the present, moves in class opposition, so morality is, from beginning to end, class morality. Perhaps it can be spoken of in terms of rule or benefit to the ruling class or when the oppressed class becomes strong enough, representing the resistance of oppressed peoples in opposition to the ruling class and their future benefits."[3]
>
> *(Zhao 2014b, 116)*

So it follows that, when considering the ethical requirements that are placed on military S&T professionals, Marxist ethics is used to define such requirements:

> Military S&T ethics is a special discipline that is carried out from a theoretical high-level regarding military S&T morality. It is guided by Marxist ethics with military theory and reality as its foundation; systematically researching the forms, natures, effects, principles, norms and developing regulations of military S&T morality.[4]
>
> *(Gao and Gong 2005, 92)*

The Party leadership also provides ethics guidance to the PLA. Tang argues that Mao Zedong's fundamental contributions to military ethics were his assertion that the PLA should "wholeheartedly serve the people" and that such a perspective is "the core principle of building a military ethics culture." Mao also emphasized that the PLA is built and controlled by the Party and he also formulated the "ethical orientation for building national defense" – factors that have both strongly influenced the PLA's understanding of military ethics (Tang 2016, 20–23).

Party influence on military ethics did not end with Mao's demise in 1976. During the next three decades, the military guidance of leaders such as Deng Xiaoping and Hu Jintao continued to influence PLA military ethics. In recent years, CCP Chairman Xi Jinping's declaration of China's "Strong Military Dream" [literally: "The Dream of a Strong Military" 強軍夢 qiangjunmeng], a military goal to "return" China's military capabilities to a level concomitant with the PRC's aspirational goal of "national rejuvenation", has provided the PLA with new objectives that have had a substantial influence on the PLA's current understanding of military ethics (Tang 2016, 29–32). Ethical considerations resulting from such direction are discussed below. In fact, Tang argues that there is an obvious synergistic relationship between the Strong Military culture that results from Xi's assertive ideology and the building of a PLA military ethics culture:

> ...military ethics culture building inevitably promotes the thorough advance of the Strong Military culture and the vigorous expansion of the Strong Military culture will certainly bring about the flourishing and development of the military ethics culture.
>
> *(Tang 2016, 22)*

Thus, the PLA's view of military ethics is principally based on Marxist-Leninist ideology, as interpreted by the Party. The following description of applied military ethics (in the form of "military science and technology moral standards"), highlights its political motivation in both content and prioritization:

> ...military S&T moral standards can be summarized as follows: obeying the Party, loving the Motherland and devoting oneself to the Motherland; revering science, stubbornly pursuing the truth; carefully seeking realistic results, having the courage to blaze new trails; academic democracy, contending for freedom; being modest and prudent, uniting in collaboration.
>
> *(Gao and Gong 2005, 93)*

Non-Political Sources

The highlighting of the Marxist roots of PLA military ethics is not intended to imply that the works of early Chinese philosophers and historical personages are not also studied for their potential contributions to military ethics. Analyses by PLA affiliated authors have studied the insights of Laozi, Sun Bin, Zhuge Liang, and other giants of Chinese culture to identify possible contemporary military ethics applications. However, it is worth noting that Tang does not include any traditional Chinese philosophers in her description of the sources of the PLA's military ethics culture; her discussion begins with Marx and is based *entirely* on historical Marxist and Party ideologies. The military ethics writings of foreign authors are also considered by PLA researchers who investigate Western philosophers as diverse as Kant and Homer. However, these types of studies are primarily theoretical and do not address practical considerations.

Lastly, PLA military ethics researchers acknowledge that military ethics insights can be obtained from foreign militaries; in particular, those of U.S. military organizations. At the beginning of the millennium, Zhang Changling, PLA Nanjing Institute of Politics, wrote two papers describing the U.S. military's ethics research effort (Zhang 1998, 2002). In particular, he praised the openness with which the research was conducted by the U.S. military and the willingness of U.S. researchers to fully integrate foreign militaries' perspectives into their work. However, while he affirmed that there was value in evaluating U.S. military ethical thought, he warned:

> Of course, in considering US military ethical thought we must use Marxist standpoints, viewpoints, and methods to conduct specific analysis regarding the degenerate components of its common capitalist fundamental institution, ideology, and lifestyle interrelationships by resolutely applying resistance and exclusion.
>
> *(Zhang 2002, 61)*

Thus, political considerations far outweigh the potential contributions of practical military ethics experience of foreign militaries.

The PLA Considers Military Ethics

As directed by the Central Military Commission of CCP, the PLA's initial effort to "advance the important tasks of moral education and the cultivation of noble morality for soldiers" began in the 1980s. Over the subsequent decades, the PLA developed a program to inculcate its personnel with the military ethical values that it considered vital to fielding a highly capable military force. While much of this process was not made public, in recent years an increasing number of articles providing insights into PLA military ethics perspectives have been published. These publications have revealed that, unsurprisingly, the PLA has defined its own version of roles and goals for military ethics, some of which are expressed in ways that are quite different from traditional Western military ethics. The following excerpt discussing the strategic value of military ethics is a case in point:

The study of modern military ethics reveals the rational and legitimate foundations of military activities themselves, thereby scientifically distinguishing the ethical conflicts and moral choices resulting from the employment and initiation of military power. This will improve the important foundation of the army's soft power, thus to effectively argue in defence of legitimate military conduct, to provide correct value guidance to the actions of the nation, the armed forces, and soldiers on the basis of scientific military theory and armed forces doctrine and to demonstrate that the army is a just army and a cultured army.[5]

(Liu and Li 2020, 72)

While the importance of considering ethical conflicts and moral choices is also certainly an important part of Western military ethics, the emphasis on using military ethics to justify military actions or improve one's soft power is less familiar.

The PLA's vision for military ethics includes the goal of developing a "military ethics culture" throughout the PLA. Tang explains its role and importance:

The military ethics culture refers to all of the moral phenomena and military organization ethical relationships that are mutually related to military activities, as well as the sum total of moral awareness of the main part of military activities, moral activities, and moral character...[it] focuses on making known the value orientation & the current spirit of the soldiers, and guiding soldiers' aspirations, convictions, & conduct. This is an important aspect of focusing the troops' morale and combat effectiveness. It is also like the cultural basis of an adversary carrying out media warfare, psychological warfare, and legal warfare.[6]

(Tang 2016, 1)

An interesting aspect of this discussion of military ethics is that it also mentions media warfare, psychological warfare, and legal warfare; referred to in the PLA as "the Three Warfares" (Kittrie 2016, 161–195). While they are key components of PRC political warfare against an adversary, because they can also affect PLA morale and combat effectiveness, they are considered relevant to military ethics.

From Sunzi to Clausewitz to Schwarzkopf there has been one absolute dictum regarding warfare: The purpose of going to war is to win. In the context of this reality, when discussing of the value of military ethics PLA writers typically avoid theoretical matters (e.g., right and wrong). Instead, they highlight the ways that defining and implementing a PLA military ethics plan will improve military readiness, enhance warfighting capabilities, and support the progress of "socialism with Chinese characteristics." Without such an orientation, it would probably be difficult for military ethics advocates to obtain a "buy-in" from operationally focused military decision makers.

An example of such an argument on behalf of military readiness can be found in a monumental essay on the importance of military S&T ethics research:

> Military S&T ethics research must suit the requirements of military combat readiness, make the most of the moral self-confidence of our nation's development of military S&T and advancement of military combat readiness, and develop a military science & technology ethical system in the service of military combat readiness.
>
> *(Zhao 2014b, 112)*

Also discussing the practical value of building a military ethics culture, Tang presents several scenarios that have obvious ethical implications, including:

- When directed against the pluralistic, diverse, changing thought culture of society and the corroding influences of modern lifestyles, how can military ethics culture better guide officers and men to, from start to finish, maintain the pursuit and good moral practices of the revolutionary soldier's lofty spirit;
- In response to a long-term peaceful environment which causes officers and men to easily produce conditions of spiritual sluggishness and slack & numbing thought, how can military ethics sound the battle cry, enhance the revolutionary heroic spirit of the connotations of the accumulated ages, to preserve the troops' revolutionary enthusiasm and exalted will to fight;
- In response to the realities of the high psychological quality requirements of officers and men regarding their duties, how can military ethics culture strengthen humanistic care, on the basis of improving the moral psychological character of officers and men promoting the mental and physical health of officers and men;
- In response to the profound influence of information networks and mass media expansion on social life and people's ideological views, how to develop the networking battlefield, practically enabling networks to become vast platforms to propagate an advanced military ethics culture; and
- In response to the new trends in military morality development, how to promptly establish, revise, and refine political work relevant system rules and increase the level of military ethics culture systematization building.

(Tang 2016, 26)

In these and other examples, Tang emphasizes political solutions for significant problems that reduce military effectiveness and identifies three general issues that demonstrate "the necessity and urgency of military ethics culture building.": 1) The challenge of PRC societal changes, 2) Ethical considerations resulting from PLA military transformation, and 3) The influence of international ideological warfare (Tang 2016, 2–4).

I The Challenge of PRC Societal Changes

In the seven plus decades since the founding of the PRC, its people have experienced tremendous improvements in quality of life. However, not all changes have had a positive impact on Chinese society and, in particular, the PLA. While material conditions may be markedly better, many traditional societal norms have been adversely affected. The PLA is currently facing such challenges and must determine how best to deal with them:

> ...valuing individuality and advocating independence are replacing the sense of personal relationships. Independent individuals are gradually being produced...these are the new generation revolutionary soldier's sense of benefit, the sense of competition, the sense of time, and the sense of prestige that are constantly being increased. To bring about the wish for the full-scale development of oneself is utterly intense... the rapid transformation of the societal ethics culture presents the building of a military ethics culture with many negative influences. For instance, moral standards such as following orders, being united in action, selfless devotion are military values that are emphasized, but in the general population they are degraded... Ideologies of individualism, hedonism, money worship, etc., are what is displayed in the barracks.
>
> *(Tang 2016, 2)*

In identifying such challenges, Tang contends that this is precisely the sort of situation that argues for the development of a military ethics culture in order to achieve the Party's military objectives. That the author addresses this issue with such candour seems noteworthy as few, if any, other sources acknowledge the existence of such endemic societal challenges faced by the PLA.

> How to assimilate the essence of societal ethics culture, to discard the dregs, to inculcate contemporary revolutionary soldier core values to build a Strong Military culture is a problem that must urgently be resolved.
>
> *(Tang 2016, 2)*

In Tang's opinion, these are essential problems that the PLA must deal with if it is to become the world-class military to which it aspires.

II The Challenge of Military Transformation

Military ethics challenges that are created by military technological developments is a theme that is addressed in PLA journal articles from a variety of perspectives. For example, Tang reviews the issue under the title "Military transformation with Chinese characteristics requires a military ethics culture that provides ethical support" (Tang 2016, 2–3). Also referred to as "military S&T ethics," it is an important element of Chinese military ethics. Initially developed to address the unique roles and responsibilities of the personnel in PRC defence industries, this topic is also used to address the ethical issues faced with the development and application of S&T to national defence:

> ...the relationship between men and weapons are again being developed from a new starting point. The face of warfare is becoming increasingly vague. In modern troop building, military activities, and combat the factor of morality is becoming greater and greater and the matter of military ethics culture is receiving extensive interest. On one hand, the modernization construction of our country's national defence and troops is generating a large number of ethical questions...ethical questions in military training and education, ethical questions in high tech weapons development, ethical questions in military systems, ethical questions regarding military and civilian relationships, knowledge questions regarding the law of war and warfare ethics, etc. They all become questions that must be confronted and settled when reforming a Strong Military.
>
> *(Tang 2016, 2–3)*

Explaining the need for military ethics to conform to the developments in military combat readiness, Zhao Feng explains that dedicated ethics research is needed to allow military ethics to maximize the unique ability of ethics to provide vital contributions to several aspects of national defense:

> Military S&T ethics research must vigorously adapt to the requirements of military combat readiness, fully displaying our nation's development of military S&T, must advance the self-confidence of the morality and justice of military S&T, and while serving military S&T readiness must bring forth a new military S&T ethics system.
>
> *(Zhao 2014b, 112)*

Tracing the historical development of China's military S&T ethics research, he concludes with an appraisal of the contributions made to military ethics by Gao Xuemin in his *An Outline of Military Science and Technical Ethics* 《軍事科技倫理學教程》:

- First, it explored the goal of "military S&T ethics." ... Military S&T ethics concerns the theory of professional ethics in military S&T circles; it is professional ethics of researching military S&T morality."
- Second, taking Marxist military ethics basic principles as the guide, it systematically researched the relationship between military science & technology and morality, military S&T moral principles, military S&T moral standards, the ideal character of military S&T workers, the moral

responsibilities of military S&T workers, military S&T moral evaluation, the value of military S&T, moral principles of military S&T research, the ethics of weapons technology development, and other issues.

• Third, it proposes a system of moral principles and standard. The author considers that military technical moral principles include devoting oneself to national defense S&T pursuits/careers, a scientifically and technologically strong military, loving the Motherland and devoting oneself to the Motherland, defending world peace, etc. (Zhao 2014b, 114).

In a contemporaneous article Zhao explains the scope of military S&T ethical thought:

> The knowledge of military S&T workers regarding military S&T ethics and consideration of how to observe military S&T moral standards, the consideration of officials as they are formulating military S&T policy, as well as rational thoughts of the masses and the people regarding the application of military S&T, etc., are all part of the category of military S&T ethical thought.
>
> *(Zhao 2014a, 6)*

Zhao also presents a technology-oriented approach for dealing with military S&T ethics, highlighting an urgent need for ethics research regarding oceans, [outer] space, and cyberspace ethics. He calls for military S&T ethics research related to nuclear weapons, directed energy weapons, kinetic energy weapons, infrasonic weapons, anti-materiel weapons, meteorological weapons, genetic engineering weapons and antimatter weapons (Zhao 2014b, 115). Similarly, Shang Wei addresses the military ethics research that is urgently required to address the rapid growth of disruptive technologies (e.g., additive manufacturing, autonomous systems, etc.); a need that is complicated by the dual-use applicability of several such technologies[7] (Shang 2016, 68–69).

Others approach the ethics considerations from the perspective of the unique demands placed on military S&T professionals which are identified as 1) political accountability; 2) acceptance of one's role in destruction; 3) being a technology leader; 4) secrecy; and 5) dealing with multifaceted roles and technologies (Gao and Gong 2005, 91–92). Zhao also explains that the researchers who are responsible for military S&T ethics also have important responsibilities that will significantly influence the nation's progress:

> Military S&T ethics scholars should, by considering the advancement of China's national defense and armed forces modernization development as their personal responsibility, provide morality and justice support to China's military S&T development.
>
> *(Zhao 2014a, 7)*

As Tang argues, development of a military ethics culture when considering military reform and technical development, Tang explains:

> [With] a military ethics culture system establishing appropriate modern warfare requirements and cultivating soldiers' moral character that reflects the contemporary spirit, [we] will accelerate the advance of military reform with Chinese characteristics and bring about the inevitable requirements of the Party's 'Strong Military' objectives under the new strategic configuration of forces.
>
> *(Tang 2016, 3)*

Predictably, appropriate ethical considerations regarding military S&T development satisfy both the practical and political requirements of the Party.

III The Ideological Dimension of Warfare

A final matter in building a military ethics culture is the ideological struggles that are encountered during both in war and peace. In the battle for ideological supremacy, Tang argues that foreign powers are continually attempting to undermine China's values and actively working to denigrate China's achievements. Yet, it is important for the Party and the PLA to maintain the moral high ground:

> In recent years, the US, UK, and other nations have consecutively advanced "new interventionism", "humanitarianism", universal ethics, etc., pounding the ideological position of the ethics banner. They emphasize that human rights values exceed national sovereignty values. Moreover, advancing that as long as these [values] are beneficial to safeguarding and implementing these value concepts, they assume using all types of methods against other nations' internal affairs [in] carrying out interventions (including military operations) are reasonable on the basis of morality. Hence, with the US as the leading Western nation actively transforming military roles, [it] not only considers the armed forces as safeguarding [its] national interests, but it also considers [it] to be an instrument for propagating and peddling its moral concepts and values.
>
> *(Tang 2016, 3)*

To make matters worse, Western forces are targeting the PLA:

> ...the [PLA] military ethics culture has...inevitably become a primary target of the attacks by hostile Western forces. Many officers and soldiers accepting the rot of Western values, begin to doubt our army's requirements for patriotic thoughts and loyalty to the Party. They have lost

confidence in our army's ability to win future wars. This presents a tremendous challenge to our army's revolutionary forces to preserve their intrinsic qualities and way of work. In addition, US and other Western nations do not hesitate to [spend] huge amounts of money...to use their own ideology to influence the world, to greatly weaken the attractiveness of China's traditional military ethics culture.

(Tang 2016, 3–4)

Note that this malady is affecting both "**officers** and soldiers"; even the PLA's emerging leaders are being seduced by Western values. Given such a situation, it is incumbent on the PLA to build an effective military ethics culture to respond to and resist such Western actions.

Another application of military ethics in the ideological domain can be observed in the concept of "moral warfare" (道德戰 *daode zhan*); the moral equivalent to ways that 'the law' is used in so-called "legal warfare" or "lawfare."

Moral warfare refers to the various uses of morality that are employed to influence the spirit and wisdom of participants in military activities. a special type of military combat to strengthen and expand one's own power while weakening, dividing, and disintegrating that of the enemy.

(Zhao 2003, 88)

A more in-depth discussion of this concept is provided in the chapter "Moral Warfare: Weaponizing ethics to 'weaken, divide, and smash the enemy'", elsewhere in this volume.

Ideological warfare is certainly the most highly politicized element of the three challenges delineated by Tang that can be remedied by military ethics. As always, it is the Party that provides the guidance for conducting such activities.

IV Refuting International Objections to a PLA Aircraft Carrier: A Military Ethics Case Study

And what might the PLA accomplish with an effective military ethics culture? Several of Zhao's articles discuss the role of military ethics in preparing the PLA Navy to receive its first aircraft carrier, the *Liaoning*, and also its importance to gaining international acceptance, albeit begrudgingly, of such a potentially significant change to the PRC's strategic capabilities. The PLA's first order of business was to develop a series of "aircraft carrier ethics":

In maritime military S&T ethics research we should place emphasis on the current resolution of the issue of aircraft carrier ethics. At present our nation's navy is equipped with an aircraft carrier. In order to fully bring the aircraft carrier into military combat readiness actions, military S&T ethics must develop aircraft carrier ethics research. Researching aircraft carrier ethics must then comprehensively analyse ethics relations from a military S&T viewpoint concentrating on the aircraft carrier formation, revealing our development, using the moral quality of the aircraft carrier, investigating general aircraft carrier ethical requirements and their moral alternatives and evaluating their criteria, considering the proper moral consciousness and the type of education and training of shipboard personnel, and other issues. The basic goal of such research is the requirement to ensure the development and employment of our nation's aircraft carrier on the basis of morality and to cause the ship's crew to form with their material aircraft carrier equipped with advanced weapons equipment a corresponding spiritual aircraft carrier. Afterwards, by means of the aircraft carrier crew being of one mind and upholding a concentrated, complete, comprehensive spirit, this is the aircraft carrier combat effectiveness force multiplier, an ever-victorious spiritual fortress...

(Zhao 2014b, 115)

As evidenced in this passage, and as previously discussed, the PLA's concept of military ethics encompasses a much broader scope of activities and influence than is typically considered in the West. In fact, military ethics seems to encompass every conceivable technical and operational aspect of preparing the *Liaoning* for service in the PLA Navy. And, in this instance, military ethics have a strategic role for the PLA. It is also interesting to note that military ethics are identified as the catalyst for the creation of a cohesive and effective crew, allowing the *Liaoning* to become the "ever-victorious spiritual fortress" that it was meant to be; all under the leadership of the Party.

In a journal article written after the *Liaoning* was preparing for service with the PLA, Zhao describes how military ethics were used to gain the desired international acceptance. Arguing that "a nation's military S&T ethical thought always bears the burden of proving the important task of the righteousness of one's own development of military S&T, he explains that how such an ethics-driven methodology was used to justify the decision to add the *Liaoning* to the PLA Navy. And their conclusion after completing the evaluation process?

China's development of an aircraft carrier was in accordance with China's national interests and the peace and development of the international community. And the few nations that attacked China by means of the "China Threat" were unjust.

(Zhao 2014a, 8)

In this example, the value of military ethics to the PLA was its ability to provide both a framework for evaluating key military decisions and also to provide an effective moral confidence to rebuff the criticism of those who didn't have China's best interests at heart. In contrast to this example of "military ethics with Chinese characteristics", it seems doubtful that Western nations would consider that using this process to rationalize this sort of strategic decision-making was anything other than a "creative" application of military ethics.

Conclusion

Like the militaries of many other nations, the PLA has a great interest in furthering its military ethics and building a "military ethics culture". As this goal is politically driven and subordinated to the Chinese Communist Party, however, this is where the similarities with most other militaries end. The PLA is directly controlled by the Party. Military ethics in the PLA is a political matter; an ideological means of dealing with what the Party views as political challenges. As a result, in the PLA, military ethics is the purview of the Party (and the political hierarchy within the PLA) and ethics-related decisions are ultimately made on the basis of Marxist ideology. While certain traditional Chinese values may influence Chinese society, when compared with Marxist precepts they have relatively little influence on how the PLA considers military ethics.

The PLA is currently facing a number of significant challenges (e.g., eroding PRC societal values, undesired Western influence) that must be dealt with in order to maintain and improve its operational capabilities. As these problems are viewed as political matters that require political solutions that are based on the guidance of the Party, it is not known how they will affect the PLA's understanding of military ethics. To its credit, the PLA continues to investigate ways of defining such distinctions, using military ethics to deal with its challenges, and build a "military ethics culture"; or, more accurately, a "military ethics with Chinese characteristics". The PLA has experienced successes (e.g., *Liaoning*) with its approach to military ethics and, as it further defines and refines the PLA military ethics culture, will undoubtedly achieve more.

This chapter has presented a brief overview of the writings by PLA-affiliated authors that address PRC military ethics. Much more has been written in recent years and, given the ongoing improvements to PLA capabilities, there is likely an abundance of new insights regarding PLA military ethics that will be of value to comparative military ethics scholars. Given the importance that the PLA is placing on military ethics, as China's military capabilities continue to grow, its military ethics culture will continue to mature. Accordingly, it is vitally important to continue to investigate this topic in order to better understand the role that military ethics will play in the PLA of the future.

Notes

1 Unless otherwise noted, all translations are my own. Throughout this chapter 倫理 *lunli* is translated as "ethics" or "ethical". A similar term, 倫理學 *lunlixue*, is frequently used as a synonym that highlights the academic aspects of the subject. In contrast, 道德 *daode* refers to "morals" or "morality".

2 Dr. Tang is a graduate of the PLA Nanjing Institute of Politics and Lecturer in the Political Work Department at the People's Armed Police University.
3 Dr. Zhao is a professor in the PLA University of Science and Technology's Research and Teaching Center for Political Theory and Army Political Work. 軍事科技倫理學 *junshi keji lunlixue* is translated as "military science and technology ethics" while 軍事科技道德規範 *junshi keji daode guifan* is translated as "military science and technology moral standards."
4 Rear Admiral Gao is Political Commissar at the Naval Institute of Engineering. Dr. Gong is a philosophy professor at the Naval Institute of Engineering.
5 Dr. Liu is a professor in the Political Work Department of the PLA Army Engineering University. Li is her student.
6 The term 軍事倫理文化 *junshi lunli wenhua* is translated as "military ethics culture."
7 Dr. Shang is an ethics researcher at the Political Work Research Center at the PLA Academy of Military Science.

Bibliography

Fu Chao. 2010. *Sunzi's Art of War: A Structural Analysis*. Beijing: PLA Press. [Published in Chinese: 付朝《孫子兵法結構研究》北京；軍事書店].

Gao Xuemin and Yun, Gong. 2005. "An Outline of Military Science and Technology Ethics." *Journal of PLA Nanjing Institute of Politics* 21, no. 2: 91–94. [Published in Chinese: 高學敏、龔耘〈軍事科技倫理學論綱〉《南京政治學院學報》].

Kittrie, Orde F. 2016. *Lawfare: Law as a Weapon of War*. New York: Oxford University Press.

Liu Shuping and Lei, Li. 2020. "The Strategic Value of the Development of Contemporary Military Ethics." *Journal of Nanjing University of Science and Technology (Social Sciences Edition)* 33, no. 4(August): 72–76. [Published in Chinese: 劉淑萍、李磊〈當代軍事倫理學發展的戰略價值〉《南京理工大學學報；社會科學版》。. doi:10.19847/j.ISSN1008–2646. 2020.04.012

Lo, Ping-Cheung and Twiss, Sumner B. 2015. *Chinese Just War Ethics: Origin, development, and dissent*. New York: Routledge.

Shang Wei. 2016. "Additive Manufacturing Technology and Military Ethical Issues." *Journal of Yunmeng* 37, no. 2(March): 68–69. [Published in Chinese: 尚偉〈增材製造技術與軍事倫理問題〉《雲夢學刊》]. doi:10.16740/j.cnki.cn43-1240/c.2016.02.012.

Tang Fang. 2016. *An Investigation of Modern Chinese Military Ethics Culture Building*. Shanghai: Shanghai World Library Publishing Company. [Published in Chinese: 唐芳《當代中國軍事倫理文化建設研究》上海；世界圖書出版公司].

Zhang Changling. 1998. "Interactions Between the Core Values of Contemporary American Military Personnel and Core Social Values." *Journal of the Air Force Political Academy*4: 78–79. [Published in Chinese: 張長嶺〈當代美國軍人核心價值與社會核心價值的互動〉《空軍政治學院學報》].

Zhang Changling. 2002. "A Few Suggestions Regarding US Military Ethics Research and Our Nation's [China's] Military Ethics Research." *Morality and Culture*1: 59–61. [Published in Chinese: 張長嶺〈美國軍事倫理研究對我國軍事倫理研究的幾點啟示〉《道德與文明》].

Zhao Feng. 2003. "Military Ethics Should Investigate the Scope of Moral Warfare." *Journal of PLA Nanjing Institute of Politics* 19, no. 4: 86–89 [Published in Chinese: 趙楓〈軍事倫理學應當研究道德戰範疇〉《南京政治學院學報》].

Zhao Feng. 2014a. "Exploratory Analysis of the Characteristics of Military Science and Technology Ethical Thought." *Journal of Naval Engineering* 11, 2(June): 5–9. [Published in Chinese: 趙楓〈軍事科技倫理思維特點探析〉《海軍工程大學學報》.] doi:10.13678/j.cnki.issn1674-5531.2014.02.002.

Zhao Feng. 2014b. "Regarding the Development of Innovation in Military Science and Technology Ethics Research Conforming to Military Combat Readiness." *Journal of PLA Nanjing Institute of Politics* 30, no. 2: 112–118. [Published in Chinese: 趙楓〈論軍事科技倫理研究適應軍事鬥爭準備的發展創新〉《南京政治學院學報》].

13

MORAL WARFARE

Weaponizing Ethics to Weaken, Divide, and Smash the Enemy

Mark Metcalf

Introduction

A perennial challenge when engaging in comparisons across cultures is having to deal with situations in which a particular word may have different meanings in different cultures. An apocryphal event from a Chinese Warring States [5th century–3rd century BCE] remonstration provides a humorous example of just such a difficulty:

> Marquis Ying said, "In Cheng they call jade that has not been worked 'pure'; in Chou they call fresh-dressed rats which have not yet been preserved 'pure'.
>
> A man of Chou carrying fresh-dressed rats passed a Cheng merchant and asked him if he wanted to buy some 'pures'. The merchant replied that he did. But when he was shown dressed rats, he declined them.
>
> Now Lord P'ing Yüan is busy getting himself a name for virtue throughout the empire. It was he who banished his own ruler...in order to become minister, yet rulers everywhere still respect him. This merely proves that rulers are less intelligent that the Cheng merchant. They are so dazzled by the word 'pure' that they do not trouble to discover what reality lies behind it.
>
> *(Crump 1996, 130)*

More than two millennia after this encounter we continue to encounter this type of issue as we discuss comparative military ethics. While we aren't usually talking about rats or jade, we are encountering situations where 'well-known' or 'generally accepted' military ethics terminology may have significantly

DOI: 10.4324/9781003336372-15

different meanings in different cultures or political environments.[1] In order to more thoroughly comprehend the perspectives and intentions of our cross-cultural counterparts, it is crucial that we clearly understand the meanings and uses of subject-specific terminology.

Just such a situation occurs in a journal article written by Zhao Feng, a professor at the Nanjing People's Liberation Army (PLA) Institute of Politics, entitled "Military Ethics Should Investigate the Category of Moral Warfare" (Zhao 2003, 86–89).[2] Issues addressed in Zhao's paper highlight distinct differences between contemporary Western and Chinese views regarding the role and scope of military ethics in modern warfare. This chapter analyses Zhao's argument with the goal of better understanding how the concept of moral warfare is construed by the PLA as a vital element of military ethics.[3]

"Military Ethics Should Investigate the Category of Moral Warfare"

In his article, Zhao describes the elements of moral warfare, provides historical examples of moral warfare, and explains how moral warfare will enhance the ability of the People's Republic of China (PRC) to conduct high-tech warfare. Most interestingly, as the English language abstract explains, such factors comprise a component of military ethics:

> In military area, morale plays an important role in strengthening, exploiting one's own side, crippling, disintegrating the enemy by influencing servicemen's soul and intelligence. In military struggle, opposing sides develop trials of strength one another by moral function, which results in moral war, forming the pattern of this special military struggle. Moral war should be made a deep study in military ethics, giving full play to the morale serving the function of war."
>
> *(Zhao 2003, 89)*

Before analyzing Zhao's essay, it is important to explain why, although the crucial term *daode zhan* 道德戰 is translated as "moral war" in the essay's English language abstract, I have chosen to translate this term as "moral warfare". As used in the essay, *daode zhan* is not a synonym for "just war," however it is difficult to draw such a distinction from the abstract. In fact, there is a clear difference between the ways that "moral warfare" and "just/ unjust war" are used in the text. "Just war" and "unjust war" are terms that are described as "two categories of combat" and, in explaining the ways of defensively applying moral warfare, the reader is encouraged to "boldly attain the 'strategic advantage' of just war" (Zhao 2003, 87). In the PLA's view, "just war" *zhengyi zhan* 正義戰 is "consistent with interests of the people and promotes social development," while an "unjust war" *feizhengyi zhan* 非正義 戰 "runs counter to the interests of the people and impedes social development" (All-Army Military Technical Terminology Management Committee 2011, 47). In contrast, throughout the text the expression "moral warfare" is used to describe specific actions (i.e., warfare) that are taken to support military operations. So, while contemporary Western and Chinese understandings of just/unjust war may differ, in the context of this essay *daode zhan* is a type (e.g., psychological, economic, etc.) of warfare and not a characterization of the morality of a conflict.

The Elements of Moral Warfare

Zhao begins his discussion with a concise definition of moral warfare: "Moral warfare refers to various types of morality used in military combat to influence the hearts and minds of participants in military activities" (Zhao 2003, 86). He emphasizes that moral warfare is a *legitimate*, yet unique, type of warfare that can be used to support military objectives. He explains that moral warfare can be viewed from two perspectives. First, it is an aspect of military conflict:

> military conflict is…manifested as antagonistic powers of military, economic, political, and thought cultures, including morality, in a contest of overall strength. The morality dispute is what both sides employ to use morality to strengthen themselves and weaken the enemy…military conflict intrinsically includes moral warfare. What needs to be pointed out is that moral warfare is a military conflict application of morality…As morality is not the equal of military conflict, moral warfare is thus a part of military conflict.
>
> *(Zhao 2003, 86)*

Second, it is also a distinct element of military operations:

> [Moral warfare is] carried out in the moral domain to achieve fixed objectives. By means of specific information channels it transports and expands the influence of morality; using morality as a method, taking the hearts and minds of military operation participants as the target; strengthening myself while weakening, dividing, and smashing the enemy as the purpose of conducting combat; it is equipped with the key elements of the military conflict model…Accordingly, it is a comparatively complete model of military conflict. In time of war this conflict model has an operational support form, while in peacetime it is the primary form of hostile military power struggle.
>
> *(Zhao 2003, 86)*

While moral warfare has an important support role during armed conflict, it has an even more significant role during peacetime as the "primary form of hostile military power struggle" – ongoing war by other means. It is worth noting that the expression "weaken, divide, and smash the enemy" (*xueruo*削弱, *fenhua*分化, *wajie difang*瓦解敌方) is repeated five times throughout the paper.

Zhao next discusses the implementation of moral warfare, describing its "weapons" and their employment.

> The substance of confrontation between moral warfare and weaponry are not the same... [Moral warfare is] inherently a type of mental warfare or psychological warfare that uses various types of information channels as intermediaries and brings morality into play with the influence of moral psychological and intellectual capabilities toward those engaged in battle. Its success is also chiefly manifested as bringing the military combat zeal of our forces into play.
>
> *(Zhao 2003, 86)*

While much of what Zhao describes may seem like psychological warfare, he explicitly explains the significant differences between moral warfare and psychological warfare:

> Moral warfare is carried out by means of morality. Its effective techniques and channels possess moral characteristics... [it is] also different than other ideological warfare in psychological warfare... [it is] maintained on the basis of public opinion, conscience, and man's innermost convictions... moral warfare operations are accomplished by means of regulating and controlling public opinion, conscience, etc..... a nation's political democracy consciousness, cultural education standards, degree of public opinion development...affect how moral warfare actions are brought into play.
>
> *(Zhao 2003, 86)*

Concluding this theoretic discussion, Zhao reemphasizes the military orientation of moral warfare and highlights its goals, purposes, objectives, and essence

> The goals of moral warfare are to strengthen one's own power and to weaken, divide and smash the enemy, to win a military conflict victory, and to realize a specific level of political will... The goal is always to achieve a military combat victory...[The] immediate purposes of moral warfare are to strengthen one's own power and to divide and demoralize the enemy...The unanimity of military goals and moral values is the prerequisite to attain military conflict victory...Thus, all parties in military conflict conduct moral warfare, allow one's military personnel to form a suitable required systems of values for military conflict, to restrain the value demands of the enemy, and to shake and destroy the enemy value system... The fundamental objective of moral warfare is to bring about economic and political benefits that are served by morality...the essence of moral warfare is a moral response to class conflicts...
>
> *(Zhao 2003, 86–87)*

The argument that "all parties in military conflict conduct moral warfare" is significant and provides an initial indication of why "moral warfare" is considered to be an ethical issue.

From the PLA's perspective, one important purpose of military ethics training is to morally steel one's forces on the basis of shared values, a goal that is also an important objective of moral warfare.

As a type of warfare, moral warfare is subdivided into two categories, defensive and offensive, each with its own operational goals and implementation methods.

The goal of defensive moral warfare is to improve a state's moral standing and is "[p]rimarily aimed at states, political parties, and inside the military alliances...its goal is to allow one's side to be the first to become 'invulnerable'."[4] Specific methods for implementing defensive moral warfare include:

1. By means of morality, promoting economic, political, national defence, ideology, and building scientific culture development. To advance the nation's overall strength and military combat effectiveness;
2. Strengthening national moral education and promoting national moral character and patriotic & 'love the army' enthusiasm in order to supply a moral foundation from the masses for current military struggles and future conflicts;
3. Strengthening military moral construction, improving military moral accomplishments, causing the troops to become a "cultured army" [*wenming zhi shi* 文明之師] and a "righteous/just army" [*zhengyi zhi shi* 正義之師];
4. Boldly attaining the "strategic advantage" of 'just war', strengthening our confidence, boosting morale, strive for as much international sympathy and support as possible; and
5. Denying the moral rationale of enemy states and potential enemy states. Resisting their moral offensives

(Zhao 2003, 87)

The phrase "the 'strategic advantage' of just war" implies that if a state can show that its military actions are "just", it has significantly tipped the scales of public opinion in its favour. This is consistent with the observation that "the language of 'just war' is employed more for political effect rather than an expression of any genuine concern for ethics" (Lo 2015, 157–162). In summary, the purpose of such defensive methods is to improve the state's moral stature, both domestically and internationally, in order to make it less vulnerable to an adversary's moral warfare attacks. It is also interesting to note the assertion that the use of defensive moral warfare will improve an army's combat effectiveness, a claim that is frequently used in PLA writings to justify the value of military ethics.

Unsurprisingly and in contrast, offensive moral warfare is "[p]rimar ly aimed at an enemy state, its military forces, and the enemy's alliances. To smash the enemy's moral [credibility]..." Zhao provides a lengthy list of ways that offensive moral warfare can be employed:

1. Strengthening our moral development. Using our side's moral superiority to give the enemy side a huge amount of moral [literally: "spiri-tual"] pressure;
2. From an ethical perspective carrying out a multi-faceted negation of the enemy state's politics, economy, military and ideological culture. Destroying the military moral basis of their masses;
3. Unmasking the unjustness of the enemy side's waging of war, inactivat-ing their driving spirit, slackening their military morale, breaking up their alliances, and sticking them in isolation;
4. Attempting to win them over politically ["attacking their hearts with policy"], such as the policy of our army to give preferential treatment to POWs, owing to their having a revolutionary humanistic moral perspec-tive, to reform many enemy officers and men by means of persuasion;
5. Attempting to win them over with emotion ["attacking their hearts with emotion"], as during wars of resistance against aggression when our Party and our army shouted the people's patriotic slogan "Chinese do not attack Chinese", thereby winning over many Kuomintang force patriotic soldiers;
6. Making use of "spreading rumours and confusing people", honey traps, and other means to sow discord between the enemy side's internal and external moral relations. Our army is a righteous army and is above adopting of these sorts of methods. Yet, as luck would have it, in reality by no means have they been eliminated in nations and political groups; and
7. Targeting the enemy military commanders' characteristic stereotyped moral views, perturbing their temperaments, taking advantage of oppor-tunities, and defeating them. As in the *Sunzi* chapter [*Sunzi* 8] "Nine Contingencies" which says of high-ranking officers "the incorruptible can be insulted, those who care for their troops can become troubled." [Ames 1993, 136] Although an enemy general is firmly entrenched in incorruptible moral excellence and is inflexible, [we] can still use meth-ods of "insulting" and "troubling" to smash the enemy."

(*Zhao* 2003, 87)

When read in the context of the earlier comment that moral warfare is the primary form of military conflict employed during peacetime, such methods are also consistent with the concept of "media warfare" (or "public opinion warfare") that was promulgated by the PRC government and PLA in 2003 as part of the "Three Warfares" concept and comprises three elements:

1. *Psychological Warfare*: the use of propaganda, deception, threats, and coercion to affect the enemy's ability to understand and make decisions;
2. *Media Warfare*: the dissemination of information to influence public opinion and gain support from domestic and international audiences for China's military actions; and
3. *Legal Warfare*: the use of international and domestic laws to gain international support and manage possible political repercussions of China's military actions.

(Kittrie 2016, 161)

Moral Warfare Throughout History

Zhao next explains that moral warfare has been employed by nations for centuries and presents historical examples of its use by both the China and the West. This section is particularly useful because it provides additional insights into the types of activities that the PLA considers constitute moral warfare.

Arguing that "China is a nation that attaches great importance to morality and throughout history there have been statesmen and strategists who have all paid attention to moral warfare" the article presents examples from ancient Chinese military history to demonstrate that "[o]ur army has a tradition of moral warfare."

The text explains that Western use of moral warfare originated during the Greco-Persian Wars (5th century BCE) when Darius I commanded that a servant should remind him to "Remember the Athenians!" three times before he ate dinner, to strengthen his resolve. During the Crusades, moral warfare was established on the basis of the "religious ethics." Clausewitz was deemed to have been a proponent of moral warfare by his emphasis on morality in his contention that "Material causes and consequences are no more than the knife handle; spiritual causes and consequences are actually the precious metal and are actually the really sharp knife edge." And moral warfare was also in evidence during World War II when "the German fascists carried out racist and revanchist propaganda and the Soviets exposed the unjust nature of the German fascist war of aggression" (Zhao 2003, 87).

Since the PLA's establishment in 1927, moral warfare has also an important aspect of the way that it conducts its operations:

> [During World War II, our army] used every possible condition to carry out moral propaganda against the enemy, fully demonstrating the effect of morality in disintegrating the enemy...during the War of Resistance against Japanese Aggression, our army used the method of giving preferential treatment & release of captives and repatriating seriously injured personnel & bodies. This exposed the lies of Japanese troops...
>
> [During China's Civil War] our army organized POWs and family members to develop the impressive "firing line calling friends", "firing

line calling spouse", and "firing line calling children" activities. They put to use a method of propaganda shells and playing cards to disintegrate and slacken enemy [i.e., Kuomintang] morale…

[During the Korean War] our nation began to develop "look with hatred, look with distain, look with contempt on US imperialism" as the core contents of various educational activities, greatly arousing all of the people; especially the patriotic zeal of the army troops. On the battlefield, our volunteer army by means of distributing various propaganda materials, forwarding POW letters, extensively developing battlefield propaganda aimed at the enemy, and broadcast recordings from POWs, forcefully exposed the intrinsic aggressive character of US imperialism, expanded our army's moral influence, and smashed the fighting spirit of the enemy…

(Zhao 2003, 88)

All of these examples demonstrate the important role that moral warfare plays in the PLA's approach to military operations and highlights the range of activities that are included in such activities. And, based on such examples, Zhao reiterates the military ethics role of moral warfare:

From the preceding we can seethat moral warfare is an important tradition of our army. Continuing and developing our Party's and our army's moral warfare tradition is necessary duty of our army's military ethics.

(Zhao 2003, 88)

Zhao then addresses more recent events, citing the importance of military ethics (in this context, moral warfare) in so-called "high-tech warfare."

High-tech warfare calls for military ethics:

- Firstly, battlefield conditions change in the twinkling of an eye. This requires soldiers and civilians of belligerent states to have a high degree of spiritual character and moral character.
- Secondly, the area of the high-tech war battlespace is unprecedentedly huge with combatants sustaining attack and deterrence that has a comprehensiveness. Combat effectiveness is unprecedentedly high… [there is] all-weather, day-or-night continuity…lethality already approaches that of tactical nuclear weapons…These all influence the morale of the officers and men, requiring them to respond with a strong sense of warfare righteousness and a lofty moral sense of duty.
- Thirdly, the large-scale use of high-tech weapons, the high speed of army operations, the intelligentization [i.e., capabilities enhancement by means of artificial intelligence] of command and control, and the extraordinary consumption of materials [all] cause the duration of conflict to

be greatly reduced…This shortness of duration causes further magnifies the role of moral warfare.

(Zhao 2003, 88)

To demonstrate the importance of moral warfare to high-tech warfare, Zhao highlights its use by U.S. military forces during the two Gulf Wars. While rather brief, this section is important because it provides specific examples of activities that are characterized as moral warfare; activities which, for the most part, might also be characterized as psychological warfare:

> During the [1st] Gulf War, the US criticized the unjustness of the Iraqi invasion and occupation of Kuwait and the Iraqi government's lack of concern for human rights. They taught US officers and troops to uphold the honour of military personnel; to devote oneself to justice. During the 2nd Gulf War, before the war the US criticized the Iraqis for having weapons of mass destruction, for supporting international terrorist organizations, and for carrying out a brutal government within Iraq, making every effort to send troops to Iraq in the name of the United Nations. After the War broke out, they declared that the attack was for the freedom of the Iraqi people, disseminating news that Saddam was to be killed and that his forces would capitulate, protesting that the Iraqi military's treatment of POWs violated the Geneva Convention, etc. Thus, it is clear that the US, having ascended because of high-tech weapons, was fully aware of the importance of moral warfare.
>
> *(Zhao 2003, 88)*

While characterized as moral warfare, it can be argued that many of these actions could also be categorized as military ethics or psychological warfare. Yet, from the PLA perspective, what is being described is moral warfare. Alternatively, many of these examples can also be characterized as "dissemination of information to influence public opinion and gain support from domestic and international audiences for…military actions" – the textbook definition of media warfare or public opinion warfare, as previously discussed (Kittrie 2016, 161).

The successful use of moral warfare by U.S. forces in concert with high tech capabilities, the author argues, means that if the PLA wants to win high tech wars, then it must increase its research to improve its own moral warfare capabilities. Although, Zhao argues, the current state of the PLA's moral warfare capabilities is poor, additional research will certainly provide improvements.

- Our nation's actual military affairs struggle urgently needs to research moral warfare.
- In the first place, during the several most recent high-tech wars, the importance of moral warfare to Western nations should ring alarms bells

for us: during future high-tech wars, to underrate moral warfare [means that] we will lose the initiative in military affairs.

- Secondly, in the realm of high-tech equipment, our army does not occupy a dominant position. One day when there's a clash with a strong enemy we must oppose [the enemy] by means of inferior weapons and defeat a superior enemy. There is a strong and powerful moral warfare compatibility with this requirement.
- Third, these types of hostile forces unceasingly conduct moral warfare against us, so we cannot not take active steps to deal with the situation.

(Zhao 2003, 88)

Concluding his appeal for additional moral warfare research, Zhao presents numerous examples of ways that ineffective PLA moral warfare capabilities are hurting the PRC's interests:

> ... hostile forces criticize our human rights situation, haranguing about "the China threat", etc. and repudiate our nation from a political morality perspective. They use cultural exchanges, personnel movements, and media channels to carry out the infiltration of Western moral value concepts against us. On Taiwan, independence advocates also continuously carry out moral warfare against the Motherland; emphasizing Taiwan's indigenous culture and local morals, denouncing the patriotism of the Chinese people, and denying a moral rationale for the cross-straits people's demand to unify. These facts make clear that we must, in a complicated government and military struggle, establish ourselves in an unassailable position. We must increase moral warfare research and do all we can to take the moral warfare initiative."

(Zhao 2003, 88)

Moral Warfare as a Force Multiplier

The final section of Zhao's essay discusses the theoretical and practical value of moral warfare to PRC military forces.

Discussion of the theoretical value of moral warfare emphasizes its contribution to military ethics, providing additional insights about the PLA's view of the scope of military ethics. First, Zhao argues, a better understanding of moral warfare will lead to a better understanding of military ethics. Defining military ethics as "ethics that investigate ethical relationships in the domain of military activities," the author argues that the current emphasis on military personnel ethics should be broadened to include "social-military ethics relationships":

> ...military ethics that already must research military personnel morals, now must research the function of social public morality in military conflict, and must also research how the different morals influence each other and influence restricted relationships during military conflict. Yet moral

warfare affects the entirety of social morals that are affected during military conflict. Therefore, research of this type of military conflict contributes to a comprehensive grasp of the scope of military ethics.

(Zhao 2003, 88)

Zhao then explains that by increasing the level of moral warfare research, the field of military ethics will receive many new insights as morality is "brought into play":

Researching moral warfare, analysing and summarizing the experiences & lessons of one's forebears, is suitable to the moral warfare requirements of current military conflict and future high-tech localized wars, forms the concepts and theory of moral warfare, and can enhance and deepen the substance of military ethics...

- First, from a military conflict moral perspective, original military ethics point out that there are two categories of conflict – just and unjust. The sense of justice in conflict can directly influence the morale of the participants. To research moral warfare, considering morality by regarding [its] influence on morale promotes it to a type of military conflict level toward understanding and can make the original understanding even more generalized to express even more accuracy.
- Second, in class-based societies, differences between the morals of different classes, ethnic groups, and political groups already exist...Yet, these by no means eliminate the regulations of the jointly recognized Geneva Convention and its additional protocols regarding humane treatment of POWs. For the purpose of bringing morality into play in military conflict, we must research the moral differences and connections of different nations, ethnicities, and military forces. This would obviously be beneficial to deepening the understanding of different moral relationships.
- Third, by researching moral warfare [one] can profoundly understand the importance of moral teaching and self-cultivation to military conflict, thus extravagantly interrelating theory and substance.
- Fourth, moral warfare involves method-based questions about the various ways of weakening, dividing, and smashing in military conflict. Researching moral warfare reveals rules about deciding methods for moral warfare goals, understanding the negative influences of terrorist attack methods, and is helpful to deepening the discussion regarding the mutual relations between military ethics concerning military moral goals and methods.
- Fifth, moral warfare involves the substance of psychological warfare, political warfare, information warfare, propaganda warfare, etc. To research moral warfare and their relationships can make military ethics and other disciplines jointly form a disciplinary group for realistic military conflict service.

(Zhao 2003, 89)

While the first three arguments address aspects of military ethics that are also addressed in the West (albeit, without the tinge of moral warfare), the confluence of military ethics and moral warfare in the final two sections consider the scope of military ethics from a distinctly PLA perspective. The fourth recommendation describes the importance of thoroughly understanding the moral warfare decision-making process, particularly the details of how such plans are implemented. Zhao concludes his recommendations by explaining that integrating high-tech conflict research with moral warfare research will make the results more relevant to combat practices.

The paper concludes by addressing the practical value of increasing military ethics research regarding moral warfare. First, Zhao argues that such research will:

> Contribute to highlighting the military combat service functions of military ethics. The mission of ethics is to realistically serve. This stipulates that military ethics should be for realistic and future military conflict service, for "winning the battle" and "not degenerating" service. To research moral warfare, taking the direct relationships between morality and military conflict, taking moral warfare and turning it into an intrinsic component part of military conflict that is treated as an effective form, to explore how to effectively serve on behalf of current military conflict and winning future high-tech regional wars, this then emphasizes the function of morality, especially military morality for war service. Moreover, it will cause our army military ethics to preserve its great vitality.
>
> (*Zhao 2003, 89*)

Secondly, this type of research will result in moral warfare lessons learned that will enhance practical warfighting capabilities:

> [It] contributes to the summarization and explanation of historical and practical moral warfare provision theory, such as the development of scientific analyses and reviews of available military ethics related to theory regarding Chinese historical moral warfare. It can also, from military ethics, expound on our nation's reasoning for requiring the US to apologize during the Chinese/American aircraft collision and from it can fully display military ethics regarding realistically recognized abilities.
>
> (*Zhao 2003, 89*)

It is noteworthy that the PRC's demand for a U.S, apology in the aftermath of the April 2001 collision by a PLA Air Force interceptor with a U.S. Navy reconnaissance aircraft over the South China Sea is used as an example of moral warfare. Not only was the U.S. action a result of the relentless moral warfare activities carried out by the PRC, but in the author's opinion the outcome of the entire process resulted from the appropriate application of military ethics by the PLA.

Finally, Zhao argues that such research will provide new insights regarding the ways that adversaries are using moral warfare to their benefit. Such insights will undoubtedly be helpful as the PRC considers new concepts for both offensive and defensive moral warfare:

> [Additional research will] contribute to practical moral warfare. Military ethics from theory summarizes the history and reality of moral warfare, researches the methods of the origins, essence, rules, and effects of moral warfare, and can provide compelling military ethics ideological guidance to current military conflicts and future high-tech wars. Speaking specifically, to research moral warfare regarding enemy moral warfare motives, goals, commonly used methods, etc., achieves a pretty good idea of what is going on and an ability to carry out targeted counter-attacks. Exposing the substance of enemy moral warfare enhances the abilities of the masses and the army with an ability to distinguish enemy moral warfare and heightens vigilance against the "Westernization" and "differentiation" attempts and ways of doing things by Western hostile forces. [We are] aware so as to not listen to, not watch, not believe, and not promulgate enemy moral warfare propaganda. To seek shortcuts for preventing and striking back against enemy moral warfare [we] can employ decisive measures when necessary, jamming and destroying enemy moral warfare facilities and technical equipment, and cutting off the enemy moral warfare channels of influence. To fully understand strengthening moral construction and to resist the relations between enemy moral warfare will enable us to enhance the Chinese people's excellent moral tradition…, and to strengthen the construction of civic virtues and the moral construction of the army.
>
> *(Zhao 2003, 89)*

Conclusion

It is obvious from Zhao's essay on moral warfare that there are many differences between the ways that military ethics are perceived in the West and the PRC. While conceptual intersections exist, one must also be conscious of identical terms (e.g., "pure") with different meanings (e.g., jade and rats). The value of an article like Zhao's to comparative military ethics research is that it presents both theoretical and practical aspects of moral warfare (and military ethics), which allow us to more clearly understand not-so-obvious differences. His article also explicitly explains why moral warfare is considered to be an element of military ethics.

In a practical sense it is useful to understand the Chinese perspective regarding the role of moral warfare and military ethics when evaluating PRC military actions as such an understanding may provide novel insights into the situation. Similarly, as with most cross-cultural interactions, it can be useful to remember that this (i.e., moral warfare) is one of the cultural 'filters' that PRC analysts and policymakers will likely be using when evaluating the actions of Western military forces.

Writing a decade later, Zhao gives yet another practical application of moral warfare. He explains the importance of morality and ethics in the PRC's decision to develop an aircraft carrier. While the term "moral warfare" does not appear in the article, both moral warfare and military ethics considerations are evident in his explanation:

> Whenever a nation's military S&T ["science and technology"] ethical thought always bears the burden of proving the important task of the righteousness of one's own development of military S&T, exposes the unrighteousness of an adversary's development of military S&T, and returns fire on the adversary's righteous criticism, it always includes confrontational details with the opposing military S&T ethics viewpoint.
>
> For example, when our nation was considering whether or not to develop an aircraft carrier, we experienced this sort of course of events. During this time, we had knowledge in response to the aircraft carrier significance, rebuttals in response to different points of view, and we had evaluations in response to the "China Threat". Ultimately, only then did we reach a consensus: China's development of an aircraft carrier was in accordance with China's national interests and the peace and development of the international community. And the few nations that attacked China by means of the "China Threat" were unjust.
>
> *(Zhao 2014, 8)*

In this example, the value of military ethics was its ability to provide both a framework for evaluating key military decisions and also to provide an effective moral confidence to rebuff the criticism of those who didn't have the PRC's best interests at heart.

In conclusion, the term "moral warfare" hasn't appeared in any PLA publications since Zhao's 2003 article. However, it would seem that the PRC and PLA have taken Zhao's recommendations regarding moral warfare to heart. As thoroughly documented in recent research by IRSEM scholars Charon and Vilmer and PRC political warfare expert Gershaneck, over the past two decades the PRC has deployed powerful political influence capabilities that transcend both borders and professional domains. Relying on the triumvirate of psychological, media, and legal warfare (i.e., The Three Warfares), the PRC is currently capable of efficiently delivering its political and military narratives to a global audience by means of diversified communications channels, while concurrently denigrating competing sources of information. While it's not possible to directly attribute such developments to Zhao, the remarkable similarities between the PRC's influence capabilities and Zhao's recommendations that are discussed in this chapter, certainly suggest that "moral warfare" capabilities continue to be actively developed in the PRC.

Notes

1 See my chapter, "A Survey of 21st Century PLA Scholarship on the Role of Military Ethics in Warfare", elsewhere in this volume, for a discussion of differences between Western and PRC concepts regarding military ethics. Ping-Cheung Lo also addresses this topic (Lo 2015, 157–162).
2 Except as otherwise noted, all translations of Chinese language sources in this paper are my own.
3 At the November 2016 conference "Military Ethics: China and the West in Dialogue" at Florida State University, I briefly explained the PLA's view of moral warfare. After the conference, the late Professor Sumner "Barney" Twiss strongly recommended that I continue to investigate the topic. This chapter is the direct result of his encouragement.
4 This is reminiscent of the advice given in chapter 4 of *The Art of War* (attributed to Sunzi [Sun Tzu]): "Of old the expert in battle would first make himself invincible and then wait for the enemy to expose his vulnerability. Invincibility depends on oneself; vulnerability lies with the enemy" (Ames 1993, 115).

Bibliography

All Army Military Technical Terminology Management Committee. 2011. *China People's Liberation Army Military Terminology*. Beijing: Military Sciences Press. [Published in Chinese: 全軍軍事術語管理委員會《中國人民解放軍軍語》北京；軍事科學出版社. 2011].

Ames, Roger, trans. 1993. *Sun-Tzu: The Art of Warfare*New York: Ballantine.

Charon, Paul and Jeangène Vilmer, Jean-Baptiste. 2021. *Chinese Influence Operations: A Machiavellian Moment*. Paris: IRSEM.

Crump, James, trans. 1996. *Chan-kuo ts'e*. Ann Arbor, MI: University of Michigan.

Gershaneck, Kerry K. 2020. *Political Warfare: Strategies for Combating China's Plan to "Win Without Fighting"*. Quantico, VA: Marine Corps University Press.

Kittrie, Orde F. 2016. *Lawfare: Law as a Weapon of War*. New York: Oxford University Press.

Lo, Ping-Cheung. 2015. "Three Synoptic Views of China's 'People's Liberation Army' on Military Ethics and Justified War." In *Routledge Handbook of Military Ethics*, pp.157–162. New York: Routledge.

Zhao, Feng. 2003. "Military Ethics Should Investigate the Scope of Moral Warfare." *Journal of PLA Nanjing Institute of Politics* 19, no. 4: 86–89. [Published in Chinese: 趙楓〈軍事倫理學應當研究道德戰範疇〉《南京政治學院學報》第19卷；第4期; 2003].

Zhao, Feng. 2014. "Exploratory Analysis of the Characteristics of Military Science and Technology Ethical Thought." *Journal of Naval Engineering* 11, no. 2(June): 5–9. [Published in Chinese: 趙楓〈軍事科技倫理思維特點探析〉《海軍工程大學學報》第11卷；第2期; 2014]. doi:10.13678/j.cnki.issn1674-5531.2014.02.002.

PART III

New Comparative Horizons on Just War and Peace

14

ADJUSTING AUTHORITY

Legitimacy and War in Muslim and Christian Traditions

John Kelsay

Introduction

I will begin with a few lines from Jim Childress' oft-cited essay on "Just War Criteria."

> ... the first criterion ... is right or legitimate authority ... it determines *who* is primarily responsible for judging whether the other criteria are met ... Answering the authority question is a precondition for answering the others; it thus cannot be dismissed as a secondary criterion ... .Even in the use of these criteria for justifying and limiting revolution, surrogates for the established authority are often found in the revolutionary elite or "the people."
>
> *(Childress 1982, 74)*

If I may, I would like to add to Childress' statement on this matter: The importance of right authority, in the sense that it has a certain pride of place–first among equals with respect to the just war criteria—this is what makes the just war tradition a political ethic. Along with its Muslim and Confucian analogues, the just war tradition constitutes an attempt to marshal the human propensity for violence, and to place it in the service of a group. In so doing, the idea is to push back against more anarchic forms of violence—feuds, duels, vendettas, and the like—and to increase the chances for at least a modicum of peace, order, and justice in human societies.

Some of the contributions made in previous discussions of Western and Chinese military ethics lay stress on the importance of the authority criterion. At the same time, they suggest that locating authority—that is, promoting some sort of workable consensus regarding who should make decisions

DOI: 10.4324/9781003336372-17

regarding resort to war—can be hard work. Changing political conditions, new technologies, and other factors make it so that authority for war is often in need for adjustment. For example, contributions by Twiss and Chan note that some Confucian scholars give some degree of legitimacy to a "lord protector." (Lo and Twiss 2015, 101) This seems to designate a ruler who, while not possessed of the degree of virtue characteristic of a "true king," may nonetheless authorize armed force in the service of "political order and humane rule." Again, the same authors note that the neo-Confucian Wang Yang-ming suggests that a wise ruler might decentralize authority by designating "local or provincial commanders" as his deputies. (Lo and Twiss 2015, 157)

Both of these examples suggest a kind of pragmatic adjustment to the criterion of right authority. In a more recent essay on Mao Tse-Tung's military ethics, Twiss deals with a rather different sort of adjustment, more akin to Childress' description of the ways revolutionaries find "surrogates for the established authority."[1] For reasons I hope to make clear, I would speak of this as a matter of "relocating" authority—something is awry or amiss, so that those who ought to authorize force in the service of the common good are either unable or unwilling to do so, in which case the right of war devolves to others.

In what follows, I want to expand our discussion of right authority by way of examples from Muslim and Christian traditions. Some of these may be understood as analogues to the adjustments I've described as "pragmatic," while others will involve "relocation." In either case, I want to suggest that we move forward in the discussion of right authority by attending not only to the fact that the various types of just war thinking under scrutiny assign priority to this criterion, but also to the ways people try to give it meaningful content in particular social and historical contexts. I'll begin with the Muslim jurist al-Mawardi, then move in turn to the Protestant Reformation and Martin Luther, to John Brown and his war against slavery, and finally to the jihadist phenomenon as represented by Osama bin Laden.

Adjusting by Redefinition: Al-Mawardi and *The Ordinances of Government* (Al-Mawardi 1996)

Abu al-Hasan al-Mawardi, born 974 in Basra, in the south of Iraq, rose from humble beginnings to the first rank of scholars. Connections at the court of the Abbasid caliphs al-Qadir (in office from 991–1031) and al-Qa'im (1031–1075) brought appointments as the chief *qadi* or judge in Nishapur (in present day Iran) and then in the capital at Baghdad, as well as several sensitive diplomatic assignments—more on these later. As the judicial appointments suggest, Mawardi excelled in those aspects of Muslim tradition dealing with *Shari'a* and *fiqh*—that is, the attempt to provide judgments pertaining to particular questions about God's directives, and thus to the practice of true religion. While his responses to such questions proved important enough for subsequent generations to preserve, the most noteworthy of his writings for our purposes is a treatise bearing the title *The Ordinances of Government*. Writing at the request of an unnamed official—probably one of the caliphs under whom he served—Mawardi here lines out a number of matters related to the qualifications, mode of selection, and duties of the leader of the Muslim community. The text stands as a classic of Sunni political thought, not least for the way it applies one of the most basic themes of that trajectory of reasoning, viz., that adherence to the *sunna* or practice of the Prophet Muhammad be joined with a focus on the well-being of the Muslim community, particularly by way of ensuring a continuity of institutions. To put this another way, Mawardi represents the sensibility of scholars for whom *fitna* or the kind of civil strife by which Muslims fight one another constitutes a great evil, to be avoided wherever possible.[2]

In Mawardi's time, the threat of *fitna* was high—present on all sides, one might say. Established following a revolution in the 740s, the Abbasid rulers came to power on the basis of wide-ranging dissatisfaction with their predecessor regime, the Umayyads. Placing their capital in Baghdad, the new ruling clan found itself in charge of a far-flung and diverse imperial state. Their most important challenge was, and remained, the maintenance of a consensus regarding legitimacy. One might say that the various factions participating in the Abbasid revolt actually shared only one value. They hated the Umayyads. When it came to defining an alternative policy, however, agreement proved hard to come by. Even Harun al-Rashid, whose rule from 786–809 may be described the high point of Abbasid power, dealt with numerous challenges from the descendants of the family of the Prophet, many of them motivated by the notion that only a designated member of that clan should exercise leadership among the believers.

Given this, the Abbasid rulers usually relied on a sizeable army to maintain control. In itself, this presented problems, not least financially, and the fortunes of successive rulers waxed and waned in large part on whether they could secure sufficient funds to pay their soldiers. Nevertheless, Harun and his successors maintained control of the central areas of the empire until the mid to late tenth centuries, when a number of substantial challenges to their power emerged. On the western front, a military clan based in Tunisia advanced and took control of Egypt and Syria. Claiming a connection with the family of the Prophet, these "Fatimids"—the name tied the new dynasty to the Prophet's daughter—exercised considerable religious, as well as political influence. Meanwhile, in the east, the Buyids, another military clan, this time based in northern Iran, took control of the cities of Shiraz and Isfahan before advancing on and ultimately subduing Baghdad itself, in 945 C.E. The Buyids also claimed a connection with the Prophet, albeit here the mantle of leadership was said to pass through the son-in-law, `Ali ibn `abi Talib. In either case, the problem for the Abbasids seems clear. Unable to mount an effective military resistance, their legitimacy in question, this dynasty of "commanders of the faithful" seemed ready to go the way of all things.

And yet, they did not. Despite their dominance, and the emergence of their distinctive brand of Shi`i Islam, the Buyids were not inclined to abolish the caliphate. Instead, they sought the blessing of those holding the office. Full recognition was not forthcoming, but Abbasid diplomats—including al-Mawardi—achieved a partial rapprochement. And then, beginning in 999, a Sunni alternative emerged. Mahmoud of Gazna, in the process of building a considerable empire of his own in Afghanistan and the Indian subcontinent, took advantage of weaknesses in the Buyid's political and military organization, driving through Iran into Iraq. Proclaiming his allegiance to the Abbasids, Mahmoud received in turn an official appointment as *sultan*, meaning the "power" by which the caliph's religious authority might be maintained.

Now to al-Mawardi's treatise, which is clearly motivated by the way the Abbasids' situation had changed. Thus, even though he begins with a rehearsal of the duties of the caliph in which military affairs figure prominently, we know—as he did—that this is a kind of fiction. Of neither al-Qadir nor al-Qa`im—to recall the rulers Mawardi served—could it be said that they were able to "protect the country and the household, so that all may go about the business of living and travel anywhere unworried by deception or loss of life or property ... " or to "strengthen border posts by deterrent equipment and fighting force so that the enemies may not gain the chance to violate what is sanctified or shed a Muslim's or a protected non-Muslim's blood." (Al-Mawardi 1996, 16) Indeed, given the Fatimids and the Buyids, even the duty of guarding the faith, "upholding its established sources and the consensus of the nation's ancestors, arguing with emerging heretics or suspicious dissenters ... so that the faith should remain pristine and the nation free from error" seemed in question. (Al-Mawardi 1996, 16) In the kind of condition outlined above, how could anyone think the Abbasids capable of actions like these?

By way of an answer, al-Mawardi builds on the ruler's duty to make appointments. In naming ministers, governors, judges, military commanders, tax collectors, and the like, one who holds the office of caliph may exercise power and fulfill his duties, even under conditions of restraint. Considering the example of the Buyids, we may understand Mawardi's claim as follows—as political and military actors, they are usurpers of authority. As advocates of an irregular form of Muslim practice, they are heretics. Yet they wish to cooperate with the Abbasids, the rightful leaders of the community of the faithful. The caliph may choose to acknowledge them by granting some, though not all of their demands for recognition. In doing so, he functions as "the implementer of the dictates of religion ... transforming unlawfulness into legality, and the forbidden into the legitimate. *Although by doing so he does not adhere closely to the conventions of appointment in respect of conditions and procedures, his action upholds the canon law and its provisions in ways that are too important to be disregarded*." (Al-Mawardi 1996, 36)[3]

Through skillful use of the power of appointments, a ruler thus sustains a continuity of the institutions of governance and—it should be said—lives with the hope of better days to come. With Mahmoud of Ghazna's defeat of Buyid forces, and his subsequent appointment as sultan, al-Mawardi could say that hope was on the way to fulfillment.

Of course, it must be said that these arrangements with Mahmoud did not actually restore the Abbasid caliphs as a military force. In that sense, Mawardi's treatise represents not only a pragmatic adjustment in response to political realities. He has also effectively redefined the office, so that it becomes the symbol of Muslim unity. To put this another way, the emphasis from Mawardi forward would be on the caliphate as a *religious* office. Effectively, authority for war is transferred to the sultan. At the same time the legitimacy of those exercising that power depended on the caliph and his power to make appointments.

Adjusting by Restoration: Luther On War Against the Turk

Mawardi died in 1058.[4] His understanding of the relationship between caliph and sultan endured for two centuries, until the Mongol forces led by Genghis Khan and his successors overran the capital and effectively brought the Abbasid dynasty to an end in 1258. At that point, however, a new force began to assert itself, particularly in Anatolia. Ultimately, the Ottoman Turks would control most of North Africa and the Middle East, along with portions of southern and central Europe. With respect to the latter, the conquest of Constantinople in 1453 provided the Ottomans with a convenient launching point for further expansion. Establishing a pattern of annual expeditions into the area, the forces of the Ottomans raised fears at every level of European society. At the popular level, stories of Christians taken into captivity by Turkish forces circulated; among officials, worries about their ability to marshal sufficient forces and funds to resist led to attempts by the Vatican to impose new taxes in support of a new crusade.[5]

Born in 1483, Martin Luther thus lived his entire life in the shadow of the Turkish threat. Indeed, in 1518, one year after posting his Ninety Five Theses on the door of the cathedral at Wittenburg, Luther extended his critique of the Pope's program of raising money through the sale of indulgences so as to include the Crusade tax. In part, the Reformer's complaint focused on corruption. As various biographers point out, on this point Luther reflected widespread suspicion that funds collected by the Vatican might not be used for the stated purpose. (Roper 2016) At the same time, Luther's critique went to broader theological concerns. As he would put it repeatedly in political writings published between 1520 and 1532, the so-called Christian society overseen by the Pope and the Holy Roman Emperor involved a hopeless confusion between the spiritual and temporal realms, with church officials taking responsibility for the latter and political leaders construing their role in terms of the former. For Luther, the very idea of a crusade or war of religion constituted a violation of Christian faith. If one were to approve military resistance against the Ottomans, one ought to speak of this in terms of an effort ordered by properly political authorities, and with aims consistent with the worldly concerns central to their vocation. Religious leaders ought not play a role in this; rather, they should provide instruction in the spiritual meaning of current events. So Luther suggested that Christian leaders ought to interpret the Turkish advance by way of precedents from the Old Testament. The Sultan would thus be God's instrument for disciplining an unruly and disobedient church, to which the proper response of believers would be repentance and prayer rather than armed force.

Following his comments in 1518, many of Luther's contemporaries understood him as a sort of political quietist, at least with respect to the Turkish advance. With the publication of his 1526 book on the question *Whether Soldiers, Too, Can Be Saved*, Luther considered that he had outlined the way that Christians might consider the propriety of armed force.[6] The work ends, however, with an acknowledgement of the need to say something in particular regarding resistance to the Turks. In view of the victory of Ottoman forces in Hungary at the Battle of Mohacs in 1526, just prior to release of the tract on soldiers, this seems unsurprising. And so in 1529, with the forces of Suleiman the Magnificent surrounding Vienna, Luther returned to the topic. With respect to our interests, the following lines are particularly significant.

> In the first place, the Turk certainly has no right or command to begin war and to attack lands that are not his ... The Turk does not fight from necessity or to protect his land in peace, as the right kind of ruler does; but, like a pirate or highwayman, he seeks to rob and ravage other lands which do and have nothing to him. He is God's rod and the devil's servant ...
>
> In the second place, we must know who the man is who is to make war against the Turk so that he may be certain that he has a commission from God and is doing right. He must not plunge in to avenge himself or have some other mad notion or reason. He must be sure of this so that, win or lose, he may be in a state of salvation and in a godly occupation. There are two of these men, and there ought only to be two; the one is named Christian, the other, the Emperor Charles.
>
> *(Schultz 1967, 170)*

As the argument develops, we understand that the "man" Luther calls "Christian" is actually the community of believers. Acting in their calling as people of faith, these offer no armed resistance to the Ottomans, or for that matter to any others who commit aggression. They see God's hand in current affairs, and they understand that a proper response involves repentance and prayer. As well, members of the Christian community whose role includes preaching and theological reflection should focus on telling the truth about religion—their own, and that to which the Turks adhere. In that connection, Luther's treatise details the positive and negative aspects of Islam—there are more of the latter, though he does commend the Qur'an for affirming the Virgin Birth and the prophetic aspects of Jesus' ministry. In the course of this polemic, Luther finds that the negative aspects of Islam bear comparison with the errors of the Pope, not least because both consider that human beings win salvation through good works, rather than by the grace of God as revealed in those who received the gift of faith.

It is when Luther turns to the second "man" that we learn how he thinks about a military response. The

> man who ought to fight against the Turk is Emperor Charles, or whoever may be emperor; for the Turk is attacking his subjects and his empire, and it is his duty, as a regular ruler appointed by God, to defend his own ...
>
> *(Schultz 1967, 184)*

In affirming this point, Luther moves quickly to clarify the reasons for which the Emperor should act. Noting that the Pope and others call for Charles to organize the fighting as a crusade, he writes:

> Not so! The emperor is not the head of Christendom or defender of the gospel or the faith ... .The emperor's sword has nothing to do with the faith; it belongs to physical, worldly things, if God is not to become angry with us ... .The emperor and the princes should be exhorted concerning their office and their bounden duty to give serious and constant thought to governing their subjects in peace and to protecting them against the Turk. This would be their duty whether they themselves were Christians or not ...
>
> *(Schultz 1967, 185–86)*

In short, the Emperor's banner should read "Protect the good; punish the wicked." (Schultz 1967, 190) The ruler's armies fight for temporal goods. The advance of true religion is not to be numbered among these.

As we reflect on Luther's remarks, several matters are worthy of note. First, we should recall that the Pope's call for a "crusade tax" clearly identifies the struggle with the Ottomans, as with other Muslims, as a matter of religion. By raising funds for the effort, as well as issuing public statements encouraging Charles V to raise an army sufficient to repel the threat and ultimately to retake Christian lands in the Middle East, church officials encouraged a view of their current conflicts as the continuation of struggles dating back to the time of Charles Martel (668–741), whose troops managed to stop the advance of Muslim armies at the borders of what we would think of as France and Spain, or even earlier, to the initial conquests by which Arab forces took territory from Christian Byzantium in Syria, Palestine, and Iraq during the seventh and eighth centuries. I have already made this point, but it is worth stressing: Luther's approach renders such an understanding illegitimate. For him, Christian faith requires a proper delineation of the spheres within which religious and political leaders operate. The task of religious leaders has to do with the spiritual welfare of human beings, so that the religious reforms asso-ciated with Luther's program—proper preaching, the administration of the sacraments, and so on—may be carried out. By contrast, political leaders deal with the temporal welfare of their subjects. Their concerns have to do with securing property, enforcing laws, and defending their people against invaders. Luther believes this distinction is obscured by the relationship between the Church and the Holy Roman Empire. Indeed, even the qualifier "holy" is improper with respect to the political entity governed by Charles. For Luther, it would not matter whether Charles or other rulers were Christians or not. For their spiritual welfare, he hopes they are such. He does not count on this, however, and against all those calling for princes to reign as Christians, Luther's typically pithy judgment is that a Christian prince may be a possibi-lity, but experience shows there are not many such persons.

Second, we should note that Luther's judgments pertaining to holy war do not only apply to the relationship between the Pope and the Emperor. *On War Against the Turk* is directly addressed to Philip of Hesse (1504–67), who favored the Reformation and provided support for Luther and his colleagues. According to the practice of the time, Philip and several others responsible for governing provinces within the German-speaking areas served as "electors," and thus played a role in the selection of Charles V as Emperor. In one sense, this made them subjects of the Emperor. In another, they constituted an alternate center of power. It is of interest, then, that during the 1520s and 1530s, Philip tried to build a coalition of Protestant princes ready to use force in the effort to secure the Reformation, first by launching a preemptive strike against Catholic strongholds —this, because of the Pope's urging that political leaders eliminate the heretical teachings of Luther and others—and second, by taking independent action against the Ottomans. Luther's clear identification of the Emperor as the holder of authority with respect to war constitutes a rebuke to one of the Reformation's most important political assets. At the same time, *On War Against the Turk* makes the case that any military action against Catholic forces ought not be authorized for reasons of faith. Rather, as with fighting against Muslims, the causes for war must be tied to mundane, temporal matters like theft of property or self-defense. This distinction, so central to the entire corpus of Luther's poli-tical writings, would receive renewed attention when, in response to Charles' plan to counter the growth of the Reformation in the Netherlands, Luther and his colleagues moved in late 1530 toward recognition of a legitimate resistance to tyranny, understood as grounded in a right to self-defense.[7]

Finally, we should consider Luther's characterization of his political thought as a recovery or restoration of something lost or obscured by what he called the "pagan servitude" of the Church. As with the priority of grace and faith in matters of salvation, so with the designation of authority for war—through the centuries, as the collaboration between religious officials asserting the priority of the Pope and those princes able to assert power as the successors of the Roman Empire came to define a geopolitical idea known as Christendom, the resulting confusion of spiritual and temporal authority led to numerous errors. Crusades, the levying of improper taxes, corruption and the outright theft of ordinary people's property provided examples of such wrongs in the temporal realm, even as the failure to encourage proper administration of the sacraments and the preaching of the gospel in the spiritual. So Luther argued, at any rate; and while it is true that in these matters, his would not be the only way of parsing the distinction between the two realms to emerge from the turmoil of the sixteenth (and really, into the seventeenth) century, it is nevertheless a significant example of the struggles attendant to establishing a location for legitimate authority.[8]

Relocating Authority (1): John Brown's Provisional Constitution

In January of 1858, the abolitionist John Brown traveled to the Rochester, NY home of his friend, Frederick Douglass. Brown had, from 1856 on, been active in the struggles between pro-slavery and "free soil" factions in "Bleeding" Kansas. With the help of journalists like James Redpath, Brown's exploits in that conflict made him one of the best known of the "radical" abolitionists —that is, those who advocated the use of armed force in the service of the cause. The martial phase of Kansas' struggle was ending, however. By the time of Brown's visit with Douglass, it seemed clear that the numbers of settlers on the "free soil" side pointed toward victory in the territorial legislature, and thus that Kansas would enter the union as a free, rather than a slave state.[9]

Brown did not consider that his work was done, however. Rather, success in Kansas seemed to spur him on. In particular, he was contemplating a strike into the southern states, or as he sometimes put it, into the "heart of Africa." The plans for the raid that would ultimately take place at Harpers Ferry, Virginia in October, 1859 were thus being laid. Brown hoped to recruit a fighting force of fifty to one hundred; their action would liberate slaves in the area around the Ferry, who would then receive military training and join Brown's men in operations further south. Eventually, a much larger band, perhaps divided into several semi-autonomous units would so disrupt the southern economy that the "slave power" would fall.

Perhaps Brown discussed this plan during his visit to the Douglass home; that is difficult to know. The real purpose of the visit had to do with another trip Brown planned for later in the spring, where he would present his ideas to the community of free blacks and runaway slaves in Chatham, Ontario. In preparation for that meeting, Brown spent most of his time in Rochester holed up in his room, writing the draft of a document he hoped the Chatham community might endorse. The text began with these lines:

> Whereas slavery throughout its entire existence in the United States is none other than a most barbarous unprovoked and unjustifiable War of one portion of its citizens upon another portion, the only conditions of which are perpetual imprisonment and hopeless servitude or absolute extermination in utter disregard and violation of those eternal and self-evident truths set forth in our Declaration of Independence. Therefore, We ... Do for the time being Ordain and establish for ourselves the following Provisional Constitution and Ordinances the better to protect our Persons, Property, Lives, and Liberties, and to govern our actions.[10]

In connection with Brown's military plan, the purpose of this document seems clear. Brown's ever expanding fighting force would need some sort of command and control structure. The text thus outlines a system of governance, with procedures for the election of a president and other officers. Interestingly, article 46 insists that the Provisional Constitution "shall not be construed so as in any way to encourage the overthrow of any State Government or of the General Government of the United States ... " Rather, Brown's organization would seek to "amend and repeal" existing legal and political institutions insofar as these supported slavery.

There is more to say about this document, which was in fact approved by a number of members of the Chatham community in April 1858, with Brown himself appointed to constitutional office as Commander in Chief.[11] For now, however, I want to focus on what the text as an indicator of Brown's attempt to "relocate" authority for war—away from the existing government of the United States, which he deemed corrupt, and to a band of Americans committed to "liberty and justice for all."

In another context, I discuss Brown's gradual move away from the standard Garrisonian or nonviolent abolitionism of his day, and toward the position of the "radicals"—people, many of them freed or escaped slaves, who from the 1840s on began to argue that resistance to slavery required armed force. (Kelsay forthcoming) In Brown's earliest writing on the topic, the 1851 "Words of Advice" composed for a band of African Americans living in Springfield, Massachussetts, armed resistance to "slavecatchers" acting under the 1850 Fugitive Slave Act is justified as a matter of self-defense. No one in the northern states, writes Brown, will "convict a man for defending his rights to the last extremity."[12]

This appeal to self-defense as a justification for armed force continued throughout Brown's career in Kansas. In 1855, when Brown made the journey from his farm in upstate New York to the recently opened territory, he did so in response to the pleas of several of his sons. John, Jr. in particular wrote to his father about the harms done by bands of pro-slavery "ruffians" coming into the area at the urging of southern slaveholders determined to ensure that when the time came for a territorial legislature to petition Washington for recognition as a state, the majority of those framing the request would favor the south's "peculiar institution." Playing on the older man's sense of himself as a military expert, John, Jr.'s letters suggested that his father could provide leadership in the defense of the advocates of freedom and in preventing further expansion of slavery into the American west. Once in the territory, the elder Brown proved his mettle in several skirmishes, writing to his wife and their children back home in some detail about the violence of the pro-slavery side and the bravery of those fighting with him. Along the way, he also displayed a talent for cultivating journalists—men like the aforementioned James Redpath, whose accounts of Brown's exploits made sure that anyone interested in the abolitionist cause became aware of "the Captain."[13]

To this point, then, Brown construed his activities in terms of defense, not very different from the classic just war notion of *defensio*. Historically speaking, precedents in canon law distinguished that sort of action from *bellum*. When an aggressive force presents a threat to life, liberty, or property, anyone has the right to take defensive action, so long as it is proportionate or limited to the amount of force necessary to repel the enemy. That is of course an important qualification; it means that *defensio* does not include the right to pursue or punish the aggressor.[14]

In that connection, we should recall that Brown had other aims in mind. Success in Kansas gave him confidence, and he began to lay plans that would carry the struggle into the south. We do not know precisely when Brown settled on Harpers Ferry as the first target in what he hoped would be guerrilla campaign lasting months or years. He certainly shared the broad outlines of the plan with the community at Chatham in the spring of 1858. And early in 1859, he recruited a band of trusted comrades from Kansas, took them to a secure location in Iowa, and provided instructions for their training. These men would eventually make their way to Maryland, where Brown managed to rent a farm near the Virginia border; they would be joined by a few others, so that the force for the raid totaled twenty-two (including Brown). In the period between the move to Iowa and relocation to Maryland, Brown traveled throughout the northeast, raising money, purchasing weapons, and (once the Kennedy Farm had been secured) moving those arms to Maryland.

Now let us return to the Provisional Constitution and its adoption by the community in Chatham. What did this mean with respect to Brown's conception of the campaign he envisioned? In particular, did it signal a change in his notion of the type of fighting in which he would engage, from something like *defensio* to something more like *bellum*? And how does such a change affect claims about legitimate authority?

We have seen how Brown's Constitution depicted slavery as "a most barbarous unprovoked and unjustifiable War of one portion of its citizens upon another portion." And the Preamble continues by stating that the purpose of adopting this document is to secure the "Persons, Property, Lives, and Liberties" of those present in Chatham, as well as others who will join the effort. The language suggests a defensive struggle. But the fact that twenty-four of the forty-eight articles that follow the Preamble deal with military matters points in the direction of a community organized for a large-scale operation.

Then, too, Brown thought of the Harpers Ferry raid as an incursion into enemy territory. It was his purpose, he would say, to free slaves. To that end, the plan was to guide those liberated into the mountains near the Ferry, where those who wished could begin the journey to freedom in the north. At the same time, Brown hoped that at least some of these folk would join with him, receive military training, and carry out raids further to the south. For this, he continued to use the language of defending those unable to protect themselves. This is an "extended" notion of defense, however, perhaps closer to the contemporary notion of protection or rescue of victims of genocidal action. For this purpose, Brown felt the need for the kind of authority conferred by the Provisional Constitution. One might venture to say that he seems instinctively to have understood that the campaign he envisioned constituted war—a struggle in response to the aggression of slave holders and their supporters, but a war nonetheless. Thus, in an interview with senators, congressmen, and other public officials following the raid, Brown not only appealed to Scripture and to the notion of God's calling, but also to the Provisional Constitution as justification for his actions.[15]

> Senator Mason: "Did you consider this a military organization in this paper? (*Showing a copy of John Brown's constitution and ordinance.*) I have not yet read it.
>
> Captain Brown: "I did in some measure. I wish you would give that paper your close attention."
>
> Senator Mason: "You considered yourself the commander-in-chief of this provisional military force?"
>
> Captain Brown: "I was chosen, agreeably to the ordinance of a certain document, commander-in-chief of that force."

It would be a stretch to speak of John Brown as a just war thinker, at least in the sense of one who is actively engaged in conversations that include references to people like Augustine, Aquinas, or Gentili. Yet in talking with students or with colleagues who wonder about the utility of the just war tradition, I often tell them that if we did not have the just war framework, we would have to invent something very much like it. In this sense, we expect anyone concerned about the relationship between power and morality to worry about questions like "when is military action justified?" "For what ends?" "How should force be directed, in terms of targets and the means employed?"

And of course, "who decides?" One of the worries about people like John Brown has to do with the idea of "private" war, which most just war thinkers associated with a kind of violence that threatens anarchy. Particularly in connection with the raid at Harpers Ferry, Brown is often interpreted as a man whose religious convictions led him to take a sense of calling—that is, an idea that he might be an "instrument of Providence" in the fight against slavery— so seriously that he felt his actions justified, without attending to the type of question to which the criterion of right authority is a response. Indeed, in the interview with Senator Mason and others cited above, Brown said as much:

> No man sent me here; it was my own prompting and that of my Maker, or that of the Devil, which you please to ascribe it to. I acknowledge no master in human form ...

When one of those present asked whether he considered himself an instrument in the hands of Providence, Brown responded "I do." This is the line of thought that led some of his contemporaries, and many people since to think of Brown as a religious fanatic, whose zeal for the cause is so great as to justify or excuse any sort of action, so long as it is aimed at the "slave power."

In a forthcoming essay on "The Making of John Brown", I argue that this misconstrues Brown's religious convictions. Here, though, it seems important to note the lengths he went to answer the question of authority for war. To think with Brown is to say something like this: One did not need to address the matter when engaged in self-defense, as with the League of Gileadites or the struggles in "Bleeding Kansas." One did, however, need the authorization of a community and its organizational structure for something like the campaign Brown envisioned in the southern states, and in which Harpers Ferry constituted the first strike. The Provisional Constitution and Ordinances, affirmed by the community of African Americans gathered in Chatham provided Brown with that authorization. With respect to the interests of this paper, it constitutes an adjustment by way of relocating authority in an America expanded so as to include people of all colors and creeds.

Relocating Authority (2): Osama Bin Laden's Declaration of War

Brown's attempt to relocate authority involved a peculiar combination of Christian and American precedents. As he sometimes put it, Brown believed in two things: the Golden Rule and the Declaration of Independence. Having affirmed those, he would then add that he believed these meant the same thing. I turn now to another, rather different attempt to establish authority for war: the example of Osama bin Laden and his appeal to the Muslim notion of fighting as an individual duty.

In May of 1996, Osama bin Laden moved his family to Afghanistan. The story behind this is fairly well known.[16] Bin Laden, the seventeenth of twenty-three sons born to Muhammad bin Laden, grew up in a relatively privileged environment. Favored by the Saudi royal family, the Bin Laden Construction Company did very well, and the father took care of Osama, even after divorcing the boy's mother. When Muhammad died in an airplane crash in 1967, the older brothers took over the business, eventually bringing Osama on a manager for several projects.

The 1979 Soviet invasion of Afghanistan changed the young man's life. Always marked by a more serious piety than other family members, Osama quickly became involved with the Afghan resistance. From 1980 to 1983, he played the role of a courier, carrying cash from the royal family and other wealthy Saudis to support the mujahidin or "strugglers" in the path of God. In that role, he became a favorite of Abdullah Azzam, a Palestinian-born scholar who achieved fame as the "chaplain" for the resistance, and whose various publications urged believers to view the fighting as a matter of defending Muslim lands—to put it another way, people should understand that fighting the Soviets, and others occupying territory historically associated with Islam constitutes an "individual duty". In cases like this, Azzam argued, jihad—here in the sense of armed struggle—is like prayer. Even as every believer must perform the prayer for him or herself, so with participation in the struggle. Everyone must play a part, as he or she is able.[17]

At the encouragement of Azzam, Osama moved his family to Peshawar, a town located just on the Pakistani side of the border with Afghanistan. Using bin Laden's contacts and Azzam's growing international reputation, the two men founded a set of organizations designed to enhance the resistance, providing health care, stipends for widows and orphans, encouraging the provision of weapons for fighters, and—increasingly over the years between 1983 and 1987—the recruitment of volunteers from Saudi Arabia and elsewhere to make the journey to Afghanistan, where they might (as the title of one of Azzam's treatises put it) "join the caravan." As the numbers of foreign fighters increased, so did the need for places to train them. Osama's construction experience proved useful here, as he was able to arrange for the purchase, transport, and to oversee the use of heavy equipment for the task.

By 1987, however, Osama desired a more active role. Assuming oversight of one of the training camps, he set himself as the commander of a small group of "Arab Afghans." Mostly from Saudi Arabia and the other Gulf states, these saw some action toward the end of the war, and when the Soviets announced plans to withdraw by the end of 1989, bin Laden and his men spoke with pride about their part in the achievement.

Returning to his home in late 1989, Osama considered himself a Saudi patriot, ready to serve in any capacity for which the royal family thought him fit. Fighting in Yemen, at least in part caused by another instance of Soviet intervention, seemed a possibility. When Saudi authorities indicated this would not be an appropriate role, bin Laden puzzled over how his country could fight communism in Afghanistan, but not in another, closer setting. In August of 1990, however, attention turned to Iraq's invasion of Kuwait. The possibility of a further incursion into the Kingdom led to a proposal by which bin Laden would organize thirty thousand or so veterans of the Afghan struggle whose presence would enhance existing Saudi defenses. When the royal family announced its decision to rely on U.S. and other international forces to deal with Iraq, Osama joined with others in criticizing the regime. The further decision to allow the U.S. to build bases throughout the Gulf region following the Gulf War only intensified this criticism. When many of those associated with opposition to the royal family's policies were arrested and imprisoned, Osama bin Laden left—first returning to Afghanistan and Pakistan, and then in 1993 to the Sudan. From there, he made use of a London-based Office of Advice and Reform to publish increasingly strident analyses of conditions in his home country. Facing pressure from the Saudis and the U.S., the Sudanese finally told bin Laden to leave.

This brings us back to 1996, with bin Laden in Afghanistan—in exile, as he would put it. In August of that year, he declared as much in an audio recording which came to be known as his "Declaration of War Against America." The following lines give a sense of bin Laden's understanding of the situation.

> I meet with you today, after a long absence has been imposed on the scholars and preachers of Islam … we have suffered by being prevented from addressing the Muslims. We have been exiled from Pakistan, the Sudan, and Afghanistan, hence this lengthy absence. But by the grace of God, a safe base has become available in Khurasan on the summit of the Hindu Kush, this summit where by the grace of God, the largest infidel military force of the world was destroyed, and the myth of the superpower withered before the strugglers' cries "God is Greater".[18]

Like the Prophet and his Companions, bin Laden has been driven from his home, but he trusts that the base provided by God will serve him—even as Medina came to be the base of operations that would, in the end, allow Muhammad to proclaim that "Arabia is now solidly for Islam."

As noted, this 1996 speech came to be known as bin Laden's "Declaration of War against America". And in a sense, it is that. Describing conditions in his homeland, he tells the audience that

> It should not be hidden from you that the people of Islam have suffered from injustice, oppression, and aggression by the Judaeo-Christian alliance and their

collaborators, to the extent that the Muslims' blood became the cheapest and their wealth and natural resources loot in the hands of their enemies. Their blood was spilled in Palestine and Iraq. The horrifying pictures of the massacre of Qana, in Lebanon are still fresh our minds. The same is true for the massacres in Tajikstan, Burma, Kashmir, Assam, the Phillippines, Fatani, Ogadan, Somalia, Eritrea, Chechny, and Bosnia and Herzegovina, where the massacres against Muslims that took place send shivers through the body … .The latest of these aggressions on Muslims, a calamity that matches the greatest confronted since the death of the Prophet … is the occupation of the Land of the Two Holy Sanctuaries … by Christian armies …

In view of the last line, in particular, it is not strange that bin Laden continues by praising those responsible for (at the time) recent attacks on U.S. facilities in and near the Saudi capital. One of the great themes of the speech has to do with marshalling support for a military and economic campaign designed to compel an American withdrawal from the Gulf region. This is a point to which bin Laden would return again and again in the next few years—for example, in a 2002 "Letter to the Americans" published in the London Observer: "We … advise you to pack your luggage and get out of our lands. We only desire this for your goodness, guidance, and righteousness, so do not force us to send you back as cargo in coffins." (Lawrence 2005, 171) For bin Laden, the current generation of Muslims lives in an "age of humiliation," with the alliance between the United States and Israel imposing much of the pain

Much of the pain, but not all. As bin Laden makes clear, the United States and its allies are only able to carry out their injustice because Muslims cooperate. And this is where he speaks to the point of interest in this paper. Recall that bin Laden's mentor Abdullah Azzam urged believers to think of Soviet action in Afghanistan as the kind of condition in which fighting becomes an individual duty, so that each and all must participate. In that case, Osama bin Laden understood his role to involve carrying out Saudi policy. To put it another way, the royal family's support for the Afghan resistance encouraged its citizens to answer Azzam's call. Osama and others could thus join Azzam's caravan, understanding that when authorities in the Kingdom spoke of their role as "guardians" of the Muslim holy places, they did so in good faith.

The decision to allow a continuing American presence in the Gulf region called that judgment into question. Believers should not only strive for Americans to leave; they must also renounce and resist the Saudi state, which has

subdued the most prominent scholars, deriding them by issuing legal opinions founded in neither the book of God nor the pathway of the Prophet. [The state] handed over the Al-Aqsa mosque [in Jerusalem] to the Jews and exposed the Land of the Two Holy Sanctuaries to Christian armies. Twisting and throttling the holy scriptures can change truth not a wit.

By doing these things, the royal family forfeits legitimacy. Indeed, the standing of prominent officials as believers is in question. As bin Laden put it in a 2000 speech,

> ... there are ten clear ways that Islam can be legally nullified. Among them is supporting polytheists against believers ... Not only are such people loyal to Jews and Christians. They also rule by laws other than those revealed by God ... .[by inviting the United States into the Arabian Peninsula, they show their true colors, for the Americans] likened themselves to a god who must be worshipped. Now the rulers of the region no longer worship the Lord of the Ancient House [that is, the Ka`ba, in Mecca], they instead worship the lord of the White House.[19]

In this connection, authority for war no longer rests with public authorities. Rather, conditions suggest that authority now rests with the community as a whole, or with a vanguard willing and able to organize an effective resistance. The duty to fight Americans and their allies, including those Saudis whose actions turned the country into an "American colony" applies to any believer able to carry it out, in any location where such is possible. In this instance, Osama bin Laden and those with urge a relocation of authority for war, from regimes unwilling or unable to protect Muslim interests, and to those individuals and groups who recognize the true condition of the community God called to command right and forbid wrong.

Conclusion

I began this essay by noting the importance of right authority for the just war, jihad, and Confucian traditions.[20] As Jim Childress notes, this criterion is required for the others to work. To put it another way, some agreement regarding the location of and procedures for making decisions about resort to and conduct of war is necessary, if these various war conventions are to provide an effective way of marshalling the human propensity for violence and placing it in the service of group life.

One of the reasons for undertaking the kind of analysis presented in this paper has to do with illustrating the fact that in addition to its importance, the requirement of right authority also constitutes a challenge. To put it bluntly, nothing lasts. Demographic shifts, foreign conquests, religious and social change, developments in technology all have import for the history of authority as related to war. Indeed, if we follow someone like Max Weber on the point, the function of religion (or more generally, of normative vocabularies) in human societies always suggests the possibility of conflict over the distribution of power and other goods.[21] To forge a consensus regarding the location of authority for war is almost always the work of a particular group within a broader social setting, and while those involved may talk about the common good or universal justice, we should expect that over time, others will perceive contradictions between such claims and the actual order of poitical life. We might think such moments as matter of tension between the order-preserving and order-transforming aspects of discourse about war. Of the cases discussed above, the clearest examples of the latter have to do with those in which groups propose to relocate authority, arguing that the existing command and control systems do not serve the cause of right and thus opening the possibility that those ready and willing to serve the cause of justice may assert the right to make war.

The case of al-Mawardi, by contrast, represents something more of the order-preserving function of discourse about war, and that of Luther occupies something of an in-between space. Even with al-Mawardi, however—an author whose conservative tendencies are pronounced– one sees that preservation is accomplished, in part, by redefining or redescribing legitimate authority. In all, the examples discussed here serve to remind us that filling out this criterion is rarely a settled matter.

Returning to Childress' affirmation of the status of right authority—one might say as first among equals with respect to the criteria measuring justice in war—the readiness of conservatives, reformers, and revolutionaries alike to assert the importance of this matter suggests the importance people attach to order in human societies. Reading the history of just war, jihad, and Confucian traditions on these matters points in the direction of the late Paul Ramsey's assertion: wise statecraft involves an attempt to achieve a balance between the goods of peace, order, and justice; yet among these, one must attach a certain priority to order, since without it, neither peace nor justice will prevail.[22] At the same time, it seems we have to say that the priority of order has its limits. Peace and justice also matter, and the various adjustments to right authority discussed here bear witness to that.

Notes

1 See "Mao Zedong's Ethics of War (1927–1949)?", Chapter 9 in the current volume.
2 It is probably worth saying that the term caliph approximates the Arabic *khalifa*, which implies "succession" or "following". In the Qur'an, the term is sometimes used for human beings as such, so that they are to be God's covenant partners in the work of establishing right order in the world. In the tradition of Muslim political thought, the term is applied to various rulers, and is supposed to distinguish them from the more controversial term "kingship." Like their Umayyad predecessors, the Abbasids also used other terms to describe the power as rulers: a favored phrase for this designated the ruler as "commander of the faithful." A standard, though dated survey of these matters is (Rosenthal 1962); I also recommend (Kennedy 1986; Kennedy 2001).
3 Emphasis added.
4 By which time the Seljuq clan came to play the role of sultan in Iraq and regions to the west.
5 On these matters, the relevant portions of Martin Brecht's highly regarded work, (Brecht 1985; 1990; 1993), three volumes are helpful, as is the much briefer

editorial introduction to *On War Against the Turk* provided in (Schultz 1967, 46: 155–206).

6 Included in Schultz 1967, 89–137.

7 See the discussion in (Skinner 1978, 2:189–238).

8 Indeed, even in Luther's own time, the practice by which various rulers proclaimed themselves defenders of the reformation seems to have overwhelmed his attempts to clarify the distinction between spiritual and temporal realms.

9 The troubles in Kansas followed from the Kansas-Nebraska act of 1854. Designed to facilitate U.S. expansion to the west, the legislation also created the possibility that these territories might enter the Union as states in which slavery would be legal. The question would ultimately be decided by the membership of the state legislatures, and thus ultimately by the votes of whichever side might produce a majority.

10 Brown's "Provisional Constitution and Ordinances for the People of the United States" may be found on various websites, as well as in a number of anthologies. A recent example is Stauffer and Trodd 2012, 26–37.

11 The structure of governance outlined Brown's Constitution for the most part mirrored the existing U.S. system. One exception to this rule is here, where Commander in Chief is an office separate from that of President. At Chatham, the convention left the latter post unfilled, perhaps in the hope that Frederick Douglass might eventually join the organization (and thus take part in Brown's operations in the southern states). That did not happen; at a private meeting not long before the Harpers Ferry raid, Douglass counseled Brown against the plan, arguing that the most likely conclusion would be death for Brown and those with him, and harsher treatment for those in slavery.

12 Quoting from*l* (Stauffer and Trodd 2012, 8), where the line is set off from most of the text by Brown's use of italics.

13 In line with this notion of Brown's activities in Kansas as a kind of defensive operation, it is interesting to note that Stephen Oates, among others, interprets the infamous killings at Pottawatomie Creek (May 21, 1856) in ways suggestive of a reprisal. In this case, Brown and his band of eight men executed five pro-slavery advocates, dismembering them in the process, but the number of victims and the manner of their death mirrored the killings of five free soil settlers. The idea, then, would be to send a message: if pro-slavery advocates can kill in this manner, so can we. In this regard, it is of interest that some of those involved would later report that following the incident, Brown commented that "there will be no more work such as this." Of course, it also remains of interest that Brown's immediate answer to charges of his involvement at Pottawatomie was (1) to say that he did none of the killing there, though he approved; and (2) to allow Redpath and others to report that at the time of the incident, Brown was twenty-five miles away. The truth, that Brown was the leader of this little band of vigilantes, would not be a matter of public record until sometime in the late 1870s, when James Townsley revised testimony regarding his own involvement. Cf. the account in (Oates 1984).

14 As mentioned above, Luther and his colleagues moved in the direction of justifying armed resistance to tyranny in late 1530. When they did so, they invoked these canon law precedents, which appealed to a natural right to defend oneself and one's belongings. A convenient example of such precedents is provided by the selections from Pope Innocent IV included in (Reichberg et al. 2006, 148–51).

15 "Interview with Senator Mason and Others," included in (Stauffer and Trodd 2012, 44–53). The quote that follows is from this source. For a discussion of the religious aspects of Brown's responses, see my paper "The Making of John Brown" mentioned above.

16 A convenient, if somewhat polemical biography is (Scheuer 2011). More recently, the material is covered in (Miller 2015) and in (Bergen 2021); also see (Lahoud 2022).

17 For a brief introduction to Azzam and a sample of his writings, see (Kepel and Milelli 2008). Thomas Hegghammer's *The Caravan* (Hegghammer 2020) provides a more extensive analysis.
18 The text may be found in several versions online. A convenient, if partial print version is included in *Messages to the World* (Lawrence 2005, 31–43). Subsequent quotes from the Declaration are presented without further citation; where I quote from other statements of bin Laden, I indicate the source.
19 As presented in (Miller 2015,43)
20 In addition to the essays by Twiss and Chan in Lo and Twiss 2015, I should mention the following contributions to the current volume: Yvonne Chiu, "'Good Governance' as Jus Ad Bellum in Chinese Just War Theory"; S. Twiss, "Chiang Kai-Shek's Military Ethics: An Analysis of his Wartime Rhetoric"; idem, "Mao Zedong's Military Ethics (1927–1949)"; and Ping Cheung Lo, "Gratian and Mengzi: Seminal Works in the Christian and Confucian Just War Traditions." Twiss' discussion of Mao provides a suggestive avenue of comparison with my discussions of John Brown and Osama bin Laden, as does Lo's description of Mengzi's position on just revolution. A project for the future might well bring their material into the discussion of order-transforming moments in just war discourse; such would certainly expand the scope of the comparative inquiry.
21 See, for example, "The Social Psychology of the World's Religions," (Gerth and Mills 1946, 267–301).
22 I have in mind some of the essays included in *The Just War* (Ramsey 1983)

Bibliography

Al-Mawardi. 1996. *The Ordinances of Government.* Translated by Wafaa H. Wahba. Reading, UK: Garnet Publishing.
Bergen, Peter. 2021. *The Rise and Fall of Osama bin Laden.* New York: Simon & Schuster.
Brecht, Martin. 1985, 1990, 1993. *Martin Luther.* Translated by James L. Schaaf. Philadelphia: Fortress.
Childress, James F. 1982. *Moral Responsibility in Conflicts.* Baton Rouge: Louisiana State University Press.
Gerth, H.H., and C. Wright Mills, eds. 1946. "The Social Psychology of the World's Religions." In *From Max Weber.* New York: Oxford University Press.
Hegghammer, Thomas. 2020. *The Caravan.* New York: Cambridge University Press.
Kelsay, John. Forthcoming. "The Making of John Brown".
Kennedy, Hugh. 1986. *The Prophet and the Age of the Caliphates.* Harlow, UK: Pearson Education, 1986.
Kennedy, Hugh. 2001. *The Armies of the Caliphs.* London and New York: Routledge).
Kepel, Gilles and Jean-Pierre Milelli, eds. 2008. *Al-Qaeda in its own Words.* Cambridge, Mass.: Harvard University Press.
Lahoud, Nelly. 2022. *The Bin Laden Papers.* New Haven: Yale University Press.
Lawrence, Bruce, ed. 2005. *Messages to the World.* London: Verso Books.
Miller, W. Flagg. 2015. *The Audacious Ascetic.* Oxford: Oxford University Press.
Oates, Stephen B. 1984. *To Purge this Land with Blood: A Biography of John Brown,* 2nd ed. Amherst: University of Massachusetts Press.
Ramsey, Paul. 1983. *The Just War.* Lanham, Md: Rowman and Littlefield.
Reichberg, Gregory et al, eds. 2006. *The Ethics of War.* Oxford: Blackwell Publishing Ltd.
Roper, Lyndal. 2016. *Martin Luther.* New York: Random House.

Rosenthal, Erwin I. J. 1962. *Political Thought in Medieval Islam*. Cambridge: Cambridge University Press.

Scheuer, Michael. 2011. *Osama bin Laden*. Oxford: Oxford University Press.

Schultz, Robert C. 1967. *Luther's Works, volume 46: The Christian in Society*. Philadelphia: Fortress Press.

Skinner, Quentin. 1978. *The Foundations of Modern Political Thought*. Cambridge: Cambridge University Press.

Stauffer, John, and Zoe Trodd, eds. 2012. *The Tribunal: Responses to John Brown and the Harpers Ferry Raid*. Cambridge, Mass. and London: The Belknap Press of Harvard University Press.

Twiss, Sumner B., and Jonathan Chan. 2015. "The Classical Confucian Position on the Legitimate Use of Military Force" Chinese Just War Ethics: Origin, Development, and Dissent. Edited by Ping-Cheung Lo and Sumner B. Twiss, 93–116. London and New York: Routledge.

15

THE RIGHT OF SELF-DEFENSE AND THE ORGANIC UNITY OF HUMAN RIGHTS[1]

David Little

Challenge

There exists a potent two-pronged attack on human rights, one political and the other academic. Authoritarian governments, such as those of Russia, China, Iran, Venezuela, and Saudi Arabia, are making a concerted effort to undermine the credibility and effectiveness of human rights.[2] At the same time, prominent scholars in the United States, the United Kingdom, and elsewhere pose harsh challenges to the grounding, coherence, and durability of human rights.[3]

For those appalled by authoritarian assaults and unconvinced by scholarly skepticism, there is an urgent need to reexamine and reaffirm the nature and purpose of human rights. To be sure, Western academic critics do not intend to support the authoritarian campaign. But if it turns out that an essential objective of human rights is to withstand authoritarianism and offer a better way toward human flourishing, it may be worthwhile to mount a fresh defense of that objective and the grounds it stands on, lest scholarly attacks, left uncontested, give inadvertent aid and comfort to the authoritarian cause.

Since the 1940s, when the first of the human-rights instruments – the Universal Declaration of Human Rights (UDHR)[4] – was drafted and unanimously accepted by the United Nations General Assembly,[5] critics have argued that common agreement on a list of rights was possible, but no acceptable cross-cultural justification of the list could be found. As early as 1949, Jacques Maritain, a member of a group of experts working on human rights under UNESCO auspices, summarized widespread doubts over finding a common justification: "We agree about the rights but on the condition that no one asks us why."[6]

DOI: 10.4324/9781003336372-18

Certain human-rights scholars, among them Michael Ignatieff, have argued that no unifying theory is needed or possible, since human rights are not intended to serve as "an ultimate trump card in moral argument" or "for the proclamation and enactment of eternal verities." Rather, they are designed "for the adjudication of conflict" and therefore constitute a quite prosaic, down-to-earth language of political "trade-offs and compromises" (Ignatieff 2001, 20, 84).[7] Others, like Samuel Moyn in *The Last Utopia: Human Rights in History*, take a more radical view. Moyn claims that human-rights language as formulated in the late 1940s was intrinsically incoherent and impractical. It consisted of words that were "victims of their own vagueness," "empty vessel [s] that could be filled by a wide variety of different conceptions." Contrary to conventional wisdom, says Moyn, the language was not composed in reference to Nazi atrocities in the Holocaust and elsewhere. It took on that connotation only years later and was thus thoroughly "utopian" in its origins for being unrelated to real-world problems.[8]

Although Charles Beitz, in his acclaimed volume *The Idea of Human Rights*, does not forswear the possibility of a normative theory of human rights, he rejects all unitary proposals, according to which the authority of human rights is based on "a single, more basic value or interest." He writes: "There is no clear reason to hold that human rights should be explicable in terms of a single master value. Perhaps the pressure to regard them in this way derives from a desire to see them all as standards of the same generic kind. But ... an account of their normativity need not be embarrassed to appeal to a variety of distinct justifying considerations."[9]

While it is no doubt possible to connect the various human-rights provisions enumerated in, say, the UDHR without referring to a "single master value," it does seem reasonable to infer, as Beitz does, that affirming such a value would lead to conceiving of the provisions as being of "the same generic kind." Conversely, positing, with Beitz, "distinct justifying considerations" or, with Ignatieff and Moyn, no common justifying considerations at all, would appear to foster intense skepticism as to whether human rights enjoy any deeply shared, let alone universal, characteristics or grounding.

It is worth recalling that debates about the coherence of the various human-rights standards laid out in the UDHR go back to the drafting period of the late 1940s. Johannes Morsink, in his definitive account of the legislative history of the UDHR, discusses at length disagreements over the "organic unity of the document," particularly in regard to the relationship between the so-called old and new human rights, namely, civil and political rights (the old) and economic, social, and cultural rights (the new), both of which are included in the final version of the UDHR (Morsink 1999, 222–38). It must be added, of course, that the two sets of rights were later separated into two different legally binding international human-rights treaties that entered into force in 1976: the International Covenant on Civil and Political Rights (ICCPR), and the International Covenant on Economic, Social, and Cultural

Rights (ICESCR). The separation made it possible for countries like China and the United States to ratify one treaty and not the other, thereby posing again the question of the coherence of the two sets of rights.[10]

Response

These criticisms are mainly mistaken. Hiding in plain sight in the Preamble to the UDHR are two passages that suggest a rationale capable of providing a compelling moral (and legal) justification and a convincing basis for the organic unity of the rights in the UDHR, and, by extension, in the ICCPR and the ICESCR.

> *Whereas* disregard and contempt for human rights have resulted in barbarous acts which have outraged the conscience of [humankind],[11] and the advent of a world in which human beings shall enjoy freedom of speech and belief and freedom from fear and want has been proclaimed as the highest aspiration of the common people, … *Whereas* it is essential, if [all human beings] are not to be compelled to have recourse, as a last resort, to rebellion against tyranny and oppression, that human rights should be protected by the rule of law.[12]

These passages leave no doubt that the defense of human rights, which are summarized as freedom of speech and belief and freedom from fear and want, is understood as a justified response to "tyranny and oppression" expressed in the multitude of "barbarous acts" perpetrated by German and other fascists before and during World War II.[13]

At bottom, these passages are about the legitimate and illegitimate use of force. Under tyranny and oppression, human beings have the right to exercise a *legitimate* use of force – "as a last resort" – in response to an *illegitimate* use of force – one that is cruel and arbitrary, the kind tyrants and oppressors depend on. Human beings thus have the fundamental right-called, the "greatest of rights" in the Western tradition (Tierney 1997, 314) – to defend themselves against "tyranny and oppression." The passages further assert that the worthy objective of force is establishment of "the rule of law" – that is, a legal and political order where cruel and arbitrary behavior is reduced by regulating force according to well-defined and established laws, impartially administered and enforced, that include, indispensably, the protection of human rights. Thus, all rights protected by the rule of law are ultimately justified by the right of self-defense against tyranny and oppression.

The Right of Self-Defense as Morally Foundational

These passages from the preamble to the UDHR assume the idea of universal subjective rights that grant all individuals, simply as human beings, entitlement to claim (or to have claimed for them) a certain performance or forbearance under threat of sanction for noncompliance. The passages also assume corresponding duties to respect rights. In the second passage, the key reference is to the *right of self-defense*, defined as the equal right of everyone to resort to reasonable or defensive force to protect oneself or others against arbitrary force, where *arbitrary force* means (at a minimum) intentional infliction of death, impairment, serious injury, severe pain or suffering, or involuntary confinement primarily for self-serving or unfounded reasons.

The relevant sanction available to an individual or group exercising the right of self-defense against such a threat is the use of force *under certain strict conditions* – necessity, imminence, proportionality, and right intention – conditions that spell out the meaning of "reasonable or defensive force" and serve to distinguish defensive from arbitrary force. The conditions work to assure that "recourse...to rebellion against tyranny and oppression" is undertaken only "as a last resort," in the words of the second passage of the Preamble. As George P. Fletcher has noted, if an individual is faced with a mortal threat, "*necessity* speaks to the question whether some less costly means of defense, such as merely showing a gun or firing into the air, might be sufficient to ward off the attack." "*Imminence* means that the time for defense is now! The defender [simply] cannot wait any longer." "*Proportionality* [prescribes that the] harm done in disabling the aggressor must not be excessive or disproportionate relative to the harm threatened and likely to result from the attack." *Right intention* means that "the defender must know about the attack and act with the intention [only] of repelling it [rather than of doing gratuitous harm to the attacker].[14] The conditions apply similarly to collectivities confronted with the "barbarous acts" of tyrants and oppressors. Arbitrary force is the opposite of defensive force and is thus unconstrained by any of the four conditions.

The term arbitrary force needs to be broadened to include features also typically exhibited by tyrants and oppressors, and of special pertinence to the World War II setting against which human-rights language was composed and adopted. The broader version may be called, *arbitrary deprivation/neglect*, meaning enforcement of laws and policies aimed intentionally at withholding or negligently disregarding – primarily for self-serving or unfounded reasons – the means necessary for basic human survival. The means in question would include the opportunity to make a living and to have adequate sustenance, housing, clothing, healthcare, education, and assistance "in the event of unemployment, sickness, disability, widowhood, old age or other lack of livelihood in circumstances beyond [one's] control."[15] If means like these are available and are willfully and forcibly withheld or disregarded primarily for self-serving or unfounded reasons, such actions are the proper target of opposition and resistance, including the use of defensive force.

Words used in the passages from the Preamble – such as "contempt for," "barbarous acts," "outraged the conscience of humankind," "tyranny," and "oppression" – are unquestionably moral utterances in the sense that they presuppose fundamental standards by which to condemn what is wrong and bad, and to justify what is right and good as regards protecting and support-ing basic human welfare. Accordingly, arbitrary force, expanded to include arbitrary deprivation/neglect, is believed to be morally condemnable, while efforts and means to protect against it, including the use of defensive force, are thought to be morally justifiable.

The important point is that force may be justified only by a certain set of reasons. Given the universally unwanted effects of force – death, impair-ment, serious injury, severe pain/suffering, or involuntary confinement, whether inflicted directly or as the result of withholding or ignoring the means to avert them – *everyone everywhere has the exact same presumed good reason to avoid or resist those effects*. Therefore, if force is ever to be used, compelling and overriding reasons must be offered. "Because I (we) want to" or "because it gives me (us) pleasure" or "because it serves my (our) interest" are *no reasons at all*. [16] Neither, obviously, are unfounded reasons, and prevaricating or offering mistaken claims about the use of force is par-ticularly egregious because so much is at stake.

Giving justifying reasons presupposes a set of shared rational standards sensitive to the common interests of all parties involved and has as its objective bringing about consensual agreement among the parties. To refer primarily to one's own interest, or to tell lies or mislead, in support of using force is bla-tantly offensive because the references in no way address the severely unwan-ted effects of force upon the victim; such actions leave the victim's injured or deprived condition altogether out of account, and they rule out any equal opportunity for informed dissent or consent. The principle of equality, in particular, is violated since applying force for self-serving or unfounded pur-poses either favors one set of interests over others, or obscures the real inter-ests served, without good reason. Accordingly, such a self-interested or misleading statement could not possibly justify a use of force. In that respect, arbitrary force overlaps with the idea of cruelty, which means callous indiffer-ence to or pleasure in causing pain and suffering. It need hardly be mentioned that what has just been described accurately reflects Nazi and other fascist practices during the World War II period. [17]

Conversely, the right of self-defense is a *moral right*, or a right justified by good reasons of a moral kind, [18] namely, the great value of acting in a mea-sured way to resist and subdue what has been called, a "*summum malum*," to be understood as the exercise of arbitrary force, whether by directly inflicting death, impairment, serious injury, severe pain or suffering, or by involuntary confinement, or by withholding or disregarding provisions necessary for basic human survival. [19]

The second passage quoted from the preamble, however, also implies a profound complication. No matter how morally justified armed rebellion might be in self-defense against tyranny and oppression, it is something to be avoided, if possible. That is because of the precariousness of the practice. On one hand, self-defense against arbitrary force, undertaken under strict limits, is morally justified. On the other hand, if practiced widely and regularly, its exercise, ironically, would likely increase the incidence of arbitrary force, since defenders take the law into their own hands. They become their own judges, juries, and enforcement officers in a "state of nature," typically subject to passion and fear and therefore to conditions that are altogether unfavorable to judicious and measured (nonarbitrary) action.[20]

According to the second passage, the complication is resolved by subjecting the right of self-defense to the control of a well-ordered government, namely, one based on the rule of law and the protection of human rights. Consequently, what is a moral right is deemed to be a *legal right* [21] as well. Monopolizing force in accord with established laws, fairly administered by impartial legislative, judicial, and enforcement officials, and aimed at protecting individual citizens against acts of arbitrary force constitutes an *enormous public good*. It represents a surer guarantee of the personal security and basic welfare of all citizens than allowing individuals to be laws unto themselves, and it can be achieved only by establishing a well-ordered government. The moral basis of the government, it must strongly be emphasized, is *the equal right of everyone in common* to personal security and basic welfare against the threat of arbitrary force. Self-defense is not egoistically grounded. It is defined as the right to use reasonable or defensive force to protect oneself *or others* because the exercise of arbitrary force constitutes a *universal moral violation*. That means, by implication, that everyone everywhere has a duty to refrain from using arbitrary force against anyone anywhere, and, to that end, to join and support a well-ordered government dedicated to equal protection against arbitrary injury and neglect by providing laws that are more settled, more impartially administered, and more reliably enforced than is true in the state of nature.[22]

Acknowledging the moral primacy of the right of self-defense, rule-of-law governments do continue to permit a remarkable, if limited, exception to their authority. Citizens retain the legal right to defend themselves, forcibly if necessary, in face of threats of arbitrary force where the police cannot protect them, although the government still reserves to itself final authority to determine whether such acts are carried out within the proper bounds of defensive force.

The crucial implication of the second passage is that governments must be held accountable according to the standards of the rule of law, insofar as those standards protect human rights, which is to say, defend the basic security and welfare of all citizens against the threat of arbitrary force, broadly understood.

Is the Right of Self-Defense a Human Right?

None of the major international human-rights documents specifically lists the right of self-defense as a human right (Hessbruegge 2017, 78). The debate among the drafters of the UDHR over whether to refer obliquely to the right in the Preamble or to list it explicitly among the other rights provides a partial explanation. There was no disagreement over the existence of the moral right of both individuals and collectivities to defend against tyranny and oppression. A majority of the drafting committee, however, worried that an unqualified endorsement of such a right might encourage rebellion and thus imperil fragile postwar democracies. As a compromise, the Preamble asserted the right of self-defense in a slightly disguised way.[23]

Nevertheless, the right is explicitly affirmed by the official body authorized to interpret and administer the ICCPR, the Committee on Civil and Political Rights (CCPR),[24] and it is recognized in the European Convention of Human Rights (ECHR)[25] and the Statute of the International Criminal Court.[26] Similarly, the personal right of self-defense is "recognized in all of the world's major legal systems," and, accordingly, constitutes a binding general principle of international law derived from domestic law. (Hessbruegge 2017, 58–59) Although applications of the principle vary, the same basic conditions defining a legitimate act of self-defense – necessity, imminence, proportionality, and right intention – "are found in virtually every legal system in the world."[27] As such, the right is consistently recognized and applied in human-rights jurisprudence affecting private citizens, as well as police and military personnel, around the world. (Hessbruegge 2017, 235–43, 91–234) In addition, Article 51 of the UN Charter recognizes an "inherent" collective right of self-defense, and international humanitarian law makes the conditions of defensive force obligatory.[28]

One reason why the international documents (and most domestic constitutions) do not explicitly identify the right of self-defense as a human right is that the social instability that might result from officially authorizing it. But there is a deeper reason. The right occupies a unique, if precarious, status in providing the moral foundation of the entire human-rights corpus. It is not simply one right among many; it rather pulls together the entire framework of rights by offering the rationale for their defense. The right to life, and, by extension, the right of self-defense, has properly been described as "the fulcrum of rights...the fountain through which other rights flow, or even the supreme right."[29] Precisely because the right of self-defense is precarious, owing to its potential for causing social disorder, it can satisfactorily be secured only by proper governmental regulation, namely, by means of a political-legal system dedicated to "human rights protected by the rule of law." It yet remains to demonstrate how it is that in its widespread, if not universal, cultural presence and in its affirmation in international human-rights documents, the right of self-defense confirms the organic unity of human rights.

The Right of Self-Defense and the Organic Unity of Human Rights

The words in the first passage quoted from the UDHR – "freedom of speech and belief and freedom from fear and want" – recall the "four freedoms" famously declared by U.S. President Franklin D. Roosevelt on January 6, 1941, and further elaborated in his "second Bill of Rights" proposed three years later. Emphasizing that traditional civil and political rights by themselves "proved inadequate," he outlined a set of economic and social rights, supplementary to civil and political rights, as the "new basis of security and prosperity" "established for all – regardless of station, race or creed."[30]

Consequently, President Roosevelt provided the drafters of the human-rights documents with a rhetorical warrant for combining "old" and "new" human rights, and then, several decades later, for elaborating them in legal form in twin international covenants, the ICCPR, and the ICESCR.[31] As he made clear, Roosevelt perceived the grave danger represented by "the new order of tyranny" to the old rights that guaranteed civil and political freedoms. Nobody, he said, "can expect from the dictator's peace ... freedom of expression, or freedom of religion," and for that reason it was necessary to commit to "the preservation of civil liberties for all." At the same time, he called attention to "the economic and social problems which are the root cause of the social revolution" enveloping the world at the time, and concluded that because "people who are hungry and out of a job are the stuff of which dictatorships are made," the new social and economic rights must also be defended with great urgency.[32]

Morsink's account of the provenance of human-rights language leaves no doubt that the drafters of the UDHR shared Roosevelt's outlook. The selection and articulation of each of the provisions, both civil/political and social/economic/cultural, can be understood only in the light of "the experience of the war." Morsink discusses other influences, such as the Cold War, the Western tradition of natural rights, and Latin American socialism, but "none of them match the Holocaust in importance."[33]

The original human-rights corpus, at least the UDHR and the ICCPR, was conceived as an indispensable effort to impose stringent restrictions on governments to prevent the recurrence of the kind of arbitrary injury and arbitrary deprivation/neglect practiced by Hitler and other fascists. The ICESCR was a somewhat different story. While the rights it contained were drawn from UDHR and thus bore the stamp of the World War II period, the interpretation and implementation of that document, for special reasons, would not achieve a comparable level of stringency and urgency until the late 1980s.

The Collective Right of Self-Defense as a Human Right[34]

In debating whether to include the right of self-defense among the other human rights or to refer to it obliquely in the Preamble of the UDHR, the drafters assumed that both individuals and collectivities possess a moral and legal right to defend themselves as a last resort against arbitrary force, and most of the opinions on the subject gathered from the cross-cultural survey discussed above consistently echoed the same understanding.

Article 51 of the UN Charter clearly reinforces the idea that, under the UN system, states, as collectivities, possess a moral and legal right of self-defense.

The wording of Article 51 – "Nothing in the present Charter shall impair the inherent right of individual or collective self-defence if an armed attack occurs against a Member of the UN, until the Security Council has taken measures necessary to maintain international peace and security" – refers, obviously, to states either individually or in concert. The wording even replicates the implication of the human-rights instruments that resorting to the right of self-defense is ultimately to be regulated by a supervening governmental authority, in this case the Security Council. The use of the words, "inherent right" or, in the French version, "natural right of legitimate defense," is clearly reminiscent of the moral predilections underlying a human-rights understanding of the right of self-defense discussed so far. Similarly, modern just-war thinking, which informed the UN Charter and particularly Article 51,[35] had come to elevate the right of self-defense above other justifications in premodern thinking as a just cause for war.[36]

Ironically, however, narrowing to self-defense the legitimate reasons for using force among states, in the way Article 51 does, released the right of self-defense from its human-rights mooring. For one thing, limiting the practice of permissible self-defense to "a Member of the UN" restricted the universal scope of the right as it is understood in the context of human rights. The definition of self-defense as the right to protect oneself *or others* from unwarranted attack attaches no restriction on the others under consideration, except for their urgent need. For another, the UN Charter redirected the use of force away from protecting human rights by the rule of law, as prescribed in the UDHR. In the categorical language of Article 2(4), "All members shall refrain in their international relations from the threat or use of force against the territorial integrity or political independence of any state," a position reinforced by Article 2(7): "Nothing contained in the present Charter shall authorize the UN to intervene in matters which are essentially within the domestic jurisdiction of any state." In effect, as the eminent international legal scholar, Louis Henkin, notes, "peace was the paramount value" of the UN Charter. The Charter was "dedicated to realizing other values as well – self-determination, *respect for human rights*... . But those purposes could not justify the use of force between states to achieve them... . Peace was more important than progress and more important than justice" (Henkin 1989, 37–69, at 38; italics added).

The implications were ominous. So long as the way a state used force did not expand beyond its territorial boundaries, no matter how badly it violated the rights of its citizens, any "threat or use of force" by outside states, or by the UN itself, seeking to intervene and curtail those violations was strictly illegal. Even more surprising, the offending state was understood to possess a "collective right of self-defense" against any such effort, should it be undertaken. The question is obvious: could a right legalizing state-sponsored armed resistance against efforts to protect the victims of arbitrary force possibly be a human right?

By 2005 and the appearance of a new doctrine, summarized in the publication *The Responsibility To Protect*, drawn up by the Canadian-sponsored International Commission on Intervention and State Sovereignty, the answer for at least some members of the international community was an emphatic No (Evans 2005, 3–9, at 5). Far from standing aside in the face of severe and extensive violations of human-rights standards within "the jurisdiction of any state," member states now had the obligation, according to the new doctrine, to override the sacrosanctity of state sovereignty inscribed in the UN Charter and to use force, if necessary, "to protect people at grave risk" (Evans 2005, 7–9). This need for new thinking was inferred from the appalling results of a policy of bystander inaction prescribed by Articles 2(4) and 2(7) that occurred in places like Cambodia, Rwanda, the Balkans, and Kosovo during the last quarter of the 20th century. Faced with the prospect of genocide, crimes against humanity, or other unthinkable human-rights violations of the kind that occurred in those places, states now had three obligations: to endeavor initially to prevent such things; to curtail them should they take place – forcibly, if required; and, afterward, to help rebuild a just and peaceful society.

In curtailing violations, intervening states should employ, where necessary, the standards of defensive force similar to the standards of just warfare enshrined in just-war thinking (Evans 2005, 7–9): *right authority*, meaning, in the first instance, approval by the Security Council, although intervening states could claim their own authority under the right of self-defense (as authorized, at least temporarily, in Article 51) should the Security Council "fail to approve military action in a case crying out for it"; *seriousness of the threat (just cause)*, understood as grave and extensive violations, especially of basic security rights, "either actual or imminent"; *proper purpose (right intent)* "to halt or avert the threat in question" (Evans 2005, 7–9); *last resort*, or the belief that "other lesser measures than military coercion will not succeed"; *proportionality*, or accepting that "the scale, duration and intensity of the proposed military action must be the minimum necessary to meet the threat"; and *balance of consequences (reasonable probability of success)*, or well-founded confidence that the mission can be achieved.[37]

These prescriptions begin to restore the connection between the use of force and a human-rights interpretation of the right of self-defense, collectively understood in the context of international relations. On this reading, a state caught seriously violating human rights could no longer claim as a human right the authority to resist forcibly a military operation against it carried out in accordance with the standards of defensive force just outlined. An offending state engaged in severe arbitrary abuse of its citizens is now in the same position as an individual aggressor found attacking a defenseless person: in neither case is the offender morally or legally entitled to resist armed efforts – where required and duly measured – undertaken to restrain the abuse.

Conclusion

The argument developed in this essay is, for better or worse, a theoretical one. The limited, if daunting, objective is to offer a theory that calls into question strongly held convictions in political and academic circles around the world that there exists no compelling, unitary, universally resonant moral and legal justification for human rights language or for its "organic unity." The theory, intimated by two overlooked passages in the Preamble to the UDHR, dwells on the right of self-defense against arbitrary force, understood both individually and collectively, and as both a moral and legal concept. It suggests that the salience of the right of self-defense as grounds for a unified understanding of human-rights language makes perfect sense against the background of the atrocious abuses of force perpetrated by the German fascists and their allies before and during World War II, and that, indeed, the language should be understood in that context.

The theory also summons arresting, if still preliminary, evidence in auxiliary support. That there appears to be widespread cross-cultural acceptance of at least the rudiments of a moral and legal right of self-defense, both individual and collective, does not, of course, validate the basic moral argument of the essay. Validation depends on a separate normative account supplied in the section on moral foundations. But such cross-cultural consensus does give reason for believing in the worldwide *comprehensibility* and *possible appeal* of human-rights language. That point of contact may become, then, a plausible cross-cultural basis for appreciating why the array of rights enumerated in the definitive human rights documents is necessary for protecting persons from arbitrary force.

Important as theorizing is, of course, it is by no means everything. In a thorough discussion, I would take up the question, regularly asserted by the critics, that, on balance, human-rights language does more harm than good.[38] In addition, we would attend to the character and effectiveness of legal remedies for violations of the ICCPR and the ICESCR, something mandated in the covenants themselves, and a subject of great controversy. We would also examine the fortunes in practice of the doctrine of the "responsibility to protect," introduced toward the end of the article and also a topic of intense debate. The doctrine is controversial precisely because – however justified, and whether applied individually or collectively – the practice of lethal self-defense is unavoidably precarious, as the article emphasizes throughout.

All that can be pled "in defense" is the importance of taking first things first. If I have succeeded in providing a compelling theory of human rights, that is perhaps a necessary, though hardly sufficient, achievement.

Acknowledgments

Special thanks to Sumner Twiss for encouraging attention to the subject matter discussed in the second section of this essay, "Widespread Acceptance of the Right of Self-Defense as Morally and Legally Justified," to John Witte,

Jr. for acute suggestions regarding refinements of the argument and the addition of pertinent material and, to David Hollenbach, John Kelsay, and John Feldmann for helpful suggestions and encouragement.

Notes

     This chapter is an abridged version of the article published in the *Journal of Law and Religion* 36, no. 3 (2021): 459–94, which bears the same title. The author is grateful to Cambridge University Press for granting permission to reprint the article in this book.

2 See the contributions of Diamond, Plattner, and Walker 2016 for a well-documented account of the attempts of authoritarianism's "Big Five" to undermine liberal democracy and human rights. See also, the excellent article by Anne Applebaum (2020, 86–93), discussing in compelling detail the challenges for the new Biden administration human rights policy posed by China and other authoritarian countries, and by the derelictions of the Trump administration. President Trump's 2019 speech to the United Nations – containing the words, "The future belongs to sovereign and independent nations who protect their citizens, respect their neighbors, and honor the differences that make each country special and unique" – "was," says Applebaum, "music to the ears of the Chinese and Iranian diplomats who want all criticism of their respective countries shut down" (Applebaum 2020, 92).

3 Such Western academic skeptics are singled out because, unlike many of their non-Western counterparts, they do not appear to support the rise of authoritarianism. For U.S. critics, see esp. note 8, below. One British critic is Stephen Hopgood (Hopgood 2013). Hopgood contends that human rights, understood as "universal, secular, and categorical" international norms, are the creation of illusory humanitarian sentiments born of parochial Christian and Western European ideas that are now in decline. That is true, he says, because "we live in a postmodern world where there are other truths supported by other authorities," all of which casts doubt on the validity of the Universal Declaration of Human Rights (66). (My thesis here is directly opposed to Hopgood's claims.) Another British critic is Nigel Biggar (Biggar 2021a), who attempts to discredit any notion of natural or human rights understood to exist independent of the institutions of law and government. For a critique of Biggar's claims, see Little 2021a. For Biggar's response, see Biggar 2021b.

4 G.A. Res. 217 (III) A, Universal Declaration of Human Rights (Dec. 10, 1948) (hereafter UDHR).

5 With eight abstentions: Saudi Arabia, South Africa, and the six members of the Soviet Bloc.

6 Cited in Glendon 2001, 77.

7 Older well-regarded studies affirm claims similar to Ignatieff's. Jack Donnelly argues that we may by now be satisfied with a purely "analytic or descriptive" theory of human rights rather than a "normative or prescriptive" one because "there is [now] a remarkable international consensus on the list of rights" (Donnelly 1989, 21–22). Paul Sieghart, in believes that there is no longer a need, if there ever was, to consider the philosophical or theological grounds of human rights, since all that is needed now is to "refer to the rules of the international human rights law as defined by the relevant instruments which have been brought into existence since 1945" (Sieghart 1983, 15). These days, in fact, it is both the usefulness and the legal and moral authority of human rights that is being widely and persistently contested in political and academic circles around the world. It surely will not do simply to ignore the growing skepticism.

8 Moyn 2010, 51, 64, 83, 89, 220–227. Moyn has extended his deconstructive efforts, asserting that human rights language is mainly the result of post-World War II conservative European Christian influence (Moyn 2015). More recently, he has developed a somewhat different line of attack, contending that human rights are largely discredited by their intimate association with neoliberalism or what he calls, "market fundamentalism" (Moyn 2018). Moyn's argument has been partially supported by Jessica Whyte (Whyte 2019). Another ally of Moyn's, Eric A. Posner, mounts his attack from a legal perspective in Posner arguing that human rights law has been generally ineffective because it "reflects a kind of rule naivete," as though "the good in every country can be reduced to a set of rules that can be impartially enforced" (Posner 2014, 7). For a critique of *The Last Utopia*, see Little 2015, chap. 2. For a critical assessment of *Christian Human Rights* and *The Twilight of Human Rights Law*, see (Little 2016, 354–66). For a critical assessment of the central claims of *Not Enough* and *The Morals of the Market*, see Little 2021b, fn. 105. Further, some American scholars readily embrace the political consequences of rejecting human rights, offering what amount to reasons against President Biden's proposal to convene a gathering of the world's democracies for the purpose of "advancing human rights in their own nations and abroad," and "defending against authoritarianism," as he put it (Biden 2020). According to one critic, the United States, to the contrary, ought to give up such ill-advised objectives and start showing respect for the great-power status of an authoritarian regime like China. It should "accept the reality that its liberal values are not universal," and stop "interfering in China's internal affairs by condemning Beijing's policies in Hong Kong and Xinjiang" (Layne 2020).

9 Beitz 2009, 128, 138. In what follows, we reject Beitz's claim that no compelling unified theory of human rights is to be found. For a related criticism, see Dworkin 2011, 474–75, note 5: "Beitz believes that human rights should be identified not through some 'top down' principle, … but through interpretation of human rights practice, guided, as it must be, by a sense of the point of the institution… . But … interpretation of that kind requires general principles that can fix the best justification of the raw data of that practice, and these must be 'top down' principles of the sort Beitz wants to avoid."

10 For a detailed discussion of the process of development from the UDHR (1948) to the two treaties, the ICCPR and ICESCR (1976), see Steiner and Alston 2000, 137–320, for a detailed discussion of the process of development from the UDHR (1948) to the two treaties, the ICCPR and ICESCR (1976). China has ratified the ICESCR, but not the ICCPR. For the United States, the reverse is true.

11 In keeping with proper, but inconsistently applied, objections among the drafters of the UDHR to gender-specific language, the words "mankind" in the first passage and "if man is" in the second should be changed as proposed. For a discussion of the objection, see Lash 1972, 70. See also Kathryn Sikkink's important comments on women's rights in the UN Charter and the UDHR (Sikkink 2017, 79–84).

12 The objection that preambular references have no legal weight in statutory law should be anticipated. The argument developed in this article is a moral one and not, finally, a legal one, although it does offer a moral foundation for law. The preambular references are not invoked as legal warrants. They are invoked as intimations of a moral perspective on human-rights language. Their status in the argument stands or falls on the merits of the moral argument, not on their legal authority.

13 12 UDHR. This is the convincing interpretation put forward by Morsink in *Universal Declaration of Human Rights*, esp. chapter 2 ("World War II as Catalyst") (Morsink 2019). Morsink's evidence conclusively refutes Moyn's denials, mentioned above, that human-rights language was composed in response to the persecution of the Jews by the Nazis. See also an updated and expanded defense of the same proposition (Morsink 2019, esp. chap. 3), "The 1940s Moment of Human Rights." Although the major emphasis in this article is on the pertinence of mid-

20th-century German fascism to the drafting of human rights language, Italian and Japanese versions are also relevant, making the fascist experience something of a universal reality. As I have noted elsewhere, "While the three versions do differ in degree, they all share certain fundamental characteristics identified by Paxton 2004. These include 'a sense of overwhelming crisis beyond the reach of any traditional solutions; the primacy of the group, toward which one has duties superior to every right, whether individual or universal, and the subordination of the individual to it; the belief that one's group is a victim, a sentiment that justifies any action, without legal or moral limits, against its enemies, both internal and external;...the need for closer integration of a purer community, by consent if possible, or by exclusionary violence if necessary;...the beauty of violence and the efficacy of the will, when they are devoted to the group's success; the right of the chosen people to dominate others without restraint from any kind of human or divine law, right being decided by the sole criterion of the group's prowess with in a Darwinian struggle'" (Little 2021c, 24–36, at 25n6), quoting Paxton 2004, 219–20.

14 Fletcher 1988, 18–29, 19–20, 23–24, and 25 (italics added). See also chapters 2 and 4 in Fletcher and Ohlin 2008, 30–63, 86–107. For a comparable discussion of the universal conditions of defensive force, see Hessbruegge 2017, 65–68.

15 Article 25 of the UDHR.

16 For a discussion of whether the "necessity defense" – the claim that it is permissible to kill or injure an innocent person where there is no realistic alternative to preserving one's own life or vital interests – represents an exception to this conclusion, see Little 2021b, note 16.

17 For supporting evidence, see Little 2021b, note 17.

18 A person "has a moral right when he [or she] has a [valid] claim, the recognition of which is called for – not (necessarily) by legal rules – but by moral principles, or the principles of an enlightened conscience" (Feinberg 1973, 67).

19 (Shklar 1989, 21–38, at 29). Shklar means by the term "arbitrary, unexpected, unnecessary, and unlicensed acts of force and by habitual and pervasive acts of cruelty and torture performed by military, paramilitary, and police agents in any regime." I would expand the reference to include forcible acts of arbitrary deprivation or neglect.

20 Nussbaum provides an eloquent account of the strong tendency of fear to produce an exaggerated and irrational obsession with self-protection (Nussbaum 2012, 20–58, especially 55–58). Nussbaum's description of the destructive potential of the "narcissistic emotion" adds weight to the phrase, "the only thing we have to fear is fear itself," uttered by Roosevelt in 1933 and lying behind the reference to "freedom from fear" in the first passage from the UDHR preamble.

21 A person "has a legal right when the official recognition of his [or her] claim (as valid) is called for by governing rules" (Feinberg 1973, 67). Elsewhere, Feinberg calls legal rights "enforcement claims," which are a state's obligation to guarantee for the protection of citizens (Feinberg 1980, 224).

22 For a discussion of John Locke's understanding of these matters, see Little 2021b, note 22.

23 (Morsink 1999, 307–12). Representatives to the UDHR drafting committee from the USSR, along with those from Chile, Brazil, and El Salvador, argued for explicit inclusion of the right of self-defense in the Declaration but were opposed by the United States, the United Kingdom, and, eventually, a majority of the committee, who countered that doing so would likely destabilize struggling democracies. One such proposal included in an early draft read as follows: "When a government seriously or systematically tramples the fundamental human rights and freedoms, *individuals and peoples* have the right to resist oppression and tyranny, without prejudice to their right of appeal to the United Nations" (Schabas 2013, 793 (italics added)); see 17, 285, 566, 744, 810, 815 for other references to the right of self-defense in drafts of the UDHR and in the debates surrounding them.

24 CCRP General Comment No. 36 (ICCPR, Article 6, the right to life) (UNDoc. CCPR/C/GC 36 (2018) states: "The Human Rights Committee recognizes that while Article 6 explicitly prevents arbitrary deprivations of life, the right to life is not absolute… . For example, deprivations of life are permissible *when individuals are exercising self-defense* or in certain circumstances when issuing the death penalty. However, the Committee emphasizes *that any permissible deprivation of life must be reasonable, necessary, and proportional to the aims sought, and must be established under the law with effective institutional safeguards to protect against potential arbitrary abuses*" (italics added).

25 ECHR Art. 2(1)(a): "2. Deprivation of life shall not be regarded as inflicted in contravention of this Article when it results from the use of force which is no more than absolutely necessary: 1(a) in defense of any person from unlawful violence.'

26 Statute of the International Criminal Court, Art. 31(1)(c): "1. In addition to other grounds for excluding criminal responsibility provided for in this Statute, a person shall not be criminally responsible if, at the time of that person's conduct: … (c) The person acts reasonably to defend himself or herself or another person or, in the case of war crimes, property which is essential for the survival of the person or another person or property which is essential for accomplishing a military mission, against an imminent and unlawful use of force in a manner proportionate to the degree of danger to the person or the other person or property protected. The fact that the person was involved in a defensive operation conducted by forces shall not in itself constitute a ground for excluding criminal responsibility under this subparagraph. '

27 Fletcher 1988, 27. For a similar conclusion, see Hessbruegge 2017, 32.

28 Art. 51 of Protocol I of the 1977 Protocols to the Geneva Conventions. (I discuss Article 51 and the UN Charter below, in the final subsection, "The Collective Right of Self-Defense as a Human Right.")

29 (Hessbruegge 2017, 103). *Forum of Conscience v. Sierra Leone*, Comm. No.223/98 (2000), para. 19; Human Rights Committee General Comment No. 6: (ICCPR, Article 6, para. 1) (UNDoc. HRI/GEN/1/Rev.1 6 (1994).

30 Franklin Delano Roosevelt, "State of the Union Message to Congress", January 11, 1944. www.fdrlibrary.marist.edu/archives/address_text.html. See also Cass R. Sunstein, *The Second Bill of Rights: FDR's Unfinished Revolution and Why We Need It More Than Ever* (New York: Basic Books, 2004).

31 In Morsink 2019, 40–41, Morsink emphasizes the connection between the Atlantic Charter telegram, issued in August 1941 by Roosevelt and Winston Churchill, and the wording of the UDHR.

32 Roosevelt, "State of the Union," January 11, 1944, fdrlibrary.marist.edu. "Without the consequences of the Great Depression, which hit Germany particularly hard, the [Nazi Party] would never have become a mass movement. And it was [Hitler] who best understood how to articulate and exploit people's desires for a savior who would inject order into chaos, create an ethnic-popular community in place of party squabbling and class warfare and lead the Reich to new greatness" (Volker 2016, 378).

33 (Morsink 1999, 37). See Morsink, *Universal Declaration of Human Rights and the Holocaust* for further elaboration. Still, without detracting from the importance of Franklin Delano Roosevelt's contribution to broadening the concept of human rights based on his reaction to the Holocaust, it is important not to overlook the influence of the Latin American experience especially on the shape of the economic and social rights. Morsink himself stresses it. (Morsink 1999, 130–39). Sikkink provides additional support, particularly as to the influence of the American Declaration of Independence on the drafting of the UDHR: (Sikkink 2017, 59–64, 74–79). This emphasis is part of a broader point that Sikink develops effectively, namely, the great significance of the Global South for the postwar development of human rights and the fact that the Global North was by no means always a dependable partner in the cause. See, for example, (Sikkink 2017, 101–04; 125–26; 131–32). For further evidence of

sustained legal, political, and religious opposition to human rights in the United States during the 1950s and 1960s, see Little 2015, 72–76.

34 The "personal" and "collective" rights of self-defense are not in every respect interchangeable. In regard to personal self-defense, the offending party is held liable both for *initiating* an arbitrary attack and for *conducting* it in disregard of the conditions of defensive force. Concomitantly, the defender, in responding, is strictly responsible to abide by the rules of defensive force. Collective self-defense, under the laws of armed combat, is different. Because the decision to use force is an act of state, state authorities are taken to be primarily responsible and liable for initiating an unlawful attack, as distinct from the combatants who carry out the attack. Accordingly, offending state authorities, if captured, are subject to legal punishment; however, combatants fighting under their orders may, if captured, only be confined as prisoners of war for the duration of the conflict and then released. As far as treatment of noncombatants goes, combatants on both sides of an armed conflict are liable under the conditions of defensive force codified in the Geneva Conventions; but in using force against enemy combatants, the conditions of defensive force are extensively relaxed, although they are not withdrawn altogether – for example, there are still limitations on weapons that cause "unnecessary suffering" to combatants (dum-dum bullets), or have massively indiscriminate effects (poison gas).

35 Johnson 2008, 550–51, fn. 42, above.

36 See Johnson 2008, 551–54, where he argues that preeminent 20th-century defenders of just-war theory, such as Paul Ramsey, Michael Walzer, and the authors of the Catholic pamphlet *Challenge of Peace* (1983), either played down the salience of the considerations of *jus ad bellum* as compared to the considerations of *jus in bello* (Ramsey, partly following Grotius), or elevated self-defense as a cause for war (Walzer and the authors of *Challenge of Peace*, also partly following Grotius).

37 Although the two standards of *jus in bello* – military proportionality and noncombatant immunity – are not specifically mentioned, one assumes they are implied.

38 Kathryn Sikkink, in "The Effectiveness of Human Rights Law, Institutions, and Movements" (Sikkink 2017, Part III, 139–221), has begun to provide persuasive answers to such charges, as registered by Posner, Moyn, Hopgood, and others (see fns. 3 and 7, above). See also Morsink 2009, chap. 5, 205–52.

Bibliography

Applebaum, Anne. 2020. "American Surrender." *The Atlantic* (November): 86–93.

Beitz, Charles R. 2009. *The Idea of Human Rights.* New York: Oxford University Press.

Biden, Joseph. 2020. "Why America Must Lead Again: Rescuing U.S. Foreign Policy after Trump." *Foreign Affairs* 99, no. 2(2020). www.foreignaffairs.com/articles/united-states/2020-01-23/why-america-must-lead-again.

Biggar, Nigel. 2021a. *What's Wrong with Rights.* New York: Oxford University Press.

Biggar, Nigel. 2021b. "Theological Critiques of WWWR: A Reply to Little & Herdt." *Canopy Forum*, April 13, 2021. https://canopyforum.org/2021/04/13/theological-critiques-of-wwwr-a-reply-to-little-and-herdt/.

Diamond, Larry, Plattner, Marc F. , and Walker, Christopher, eds. 2016. *Authoritarianism Goes Global: The Challenge to Democracy.* Baltimore, MD: Johns Hopkins University Press.

Donnelly, Jack. 1989. *Universal Human Rights in Theory and Practice.* Ithaca, NY: Cornell University Press.

Dworkin, Ronald. 2011. *Justice for Hedgehogs.* Cambridge, MA: Harvard University Press.

Evans, Gareth. 2005. "The Responsibility to Protect: Moving towards a Shared Consensus." In *The Responsibility to Protect: Ethical and Theological Reflections*. Edited by Semegnish Asfaw, Guillermo Kerber, and Peter Weiderud. Geneva: World Council of Churches.

Feinberg, Joel. 1973. *Social Philosophy*. Englewood Cliffs, NJ.

Feinberg, Joel. 1980. *Rights, Justice, and the Bounds of Liberty*. Princeton, NJ: Princeton University Press.

Fletcher, George P. 1988. *A Crime of Self-Defense: Bernhard Goetz and the Law on Trial*. Chicago, IL: University of Chicago Press.

Fletcher, George P. and Ohlin, Jens David. 2008. *Defending Humanity: When Force Is Justified and Why*. New York: Oxford University Press.

Glendon, Mary Ann. 2001. *A World Made New: Eleanor Roosevelt and the Universal Declaration of Human Rights*. New York: Random House.

Grotius, Hugo. 2005. *The Rights of War and Peace* [*1625*]. Vol. 2. Edited by Richard Tuck. Indianapolis, IN: Liberty Fund.

Henkin, Louis, Hoffmann, Stanley, Kirkpatrick, Jeane J., Gerson, Allan, Rogers, William D., and Scheffer, David J. 1989. "The Use of Force: Law and U.S. Policy." *Right v. Might: International Law and the Use of Force*. New York: Council on Foreign Relations.

Hessbruegge, Jan Arno. 2017. *Human Rights and Personal Self-Defense in International Law*. New York: Oxford University Press.

Hopgood, Stephen. 2013. *The Endtimes of Human Rights*. Ithaca: Cornell University Press.

Ignatieff, Michael. 2001. *Human Rights as Politics and Idolatry*. Princeton, NJ: Princeton University Press.

Johnson, James Turner. 2008. "The Idea of Defense in Historical and Contemporary Thinking about Just War." *Journal of Religious Ethics* 36, no. 4(2008), 543–556, 550–551, fn. 42, 551–554.

Lash, Joseph P. 1972. *Eleanor: The Years Alone*. New York: W.W. Norton.

Layne, Christopher. 2020. "Coming Storms: The Return of Great-Power War." *Foreign Affairs* 99, no. 6(2020). www.foreignaffairs.com/articles/united-states/2020-10-13/coming-storms.

Little, David. 2021a. "Nigel Biggar, What's Wrong with Rights?" *Canopy Forum*, January 20, 2021. https://canopyforum.org/2021/01/20/nigel-biggar-whats-wrong-with-rights.

Little, David. 2015. *Essays in Religion and Human Rights: Ground To Stand On*. New York: Cambridge University Press.

Little, David. 2016. "Law, Religion, and Human Rights: Skeptical Reponses in the Early Twentieth Century." *Journal of Law and Religion* 31, no. 3(2016), 354–366.

Little, David. 2021b. "The Right of Self-Defense and the Organic Unity of Human Rights." *Journal of Law and Religion* 36, no. 3(2021), fn. 105, note 6–17.

Little, David. 2021c. "Freedom of Religion: Fundamental Right or Impossibility?" In *Law, Religion, and Freedom: Conceptualizing a Common Right*. Edited by W. Cole Durham, Jr., JavierMartínez-Torrón, and Donlu Thayer. London: Routledge.

Morsink, Johannes. 1999. *Universal Declaration of Human Rights: Origins, Drafting, and Intent*. Philadelphia, PA: University of Pennsylvania Press.

Morsink, Johannes. 2009. "The Charge of Unrealistic Utopianism." In *Inherent Human Rights: Philosophical Roots of Human Rights*. Philadelphia, PA: University of Pennsylvania Press.

Morsink, Johannes. 2019. *The Universal Declaration of Human Rights and the Holocaust: An Endangered Connection*. Washington, DC: Georgetown University Press.

Moyn, Samuel. 2010. *The Last Utopia: Human Rights in History*. Cambridge, MA: Belknap Press of Harvard University Press.

Moyn, Samuel. 2015. *Christian Human Rights*. Philadelphia, PA: University of Pennsylvania Press.

Moyn, Samuel. 2018. *Not Enough: Human Rights in an Unequal World*. Cambridge, MA: Harvard University Press.

Nussbaum, Martha C. 2012. *The New Religious Intolerance: Overcoming the Politics of Fear in an Anxious Age*. Cambridge, MA: Harvard University Press.

Paxton, Robert O. 2004. *The Anatomy of Fascism*. New York: Alfred A. Knopf.

Posner, Eric. 2014. *The Twilight of Human Rights Law*. New York: Oxford University Press.

Roosevelt, Franklin Delano. 1944. "*State of the Union Message to Congress*." January 11, 1944. www.fdrlibrary.marist.edu/archives/address_text.html.

Schabas, William A., ed. 2013. *Universal Declaration of Human Rights: Travaux Preparatoires*. New York: Cambridge University Press.

Shklar, Judith N. 1989. "The Liberalism of Fear." In *Liberalism and the Moral Life*. Edited by Nancy L. Rosenblum. Cambridge, MA: Harvard University Press.

Sieghart, Paul. 1983. *The International Law of Human Rights*. Oxford: Clarendon Press.

Sikkink, Kathryn. 2017. *Evidence for Hope: Making Human Rights Work in the 21st Century*. Princeton, NJ: Princeton University Press.

Steiner, Henry and Alston, Philip, eds. 2000. *International Human Rights in Context: Law, Politics, Morals*. 2nd ed. New York: Oxford University Press.

Sunstein, Cass R. 2004. *The Second Bill of Rights: FDR's Unfinished Revolution and Why We Need It More Than Ever*. New York: Basic Books.

Tierney, Brian. 1997. *The Idea of Natural Rights: Studies in Natural Rights, Natural Law and Church Law, 1150–1625*. Atlanta, GA: Scholars Press.

Volker, Ullrich. 2016. *Hitler: Ascent, 1889–1939*. New York: Penguin Random House Books.

Whyte, Jessica.2019. *The Morals of the Market: Human Rights and the Rise of Neoliberalism*. London: Verso Books.

16

CONFUCIANISM, KANT, AND THE PACIFIST TRADITION IN THE CONSTITUTION OF JAPAN

Benedict S. B. Chan

I Debate over Article 9 of the Constitution of Japan and the Japanese Pacifist Tradition

Japan's current national constitution, the Constitution of Japan (in shot, hereafter "the Constitution"), which came into effect in 1947, is also known as the "Peace Constitution." Indeed, the word "peace" appears several times in the document. For example, the preamble states that "We, the Japanese people, desire peace for all time" (Japan, Prime Minister's Office 1947). However, the Constitution's pacifist emphasis is not without controversy. The most controversial part is Article 9, which states:

> Aspiring sincerely to an international peace based on justice and order, the Japanese people forever renounce war as a sovereign right of the nation and the threat or use of force as means of settling international disputes. In order to accomplish the aim of the preceding paragraph, land, sea, and air forces, as well as other war potential, will never be maintained. The right of belligerency of the state will not be recognized.
>
> *(Japan, Prime Minister's Office 1947)*

Article 9 is controversial because it legally prevents Japan from maintaining any military forces, whether an army, a navy, or an air force. There are two opposite attitudes to this pacifist position. On one side are those who believe that Article 9 limits the power of the Japanese government to the extent that Japan has become a country without complete sovereignty, and that the Japanese should therefore amend or even abolish Article 9. Since the end of World War II, almost every party (e.g., the Liberal Democratic Party, the Democratic Party) that has held a majority in Japan's National Diet has

DOI: 10.4324/9781003336372-19

expressed the intention to amend the Constitution (Beer and Maki 2002; Hagström 2010; Haley 2017; Hook and McCormack 2001; Itoh 2001; Mihali 2014; Winkler 2011). On the other side are those who believe that Article 9 is the most important part of the Constitution of Japan because it shows that Japan is willing to relinquish military force and become a country of peace. This attitude is usually held by civilians rather than politicians, such as members in the "Save Article 9 Movement" (Creighton 2015, 121). They believe that many Japanese still support the Constitution. To retain Article 9, they believe that we should find the supports in the civil society rather than from the government officials (Dudden 2015; Shibuichi 2017).

I argue in a previous paper that - utilitarian contingent pacifism can be developed to justify retaining Article 9 (Chan 2022). However, the debate over Article 9 cannot be discussed in full in a single paper. This chapter can be considered a continuation of my previous paper but with a different focus. Instead of focusing on utilitarian contingent pacifism, this chapter conducts a philosophical investigation of the pacifist tradition in the Constitution of Japan (in short, "the Japanese pacifist tradition"). Some argue that the Constitution is completely foreign to Japan, as Japan was forced to adopt and enact the Constitution by the U.S. government. However, this claim is not historically accurate. Some Japanese were also involved in the writing of the present Constitution, and despite some differences, the current Constitution is similar to the previous Meiji Constitution in many respects. Hence, it is difficult to say that the current Constitution is entirely foreign to Japan (Beer and Maki 2002; Hook and McCormack 2001; Moore 2002). Instead of focusing on whether the Constitution is foreign, we should ask whether there is a pacifist tradition behind the Peace Constitution and, if so, what the content of this Japanese pacifist tradition is and how it is related to global political institutions.

Some scholars argue that various pacifist ideas, especially anti-war ideas, were part of Japan's intellectual and political life long before 1947, when the current Constitution came into effect. Indeed, some even state that Japanese culture "constitutes one of the largest literatures on the subject [of pacifism] outside of Western Europe and North America, where pacifism first developed" (Bamba and Howes 1978, 1). The pacifist tradition in the Constitution is one (and probably the most important one) of many applications of the pacifist philosophy and literature in Japanese culture. A distinctive feature of this tradition is that it involves ideas not only from Japan but also from many other sources, such as Buddhism, Christianity, Confucianism, liberalism, and socialism (Bamba and Howes 1978; Dower 2000; Yamamoto 2004; Yamamuro 2010, 2017). The Japanese pacifist tradition is a set of ideas based on Japanese philosophy and its dialogues and interactions with other philosophies. In this sense, if we consider the Japanese pacifist tradition to be the theoretical foundation of the Constitution of Japan, then it is true that the Constitution has some foreign elements. However, this is true in a separate way from that insisted on by its opponents. It is not the case that the

Constitution is completely foreign because Japan was forced by the U.S. government to adopt it; rather, the Constitution has foreign elements because it is based on the Japanese pacifist tradition, which has evolved through dialogues and interactions between Japanese culture and other cultures. It is important to determine how these dialogues and interactions have influenced and ultimately shaped the Japanese pacifist tradition.

As mentioned above, this tradition involves ideas from many sources. As it would be impossible to discuss all of them in a single chapter, only two are discussed here: Immanuel Kant's philosophy of perpetual peace and Confucianism. Some historians, such as Shinichi Yamamuro (2010, 2017), argue that although neither Kant's philosophy of perpetual peace nor Confucianism originated in Japan, they have both had an important influence on Japan in different eras. Kant's philosophy is related to the Japanese pacifist tradition through two main channels. One is the works of certain Japanese scholars from the end of the Edo (Tokugawa) period to the Meiji era. The other is the relationship between the ideas in Kant's *Perpetual Peace* (2006) and international politics, including the influence of these ideas on the Constitution (Yamamuro 2010, 201–202, 215–221; 2017, 49–99). Compared with Kantian thought, Confucianism has a longer history in Japan. Although Shinto and Buddhism are the major sources of religious and philosophical thought in Japan, Confucianism has played a distinct role in Japanese history, especially in shaping the Japanese pacifist tradition (Yamamuro 2010, 199–202; 2017, 101–143). How Kant's philosophy of peace and Confucianism – representing one source of ideas from the West and one from the East – are related to the Japanese pacifist tradition is worthy of investigation.

Yamamuro discusses the thought of several Japanese philosophers from the end of the Edo period to the Meiji era, including Nishi Amane, Ono Azusa, Nakae Chōmin, Ueki Emorim Nakamura Masanao, and Yokoi Shōnan. Yamamuro argues that the anti-war principles advocated by these thinkers provide a moral grounding for the Japanese pacifist tradition (Yamamuro 2010, 196–199, 215–221; 2017, 15–47). As a historian, however, Yamamuro focuses mainly on the historical side of Kant's philosophy and Confucianism, rather than investigating their philosophical aspects in depth. To fill this gap, I focus here on the philosophical nature of the relationships between the Japanese pacifist tradition and Kant's philosophy and Confucianism. I argue that anti-war ideas are key to these relationships.

One may also wonder whether this Japanese pacifist tradition fits into the just war theory or pacifism in the traditional schema of the ethics of war and peace. Realism, just war theory, and pacifism are considered the three dominant traditions in this schema, and both Kant's philosophy and Confucianism are usually considered in relation to just war theory rather than pacifism. In this chapter, I discuss in detail how the elements of Kant's philosophy and Confucianism that have influenced the Japanese pacifist tradition are related to certain versions of contemporary pacifism. I also argue that the Japanese pacifist tradition is a special kind of pacifism. I agree with Hasebe that the Constitution permits self-defense and does not express a "pure pacifism rejecting any use of force" (2017, 125). I further argue that the Japanese pacifist tradition can be considered a close relative of or even a pioneer in contingent pacifism, a new understanding of pacifism from within the just war tradition.

In the following sections of this chapter, I discuss the above ideas point by point. In the next section, section II, I discuss and evaluate Yamamuro's ideas about Kant's philosophy of perpetual peace and the Japanese pacifist tradition. In the subsequent section III, I discuss and evaluate Yamamuro's ideas about Confucianism and the Japanese pacifist tradition. In section IV, I explain how the Japanese pacifist tradition is related to contingent pacifism as a special kind of contemporary pacifism. Although it is not possible for a single chapter to answer all the questions emerging from the debate over the Constitution, I offer some important preliminary insights into the philosophical and practical issues surrounding the Japanese pacifist tradition.

II The Japanese Pacifist Tradition and Kant's Philosophy of Perpetual Peace

Since the end of the Edo period, many Japanese philosophers have introduced elements of Western thought to Japan. One example is Nishi Amane (1829–97) in the Meiji period. Regarding Kant's philosophy of perpetual peace, Nakae Chōmin (1847–1901) is usually considered the most important Japanese philosopher to have discussed Kant and other Western philosophers in relation to abolishing war and the military. In addition to expositions of Kant's philosophy, Nakae is famous for promoting the thought of Rousseau in Japan (Kaufman-Osborn 1992). In his book *A Discourse by Three Drunkards on Government*, he discusses and compares the anti-war ideas of St. Pierre, Rousseau, and Kant (Nakae 1984). He finds "one point unsatisfying in Saint-Pierre's theory – the question of how to abolish war," and explains that "since Saint-Pierre first expressed his theory of world peace, Rousseau has praised it and Kant has developed it so that it has come to have the rational qualities suitable for a philosophy." On Kant, Nakae writes that "Kant says that if all nations wish to abolish war and achieve the benefits of peace, they must adopt democracy" and eventually abolish the military. In other words, he takes from Kant the idea that democracy is necessary for peace, and that to promote peace, states should relinquish their military power and forsake war (Nakae 1984, 82–86) (I will further discuss Kant's idea on "democracy" later in this section). Yamamuro argues that many of the ideas expressed in the Constitution, such as the right of all people to live in peace and the abolition of war and the military, did not come from nowhere. Rather, these ideas have been discussed for a long time in various international political forums and by a range of philosophers worldwide (Yamamuro 2010, 215–221; 2017, 15–100). Japan is no exception to this, as Japanese thinkers have discussed these ideas at least since the time of Nakae. This is one way to respond to the right-wing idea that the Japanese Constitution is an entirely foreign product forced onto Japan by the United States (Yamamuro 2010, 201–202; 2017, 135–143).

Given that Yamamuro focuses mostly on the historical side of the Japanese pacifist tradition, the discussion can be extended by focusing on the philosophical aspects. Combining the historical and philosophical perspectives can give us a more complete picture of the Japanese pacifist tradition. To begin, I discuss Kant's philosophy of perpetual peace and some later developments of the democratic peace theory by Michael Doyle (Kant 2006; Doyle 1983a, 1983b, 1986, 1997). A reason for discussing these ideas is that some political theorists believe that the alliance between the United States and Japan (hereafter "the Alliance") is a union of democratic countries led by the United States and that such a union can be traced back to this tradition (Henderson 1999). Some important concepts from Kant and democratic peace theory relevant to the moral justification of Article 9 are discussed below.

In *Perpetual Peace*, Kant sets out a proposal for permanent peace for the entire world, comprising six Preliminary Articles (PAs) and three Definitive Articles (DAs). Of particular relevance to Article 9 of Japan's Constitution is PA 3, which states that "standing armies (*miles perpetuus*) shall gradually be abolished entirely" (Kant 2006, 69). Kant believes that keeping armies is one cause of offensive warfare. Furthermore, as hiring people to kill is using people merely as a means but not also as an end, it violates the second formulation of his categorical imperative. For these reasons, the third Preliminary Article can be considered as supportive of Article 9.

Also relevant to this discussion are Kant's also put down three DAs, as follows:

DA 1: "The civil constitution of every state shall be republican" (Kant 2006, 70)
DA 2: "International right shall be based on the federalism of free states" (Kant 2006, 78)
DA 3: "Cosmopolitan right shall be limited to the conditions of universal hospitality" (Kant 2006, 82)

These three DAs can also be considered as three levels of law: DA 1 proposes a republican constitution, DA 2 relates to international law, and DA 3 is what Kant calls "cosmopolitan law" (Kant 2006, 82–85). Kant believes that each of these articles makes a unique contribution to his perpetual peace proposal. As DA 1 and DA 2 are the most relevant to the discussion in this chapter, I discuss these two in more detail below.[1]

DA 1 is an important reflection of Kant's belief that in any republic, there are checks and balances and a separation of power. In this situation, political leaders cannot do whatever they want. If a political leader in a republic seeks to instigate a war, citizens have the right and freedom to express their attitudes toward the war. In this case, everyone must consider the cost of the war. On the contrary, dictators do not need to care about their citizens. In brief, a dictator is more likely to start a war than a political leader in a republic. Therefore, Kant argues, republics are usually more peaceful than dictatorships (Kant 2006, 75).

DA 2 is a proposal for international law and means all republics should join in a pacific union. According to Kant's proposal, a union lies somewhere in between, and is superior to, the alternatives of a world state and a peace treaty. A world state would violate national sovereignty and have sufficient power to be potentially tyrannical (Kant 2006, 78); a peace treaty would be too weak, capable of stopping only one or two wars, while "always allowing a pretext to be found for a new war" (79). A union requires more than a peace treaty, as it stops not only several wars but also the state of war. As republic states do not go to war easily, such a union among them can ensure that they do not fight each other in the event of a conflict. In brief, Kant prefers a pacific union over the two alternatives because such a union does not have the excessive power of a world state but can contribute more to peace than a mere treaty.

Kant's proposal, and especially the three abovementioned DAs, is developed further by Doyle. Doyle focuses on the importance of liberal democratic states as representative of republican government, respect for nondiscriminatory human rights, and social and economic interdependence (Doyle 1997, 286–287). Although Kant distinguishes "republic" (separation of power) from "democracy" (direct election from citizens) (Kant 2006, 76–78), this distinction ceases to exist in contemporary democratic political systems. That is, we now use the word "democracy" for the political system with both the separation of power and universal suffrage. Whereas Kant argues only for republics but not democracy, Doyle considers contemporary liberal democratic states as having all the elements of Kant's republics, plus many other crucial attributes. A detailed discussion of Doyle's definition of liberal democratic states is outside the scope of this chapter, but it is safe to say that most of the usual elements of democracy, such as freedom of expression, universal suffrage, and a division of power, are included in his definition (Doyle 1983a, 206–209). Based on this definition, Doyle argues that people in a liberal democratic state usually respect people in other liberal democratic states, and that their attitudes and decisions affect or even determine the policies of their governments. When these liberal democratic states form a peace union, it is impossible to wage war within the union (Doyle 1997, 287). Adding all these elements together, it is more likely for peace to be maintained among liberal democratic states than non-liberal states. Nevertheless, Doyle also argues that peace can be maintained *only* among liberal democratic states. When liberal democratic states face threats from non-liberal states, they (individually or as a union) do not avoid war. Doyle's research draws mainly on historical statistics from the 18th century through to 1980. According to his analysis of this period, he argues that liberal democratic states do not go to war with each other, non-liberal states go to war among themselves, and liberal democratic states often go to war with non-liberal states; statistically, liberal democratic states also win more wars than non-liberal states (258–277).[2] Doyle concludes by naming this phenomenon "separate peace."

Doyle's democratic peace theory can be considered as a modern version of Kant's perpetual peace proposal and related to Article 9. Against the backdrop of the *Treaty of Mutual Cooperation and Security between Japan and the United States of America* (Japan, Ministry of Foreign Affairs of 1960) (hereafter "the Treaty"), Article 9 can be roughly considered as an application of democratic peace theory and can therefore be supported by the same reasons used to support that theory. The most obvious reason to consider Article 9 as an application of democratic peace theory is that both the United States and Japan are liberal democratic states, so the Treaty and the Alliance can be considered a case of the peace union promoted by Kant and Doyle. Indeed, Doyle explicitly considers the U.S.–Japanese alliance as an example of the pacific union he discusses: "[t]his separate peace provides a solid foundation for the United States' crucial alliances with the liberal powers (NATO, *the Japanese alliance*, ANZUS) … It also offers the promise of a continuing peace among Liberal states" (Doyle 1997, 265. My italics).

There are many philosophical and empirical objections to the ideas of Kant and Doyle (Cavallar 2001; Hoffmann 1995; Lutz-Bachmann 1997; Pojman 2005; Spiro 1994). However, the purpose of this chapter is not to argue for or against Kant and Doyle but to see how their ideas relate to the Japanese pacifist tradition. The Alliance differs in some respects from the pacific union suggested by Doyle and Kant. First, Article 9 forbids Japan from keeping a military force, and hence the Alliance is characterized by a single direction of protection from the United States to Japan, not vice versa. Shinzo Abe's administration re-interpreted Article 9 to allow the self-defense force to take on overseas tasks, but this interpretation is controversial (Goodman 2017; Yamamoto 2017). Second, the Treaty alone is not strong enough to constitute a pacific union as suggested by Kant. At most we can say that the Alliance is an incomplete application of democratic peace theory; it is similar to but not the same as the pacific union described by Kant and Doyle.

To better protect the pacifist tradition of the Constitution, the Alliance could be modified to resemble a pacific union more closely, which has several important characteristics. The aim of such a union is to protect human rights, especially the right of self-defense for every individual. All states in a pacific union should be democracies, and the union should continuously expand to include more democratic states. The democratic states in the union should forfeit any right to wage war against each other, and the protective force should not be at state level or under the control of any one country. This last point is the most difficult to achieve and the furthest from the current non-ideal situation. The idea of a pacific union shows why re-interpretations of Article 9 are problematic. If any change is to be made, it should not involve simply allowing the Japanese state to maintain military forces. Rather, it should be a shift in the opposite direction: giving up military force at the state level and assigning the power over collective self-defense to the pacific union. This is a further development of Yamamuro's historical discussion of the relationship between the Constitution and global political institutions.

Having explored how the Japanese pacifist tradition is related to Kant's philosophy of perpetual peace and even to contemporary democratic peace theory, there is one more point that must be discussed before moving on. Based on the above ideas, it is easy to identify a focus on anti-war concepts and practices. People focus on how to reduce military forces and eliminate war to reach the eventual goal of promoting peace. For example, Kant's six PAs promote the eventual abolition of military forces (PA 3) and prohibit certain acts during war (PA 6) in a similar fashion to the principles of *jus in bello* underlying many contemporary just war theories. Although the DAs discuss political and international institutions that can promote peace, Kant's ideas in *Perpetual Peace* cannot be categorized as traditional pacifism. For instance, none of his ideas directly supports non-violence. Many Japanese philosophers also take an anti-war but not non-violence focus. For example, Nishi Amane discusses how to apply Kant's philosophy of perpetual peace in Japan in some detail but does not totally deny the need for Japan to maintain military forces; he simply looks forward to perpetual peace in the future (Yamamuro 2017, 130–134, esp. 133–134). Of course, anti-war and non-violence are not mutually exclusive concepts, and one may develop the idea of non-violence from an anti-war position (or vice versa). However, these philosophers neither argue that violence should be forbidden in all circumstances nor deny the importance of self-defense. From the investigation so far, it is safe to say that the Japanese pacifist tradition is related to Kant's philosophy of perpetual peace but emphasizes anti-war concepts more than other principles of traditional pacifism, such as non-violence. In section IV I discuss further whether this position is closer to just war theory or pacifism.

III The Japanese Pacifist Tradition and Confucianism

Compared with Kant's philosophy, Confucianism was introduced to Japan much earlier and has had a greater influence on Japanese thought (De Bary et al. 2001; De Bary et al. 2005). My focus here is only on some important ideas regarding the relationship between Confucianism and the Japanese pacifist tradition. To begin I discuss the ideas of two Japanese philosophers mentioned by Yamamuro. The first is Yokoi Shōnan (1809–69), an active Confucian samurai and thinker at the end of the Edo period. Although Yokoi believes that Japan can learn from Western civilization, he thinks that the most fundamental political principles should be based on Confucianism, especially the Confucian politics of benevolence (*ren*) and the way (*dao*) of Yao, Shun, and Confucius. For example, he states that "nourishing the people is the main work.... This is the natural principle (*shizen no jōri*). Yao and

Shun's rule of all-under-Heaven was none other than this" (De Bary et al. 2005, 643). Historians present different interpretations of Yokoi's ideas on the military. Whereas Shin Chiba focuses on Yokoi's ideas about the link between a wealthy country and a strong military, Yamamuro interprets Yokoi's purpose as abolishing war. Nonetheless, they agree that Yokoi's ultimate goal is to promote peace (Chiba 2014, 95–100; Yamamuro 2010, 199–200; 2017, 102–107). For example, discussing the United States, Yokoi emphasizes the need for all cultures and religions to come together "to stop wars in accordance with divine intentions, because nothing is worse than violence and killing among nations" (De Bary et al. 2005, 644).

In the preceding section, I discuss how the thought of Nakae is related to Kant's philosophy of perpetual peace. Nakae is also famous for drawing on Confucianism in support for his pacifist position. In the foreword to the English translation of Nakae's *A Discourse by Three Drunkards on Government*, Jansen writes that Nakae "blended Confucian terminology and values with the thought of Rousseau" and that "Nakae's values remained explicitly Confucian, he had grave doubts about the need for burdensome military spending, and he believed in the importance of fully representative government" (Nakae 1984, 7, 9). Nakae argues that *ren*, which is one of the most important Confucian virtues, is also a moral foundation of pacifism and a clever response to violence: "Even if he gets angry and indulges in violence, what can he do if we smile and adhere to the 'way of humanity' [*ren*]?" (Nakae 1984, 52). Nakae also argues that such a pacifist tradition is not only important in Confucianism but also a consensus of different moral traditions and theories (Nakae 1984, 123).

The above two examples are of Japanese philosophers between the end of the Edo period and the Meiji era, but Confucianism was also important in the early Edo period (De Bary et al. 2005, 29–82). Neo-Confucianism, especially the philosophy of Zhu Xi, was widely spread and discussed in Japan at that time (De Bary et al. 2005, 83–142). Indeed, tracing the history of Confucianism shows that it was introduced to Japan much earlier. Confucianism was first introduced to Japan in around the 7th century, at the time of Prince Shōtoku and the Taika Reforms (De Bary et al. 2005, 29). Japan has adapted various traditions from other East Asian regions. For example, Chinese thought influenced institutions in early Japan (De Bary et al. 2001, 63–99). Of all the East Asian religions and cultures, Buddhism is usually considered to have had the strongest influence on Japanese culture (Schlichtmann 1995). Nevertheless, Confucianism is also an important East Asian tradition that has been adapted to Japanese culture.

From this understanding of the history of Confucianism in Japan, I turn now to discuss the ethics of war and peace. It is important to discuss the role of East Asian traditions in the ethics of war and peace (Lo and Twiss 2015). Some scholars discuss in depth how Confucian philosophy is related to this major topic of applied ethics (Bai 2013; Bell 2006, 23–51; Twiss and Chan

2015a, b). My focus here is on how these scholars emphasize *ren* in the Confucian ethics of war and peace, especially in relation to the legitimate authority to use force.

Confucians usually argue that in an ideal world, there is no need to wage war. In this sense, Confucianism is anti-war on at least the ideal level. In the real (non-ideal) world, however, there are situations in which the use of military force may be unavoidable. Confucians, especially Mencius, make significant efforts to distinguish the "kingly way" from the "hegemonic way" of wielding political power. The most prominent passage along these lines is *Mencius* 2A3, which states that "one who uses force while borrowing from benevolence will become leader of the feudal lords... One who puts benevolence into effect through the transforming influence of morality will become a true King" (Mencius 2003, 69). This and other passages from Mencius and other Confucians reveal their belief that the legitimate use of force must be based on humanity (*ren*) and Confucian virtues. Twiss and Chan summarize this requirement as follows:

> So, in order to be truly legitimate, a ruler must have the virtue of *ren*, practice humane governance, follow the ancient statutes or institutions, be recommended to Heaven by another legitimate ruler or inherit his position from a legitimate ruling authority, have the support of the people, as well as the support of Heaven.
>
> *(Twiss and Chan 2015b, 96)*

Some Japanese philosophers, such as Yakoi and Nakae, assent to the Confucian philosophy of humanity (*ren*) as a basic virtue or value. The most relevant part of this Confucian notion to the present discussion is that sovereignty does not necessarily constitute legitimate authority for the use of force. It is true that many Confucian perspectives on the ethics of war and peace assume that force is held by the state. Nevertheless, the most basic or intrinsic value is humanity (*ren*), not sovereignty. Therefore, some scholars argue that sovereignty is not of utmost importance in the Confucian ethics of war and peace. For example, Bai (2013) argues that humanity overrides sovereignty. As an example, he refers to *Mencius* 7B2, which states that there was no just war in the Spring and Autumn period because "a punitive expedition is a war waged by one in authority against his subordinates. It is not for peers to punish one another by war" (Mencius 2003, 211). Bell (2006, 42) uses Mencius' idea to argue that the United States had no legitimate authority to wage the Second Gulf War because it had not obtained approval from the United Nations. This idea is supported by Bai (2013) but countered by Twiss and Chan (2015b, 108–112). Whether the United Nations is a legitimate authority is open for debate, but these examples show that there are Confucian perspectives from which sovereignty has no importance in the absence of *ren*.

As in the above discussion of Kant's philosophy of perpetual peace and democratic peace theory, I do not aim here to defend a Confucian ethics of war and peace. Instead, my focus is on how the Japanese pacifist tradition can be traced back to Confucian philosophy. Confucians are opposed to war without righteousness, which means that they are against unjust war. This is part of the pacifist idea that the Japanese pacifist tradition has learned from Confucianism. Unlike Kant, most Confucians do not offer a proposal for perpetual peace; however, they do mention another important idea that is relevant to the Japanese pacifist tradition. The above discussion of Confucianism shows that Confucians do not assign intrinsic value to state sovereignty. There is no reason to insist that only a completely sovereign state has the legitimate authority to use force. Indeed, the contemporary concept of sovereignty (i.e., Westphalian sovereignty or state sovereignty) only arose with the development of the modern European system of states in the last 500–600 years. The importance of state sovereignty, especially whether it overrides other values, is questionable (Morris 1998, 172–227). Confucianism provides an explanation for why state sovereignty does not override humanity (*ren*). Although it can be argued that Confucian ethics do not relate directly to Article 9, Confucianism helps us to understand that state sovereignty is not of intrinsic but only of instrumental value in promoting peace. One practical application of the Japanese pacifist tradition is to claim that Article 9 and a pacific union are morally superior to complete state sovereignty, and Confucian ethics provide moral support for this practical application. In the concept of *ren*, Confucianism also provides a moral foundation for peace in an ideal world that is also a moral foundation for the Japanese pacifist tradition. Therefore, Confucianism is relevant to and can be considered one of the philosophical foundations of the Japanese pacifist tradition.

IV The Japanese Pacifist Tradition and Contingent Pacifism

The above sections show that the Japanese pacifist tradition is an anti-war tradition that has borrowed elements from other bodies of thought, such as Kant's philosophy and Confucianism. A question that arises from the discussion so far is whether the Japanese pacifist tradition is closer to just war theory than pacifism. This impression might arise from the absence of a veto on self-defense (which implies violence in some situations) in the Japanese pacifist tradition, as described by Hasebe:

> Pacifism maintained under the Constitution of Japan was not pure pacifism rejecting any use of force. The successive governments held that the right of individual self-defense, in other words, the right to use force in order to repel on-going or imminent, unlawful armed attack against Japan itself, could be exercised under the Constitution.
>
> *(2017, 125)*

Although people usually think that individual self-defense is important, it is unclear whether this implies that a state should have a right of national defense. The Japanese pacifist tradition accepts the right of individual self-defense but denies the right of national defense. In general, it does not look like traditional pacifism. How should we explain this position? In this section, I argue that the Japanese pacifist tradition can be considered a close relative of or even a pioneer in contingent pacifism, which explains why it can promote the individual self-defense but not national defense.[3]

To start the analysis, I return to a discussion of Kant's philosophy and Confucianism. On the assumption that pacifism implies that there is no such thing as a just war, then neither Kant nor Confucian is pacifist. In this sense, it is also hard to treat the Japanese pacifist tradition as traditional pacifism. Given the above discussion of the relationships between this tradition and Kant's philosophy and Confucian ethics, if these philosophies echo many principles of just war theory, it is tempting to think that the Japanese pacifist tradition is also based on principles of just war theory. However, the Constitution, especially Article 9, has a strong emphasis on peace; it forever renounces war as a sovereign right and hence requires the Japanese government to give up all military forces. It is doubtful that such a radical idea of peace can be based on just war theory, at least on the ordinary understanding of this theory.[4]

Many scholars consider both Kant's philosophy of perpetual peace and Confucian war ethics as specific versions of just war theories. However, this is open for debate. It can be argued that Kant's aim is to develop an international peace framework to abolish war, not to justify war (Fox 2013, 14–15). As discussed in the preceding section, many scholars use the just war tradition to interpret the Confucian ethics of war (Bell 2006; Lo and Twiss 2015), but there are also pacifistic interpretations of Confucianism, some of which are also critiques of Confucian just war interpretations (Ching 2004; Hagen 2021, 2022; Yao 2004).[5] It is beyond the scope of this chapter to decide whether Kantians or Confucians are just war theorists, but it is important to recognize that some elements of both of these traditions are embedded in the Japanese pacifist tradition. Whether these elements support pacifism is the main concern here, for which it is necessary to analyze the meaning of pacifism further.

The "traditional schema" of the ethics of war and peace (Fiala 2018, 45) is made up of realism, just war theory, and pacifism. Realism and pacifism are usually considered two unachievable extremes of a theoretical continuum. Pacifism is considered an unachievable and extreme position by its opponents, who define it as denying the use of force in any situation, including self-defense. Such a position violates the common intuition that self-defense is morally permissible in some situations. However, there is no reason to define the term "pacifism" as narrowly as in the traditional schema. As Fiala argues, "pacifism" in the traditional schema should be renamed "absolute pacifism," representing only one (extreme) position in the big family of pacifist theories

(Fiala 2018, 35–61). Pacifists generally agree that peace is a valuable goal, but different pacifists suggest different ways of achieving this goal. Among the different versions of pacifism, I use the umbrella term "contingent pacifism" to cover the ideas that sit between just war theory and traditional pacifism. These ideas are in opposition to war at state level but do not deny the right of individual self-defense (May 2015, 43). Given that contingent pacifism is a comparatively innovative approach, it remains under development and some elements are not yet very concrete. However, with further development, it can become an important theory in the ethics of war and peace. Indeed, I argue elsewhere that utilitarianism can be used to justify contingent pacifism and then to support Article 9 (Chan 2023). To control the length of this chapter and avoid too much repetition of my previous work, I do not seek to defend or justify contingent pacifism here. Instead, I focus on the relationship between the Japanese pacifist tradition and contingent pacifism.[6]

Contingent pacifism is opposed to "warism, militarism, and the military-industrial complex" (Fiala 2018, 4), but it allows for individual self-defense in some non-ideal circumstances. The ideas of Cheyney Ryan (2013, 2015, 2018) and David Rodin (2002, 2014) are especially relevant here.[7] Ryan divides pacifism into personal pacifism and political pacifism. Personal pacifism is opposed to killing or other violence as a personal act, whereas political pacifism focuses on the problems of waging war as a social practice or a practice of sovereignty. Political pacifists believe that states should never wage war because war is always morally problematic. Ryan uses the analogy of the death penalty to explain this idea. Some death penalty opponents argue against the act of killing *per se* and others argue against the death penalty as a legal practice. Personal pacifism is similar to the former and political pacifism to the latter (Ryan 2015, 28). According to Ryan, political pacifism does not imply that individuals should never defend themselves; it simply means that war usually cannot fulfill the purpose of individual self-defense. In short, political pacifists affirm the right to individual self-defense but deny that war is a practical way to promote this right.

Rodin believes that the right of individual self-defense is defensible, but the right of national defense is not. He summarizes two arguments for the right of national defense and then declares that neither of them works. The first is the analogical argument, which likens national defense to individual self-defense; the second is the reductive argument, which reduces national defense to individual self-defense (Rodin 2014, 74). Rodin argues that the analogical argument does not work because a state and an individual person are completely different (Rodin 2014, 69–79) and that the reductive argument does not work in most situations because most defensive wars waged by modern states defend only state sovereignty, not the lives and well-being of citizens, and so national defense cannot be reduced to individual self-defense (2014, 87).[8]

The Japanese pacifist tradition echoes the ideas of Ryan and Rodin. Ryan's description of political pacifism fits quite well with Hasebe's (2017) interpretation of Japan's Constitution. In Hasebe's terms, the Constitution does not express a "pure pacifism rejecting any use of force" (125) and reserves the right of individual self-defense. In Ryan's terms, the Constitution can be interpreted as following political pacifism rather than personal pacifism, as it does not deny the right of individual self-defense and only expresses the moral illegitimacy of waging war at the state level. It also denies the maintenance of a military force because this would usually assume that war is a legitimate practice for the preservation of sovereignty. Instead, it implies that security and protection take place at the Alliance level based on the Treaty. In summary, the Japanese pacifist tradition is anti-war but attempts to balance this position with the right of individual self-defense in a practical way. This is similar to Rodin's justified interdiction theory (2014, 89), which he proposes as sitting between traditional pacifism and just war theory on the grounds that we may not be able to avoid every kind of violence but the waging of war by states is usually problematic. Putting this idea into the debate over the Japanese pacifist tradition, in addition to Article 9, the Treaty and Alliance are important components of the current defense system of Japan. Accordingly, we should not define non-violence in the narrow sense of traditional absolute pacifism. The Japanese pacifist tradition is not simply a kind of pacifism that is completely different from just war theory, but it is a kind of pacifism in its anti-war stance, which denies war as a social practice, a legitimate act of the state, or a means of protecting the right of individual self-defense. This is what Article 9 of the Peace Constitution is really about. Although Japanese pacifists may not have the term "contingent pacifism" in mind when they develop their own ideas of the pacifist tradition, this investigation makes it clear that the Japanese pacifist tradition echoes contingent pacifism. This explains why May (2015, 2) suggests that the Constitution is a significant practical example of contingent pacifism.

In this chapter, I discuss in detail how the Japanese pacifist tradition behind the Constitution can be linked with anti-war ideas in Kant's philosophy and Confucianism. I also argue that the Constitution's acceptance of individual self-defense echoes contingent pacifism. This does not totally settle the debate on retaining Article 9. Nevertheless, given the observed relationships between the Japanese pacifist tradition and other Western and Eastern philosophies, we see that the Japanese pacifist tradition is quite unique. Also, the Constitution is a prominent application of the global peace institution, and it echoes many ideas in the family of contingent pacifism. It is safe to conclude that the Constitution, backed by the Japanese pacifist tradition, is not only a promising political institution for Japan but also a prominent application of contingent pacifism in the global political field.

Notes

1 DA 3 is mostly related to international trade and universal hospitality, which are outside the scope of this chapter.

2 In addition to Doyle, there are other scholars who also discuss data on the relationship between democracy and peace (Ray 1995, 1998; Rummel 1983, 1997; Owen 1994, 2000).

3 For simplicity, this chapter does not distinguish "individual self-defense' and "the right of individual self-defense" (and also "national defense" and "the right of national defense"). They will be used interchangeably in this chapter.

4 The Japanese pacifist tradition cannot be based on realism. Realism is the foundation for right-wing positions and is generally used against the pacifist ideas in the Constitution.

5 Ching suggests that early Confucians could be considered as just war pacifists, but Twiss and Chan argue that the idea of "just war pacifism" is oxymoronic. Defending Ching, Hagen argues further that just war pacifism is "just war criteria with pacifistic intent" (Ching 2004, 253; Hagen 2022, 7; Twiss and Chan 2015b, 115n25). Although just war pacifism is not a very clear concept, it is possible to theorize a middle ground between just war theory and traditional pacifism, and just war pacifism and similar concepts may contribute to such a development. Whether Confucianism can be interpreted as just war pacifism is open for debate, and I do not have enough space here to discuss this topic in detail. However, I do discuss contingent pacifism in this section, which presents some ideas that sit between just war theory and traditional pacifism.

6 The summaries of the ideas of contingent pacifism given in this chapter, especially from Fiala, May, Ryan, and Rodin, are excerpted and revised from Chan (2023).

7 Neither Ryan nor Rodin label their ideas as contingent pacifism. Nevertheless, some scholars place their ideas in the category of contingent pacifism (e.g., Morrow 2018). Whether Ryan and Rodin should be treated as contingent pacifists is largely a matter of terminology. A detailed discussion of this terminological debate is not necessary for the purposes of this chapter.

8 For further details of the ideas of Ryan and Rodin, see Chan (2023, 645–646).

Bibliography

Bai, Tongdong. 2013. "Humanity (*Ren*) Overrides Sovereignty: On Mencius's View of a Just War" [in Chinese]. *Journal of Social Sciences* 1: 131–139.

Bamba, Nobuya, and Howes, John F. 1978. *Pacifism in Japan: The Christian and Socialist Tradition*. Kyoto: Minerva Press.

Beer, Lawrence Ward, and Maki, John M. 2002. *From Imperial Myth to Democracy: Japan's Two Constitutions, 1889–2002*. Edited by John M. Maki. Boulder, CO: University Press of Colorado.

Bell, Daniel. 2006. *Beyond Liberal Democracy: Political Thinking for an East Asian Context*. Princeton, NJ: Princeton University Press.

Cavallar, Georg. 2001. "Kantian Perspectives on Democratic Peace: Alternatives to Doyle." *Review of International Studies* 27: 229–248.

Chan, Benedict S. B. 2023. "Utilitarian Contingent Pacifism and Article 9 of the Japanese Constitution." *Philosophia* 51, no. 2: 635–657. doi:10.1007/s11406-022-00566-0.

Chiba, Shin. 2014. "A Historical Reflection on Peace and Public Philosophy in Japanese Thought: Prince Shotoku, Ito Jinsai and Yokoi Shōnan." In *Visions of Peace: Asia and The West*. Edited by Takashi Shogimen and Vicki A. Spencer, pp. 85–102. New York: Routledge.

Ching, Julia. 2004. "Confucianism and Weapons of Mass Destruction." In *Ethics and Weapons of Mass Sestruction: Religious and Secular Perspectives*. Edited by Sohail H. Hashmi and Steven Lee, pp. 246–269. Cambridge: Cambridge University Press.

Creighton, Millie. 2015. "Civil Society Volunteers Supporting Japan's Constitution, Article 9 and Associated Peace, Diversity, and Post-3.11 Environmental Issues." *Voluntas* 26 no. 1: 121–143. doi:10.1007/s11266-014-9479-5.

De Bary, William Theodore et al., eds. 2005. *Sources of Japanese Tradition Vol 2: 1600–2000.* 2nd ed. New York: Columbia University Press.

De Bary, William Theodore et al., eds. 2001. *Sources of Japanese Tradition Vol. 1: From Earliest Times to 1600.* 2nd ed. New York: Columbia University Press.

Dower, John W. 2000. *Embracing Defeat: Japan in the Aftermath of World War II.* New York: W.W. Norton.

Doyle, Michael. 1983a. "Kant, Liberal Legacies, and Foreign Affairs (Part 1)." *Philosophy and Public Affairs* 12, no. 3(Summer): 205–235.

Doyle, Michael. 1983b. "Kant, Liberal Legacies, and Foreign Affairs (Part 2)." *Philosophy and Public Affairs* 12, no. 4(Autumn): 323–353.

Doyle, Michael. 1986. "Liberalism and World Politics." *American Political Science Review* 80, no. 4: 1151–1169.

Doyle, Michael. 1997. *Ways of War and Peace.* New York: W.W. Norton.

Dudden, Alexis. 2015. "A Push to End Pacifism Tests Japanese Democracy." *Current History* 114, no. 773: 224–228.

Fiala, Andrew. 2018. *Transformative Pacifism: Critical Theory and Practice.* New York: Bloomsbury.

Fox, Michael Allen. 2013. *Understanding Peace: A Comprehensive Introduction.* New York: Routledge.

Goodman, Carl. 2017. "Contemplated Amendments to Japan's 1947 Constitution: A Return to Iye, Kokutai and the Meiji State." *Washington International Law Journal* 26(1): 17–74.

Hagen, Kurtis. 2021. *Lead Them with Virtue: a Confucian Alternative to War.* Lanham, MD: Lexington Books.

Hagen, Kurtis. 2022. "Mencius and Xunzi on the Legitimate Use of Offensive Force: A Pacifistic Critique of Recent Just War Interpretations." *Philosophy Compass*, e12831. doi:10.1111/phc3.12831.

Hagström, Linus. 2010. "The Democratic Party of Japan's Security Policy and Japanese Politics of Constitutional Revision: A Cloud over Article 9?" *Australian Journal of International Affairs* 64, no. 5: 510–525. doi:10.1080/10357718.2010.513367.

Haley, John. 2017. "Article 9 in the Post-Sunakawa World: Continuity and Deterrence within a Transforming Global Context." *Washington International Law Journal* 26 no. 1: 1–16.

Hasebe, Yasuo. 2017. "The End of Constitutional Pacifism?" *Washington International Law Journal* 26, no. 1: 125–135.

Henderson, Errol A. 1999. "Neoidealism and the Democratic Peace." *Journal of Peace Research* 36, no. 2: 203–231.

Hoffmann, Stanley. 1995. "The Crisis of Liberal Internationalism." *Foreign Policy* 98 (Spring): 159–177.

Hook, Glenn D. and McCormack, Gavan. 2001. *Japan's Contested Constitution: Documents and Analysis.* London and New York: Routledge.

Itoh, Mayumi. 2001. "Japanese Constitutional Revision: A Neo-liberal Proposal for Article 9 in Comparative Perspective." *Asian Survey* 41, no. 2: 310–327. doi:10.1525/as.2001.41.2.310.

Japan, Ministry of Foreign Affairs of. 1960. "Treaty of Mutual Cooperation and Security between Japan and the United States of America." Accessed November 14, 2022. www.mofa.go.jp/region/n-america/us/q&a/ref/1.html.

Japan, Prime Minister's Office. 1947. "The Constitution of Japan." Accessed November 14, 2022. http://japan.kantei.go.jp/constitution_and_government_of_japan/constitution_e.html.

Kant, Immanuel. 2006. *Toward Perpetual Peace and other Writings on Politics, Peace, and History*. New Haven, CT: Yale University Press.

Kaufman-Osborn, Timothy V. 1992. "Rousseau in Kimono: Nakae Chomin and the Japanese Enlightenment." *Political Theory* 20, no. 1: 53–85.

Lo, Ping-Cheung and Twiss, Sumner B. 2015. *Chinese Just War Ethics: Origin, Development, and Dissent*. London: Routledge.

Lutz-Bachmann, Matthias. 1997. "Kant's Idea of Peace and the Philosophical Conception of a World Republic." In *Perpetual Peace: Essays on Kant's Cosmopolitan Ideal*. Edited by James Bohman and Matthias Lutz-Bachmann. Boston, MA: MIT Press.

May, Larry. 2015. *Contingent Pacifism: Revisiting Just War Theory*. Cambridge: Cambridge University Press.

Mencius. 2003. *Mencius*. Translated by D. C. Lau. Hong Kong: Chinese University Press.

Mihali, Alexandra. 2014. "An Overview on Japan's National Security Strategy." *Conflict Studies Quarterly*, no. 6: 50–62.

Moore, Ray A. 2002. *Partners for Democracy: Crafting the New Japanese State Under MacArthur*. Edited by Donald L. Robinson. Oxford: Oxford University Press.

Morris, Christopher W. 1998. *An Essay on the Modern State*. Cambridge: Cambridge University Press.

Morrow, Paul. 2018. "Contingent Pacifism." In *The Routledge Handbook of Pacifism and Nonviolence*. Edited by Andrew Fiala, pp. 142–153. New York: Routledge.

Nakae, Chōmin. 1984. *A Discourse by Three Drunkards on Government*. Boston, MA: Weatherhill.

Owen, John. 1994. "How Liberalism Produces Democratic Peace." *International Security* 19, no. 2: 87–125.

Owen, John. 2000. *Liberal Peace, Liberal War: American Politics and International Security*. Cornell University Press.

Pojman, Louis. 2005. "Kant's Perpetual Peace and Cosmopolitanism." *Journal of Social Philosophy* 36, no. 1(Spring): 62–71.

Ray, James. 1995. *Democracy and International Conflict*. University of South Carolina Press.

Ray, James. 1998. "Does Democracy Cause Peace?" *Annual Review of Political Science*, no. 1: 27–46.

Rodin, David. 2002. *War and Self-Defense*. Oxford University Press.

Rodin, David. 2014. "The Myth of Self-Defense." In *The Morality of Defensive War*. Edited by Cécile Fabre and Seth Lazar, pp. 69–79. Oxford: Oxford University Press.

Rummel, Rudolph. 1983. "Libertarianism and International Violence." *Journal of Conflict Resolution* no. 27: 27–72.

Rummel, Rudolph. 1997. *Power Kills: Democracy as a Method of Nonviolence*. Transaction Publishers.

Ryan, Cheyney. 2013. "Pacifism, Just War, and Self-Defense." *Philosophia* 41, no. 4: 977–1005. doi:10.1007/s11406-013-9493-7.

Ryan, Cheyney. 2015. "Pacifism(s)." *Philosophical Forum* 46, no. 1: 17–39. doi:10.1111/phil.12053.

Ryan, Cheyney. 2018. "Pacifism." In *The Oxford Handbook of Ethics of War*. Edited by Seth Lazar and Helen Frowe, pp. 277–293. Oxford: Oxford University Press.

Schlichtmann, Klaus. 1995. "The Ethics of Peace: Shidehara Kijūrō and Article 9 of the Constitution." *Japan Forum* 7, no. 1: 43–67. doi:10.1080/09555809508721527.

Shibuichi, Daiki. 2017. "The Article 9 Association, Leftist Elites, and the Movement to Save Article 9 of Japan's Postwar Constitution." *East Asia: An International Quarterly* 34, no. 2: 147–161. doi:10.1007/s12140-017-9269-y.

Spiro, David. 1994. "The Insignificance of the Liberal Peace." *International Security* 19, no. 2(Fall): 50–86.

Twiss, Sumner B. and Chan, Jonathan K. L. 2015a. "Classical Confucianism, Punitive Expeditions, and Humanitarian Intervention." In *Chinese Just War Ethics: Origin, Development, and Dissent.* Edited by Ping-Cheung Lo and Sumner B. Twiss, pp. 117–134. London: Routledge.

Twiss, Sumner B. and Chan, Jonathan K. L. 2015b. "The Classical Confucian Position on the Legitimate Use of Military Force." In *Chinese Just War Ethics: Origin, Development, and Dissent.* Edited by Ping-Cheung Lo and Sumner B. Twiss, pp. 93–116. London: Routledge.

Winkler, Christian G. 2011. *The Quest for Japan's New Constitution: An Analysis of Visions and Constitutional Reform Proposals 1980"2009.* London and New York: Routledge.

Yamamoto, Hajime. 2017. "Interpretation of the Pacifist Article of the Constitution by the Bureau of Cabinet Legislation: A New Source of Constitutional Law?" *Washington International Law Journal* 26, no. 1: 99–125.

Yamamoto, Mari. 2004. *Grassroots Pacifism in Post-War Japan: The Rebirth of a Nation.* New York: Routledge.

Yamamuro, Shinichi. 2010. "The Source and Development of Japan's Philosophies of Non-Violence." *The Journal of Oriental Studies* 20: 196–221.

Yamamuro, Shinichi. 2017. *The Thought Lineage of Article 9 of the Japanese Constitution, Chinese Translation Edition.* Translated by Jen-Shuo Hsu. Taipei: Gusa Publishing, an Imprint of Walkers Cultural Co., Ltd. Original edition: Asahi Shimbunsha.

Yao, Xinzhong. 2004. "Conflict, Peace and Ethical Solutions: A Confucian Perspective on War." *Sungkyun Journal of East Asian Studies* 4, no. 2: 89–111.

17

CONCLUSION

Benedict S. B. Chan and Ping-cheung Lo

It is with a heavy heart that we, two of the coeditors of this volume, write this concluding chapter in remembrance of our colleague and friend, Professor Sumner B. Twiss (affectionately known as Barney), who passed away on May 22, 2023. Barney's contributions to the field of comparative religious ethics and his commitment to promoting cross-cultural dialogue and understanding were invaluable. As the first editor of this edited volume, he played an important role in the planning, preparation, and final stages of this book by offering his expertise and insights in comparative ethics to shape its contents. He continued to contribute to this volume until the month of his passing by proofreading and editing the final draft. This edited volume will doubtlessly stand as one of his lasting legacies. We deeply mourn Barney's passing and will miss his presence and contributions to the field.

Given that the details of each chapter in this edited volume were discussed in the introductory chapter, which Barney contributed to, we shall briefly discuss these chapters further to conclude the works presented in this book. This volume offers broadly comparative perspectives on the ethics of warfare, with a particular focus on Chinese and other East Asian traditions, in addition to a comparison with Western perspectives. The various chapters in this volume highlight the complexity and diversity of military ethics, just war thinking, and pacifism, in addition to providing valuable insights into the ethical considerations that arise in the context of warfare. This book serves as an illustration for comparative cross-cultural dialogue by tracing the internal complexity of both Eastern and Western warfare ethics traditions within their respective cultural contexts. Furthermore, the chapters identify important areas of ethical agreement and differences between Eastern and Western traditions when considered across the cultural divide.

DOI: 10.4324/9781003336372-20

Despite the contributions in this volume, much more work still needs to be done in this subdiscipline of comparative warfare ethics. In particular, two topics could benefit from further development. The first topic concerns a deeper exploration of what East Asian traditions say about the ethics of war and peace. The chapters in this volume have examined the perspectives of different East Asian philosophical and religious traditions on just war theory, realism, and pacifism. In terms of these traditions, the chapters have explored the Confucian, Legalist, and Daoist ethics in addition to the Japanese pacifist tradition. In terms of classical works, the chapters have probed deeper into various core texts such as the *Mengzi*, the *Hanfeizi*, the *Seven Military Classes*, and the *Four Books of the Yellow Emperor*. The chapters have also engaged with famous military leaders, such as Zeng Guofan, Chiang Kai-shek, and Mao Zedong, in addition to discussing how the current military ethics used by the People's Liberation Army focus on just cause and other *ad bellum* issues, even during peacetime. Many of these research topics are appearing in English for the first time. Taken together with its companion volume *Chinese Just War Ethics: Origin, Development, and Dissent* (Routledge, 2015), these two volumes set a benchmark for future research on the ethics of Chinese warfare. There are no other volumes in any language that provide the same sustained and rigorous treatment of these topics; however, we must point out that the field is still wide open because of China's long history. In particular, many Chinese texts and traditions embody sophisticated philosophies about the ethics of war, which are not only analogous and comparable to Western just war thought but also precede the latter by many centuries. Taking Chinese military history as a good example, many historical texts that are over two millennia old include accounts and moral evaluations of battles and warfare. To date, there has been almost no effort made to analyze the moral perspectives embedded in these military history writings. Not only the Chinese but also the peoples of East Asia were strongly historically minded with great admiration for the people and events from ancient times. We cannot fully understand how East Asian people actually think about warfare ethics without engaging with how they evaluate the warfare conduct from the past. For example, although the *Romance of the Three Kingdoms* is a historical fiction written in the fourteenth century about the wars and politics around 200 CE, it contains many value-laden narratives concerning warfare and is still widely read by Chinese, Japanese and Koreans, with different people identifying with its diverse moral heroes. In the future, interested researchers could explore the assorted topics related to the warfare ethics depicted in this renowned novel.

Another topic that could benefit from further development and exploration is how the perspectives described in these chapters can contribute to the global international order. By examining traditional East Asian perspectives on warfare and politics, scholars can gain valuable insights into how these non-Western perspectives could be used to contribute to the global international order, particularly in light of the current challenges to the liberal order that was established after World War II. As mentioned in the introductory chapter, the authors of different chapters have various positions on this topic. On the one hand, some argue that the institutions of the United Nations, which embody the liberal order created after World War II, are facing challenges from different traditions, including East Asian traditions, which promote particularistic visions for the world order. On the other hand, some defend the current international order by highlighting the moral significance of international human rights as well as the importance of the United Nations' peace-keeping framework. In addition, the Japanese Constitution could be argued as a unique case among international institutions because it prohibits military action at the state-to-state level but permits collective self-defense at the international level. By investigating these diverse perspectives, we can gain a more inclusive understanding of the ethical considerations and debates surrounding warfare, which can contribute to the promotion of just peace and stability in international relations. Further research is needed to explore the arguments for and against current international laws and institutions so that we can arrive at a more well-grounded outlook on how to promote just peace and stability in the ever-changing global order.

In conclusion, this volume provides valuable insights into the comparative ethics of war and peace and highlights the importance of continued dialogue and exchange in this field. We hope that this book will inspire further research and dialogue on these critical issues and contribute to building a more peaceful and just world.

INDEX

Afghanistan 260, 270–272
Agent sent by Heaven (or *tianli*天吏) 70, 77, 83, 85
Al-Mawardi's *The Ordinances of Government* 258–261
Aquinas, Thomas 17–19, 22–23, 35, 71–73, 99, 268
The Art of War 20–21, 25, 30–31, 44, 137, 140, 142, 178, 224, 228, 244; ethic of war 39–43; view of war 33, 36–39; violence 33
Augustine 17–19, 21–23, 35, 72, 87, 268

Bin Laden, Osama 81, 258, 269–273
Boom, Irene 66, 69
Brown, John 258, 265–269
Buddhism 297–298, 304

Cajetan 73
Canon law (or canonical rulings or decisions) 19, 260, 267
Chiang Kai-shek 2, 157, 173, 183, 186–187, 200; charges against the Japanese military 210–212; in bello ethics 214; self-defense as just cause 203–204; Sun Yat-sen's three principles 201–202, 215–217; uses of Sun Tzu's (or Sunzi's) *The Art of War* 204–206
China's narrative 52–53
Chinese Communist Party 171, 173–174, 180, 183, 185–187, 191, 193–195, 201, 224–227, 230, 233–236, 244, 246

Civil war 77–78, 170; against the Taiping Army 157, 164; American 34; Chinese 183, 186–188, 200–201, 245
Clausewitz 30–33, 37, 39, 43, 228
Cold war 30, 54, 147, 285
Combat effectiveness 228, 243, 246
Confucius (or Kongzi) 39, 65, 84, 87–88, 106, 143, 146, 152, 161, 202, 303

Daodejing (or *Laozi*) 3, 5, 119–120, 122, 124, 128, 130–138, 141–142, 144, 146, 150–152
Dar al Islam and Dar al Harb 56

Edo Period 303–304

Fa (伐) 66–70
First Opium War 160
Fu, Peirong 81–82

Genocide 149, 208, 219, 287
Gratian 17, 21–22, 65, 71–75, 79, 85
Grotius, Hugo 65, 73, 75

Han Feizi: as a state consequentialist 114–115; interpretation of the Way or the *Dao* (道) 120–122; realpolitik and machtpolitik 117–122; war and peace 122–123
Huangdi's war against Chi You 143, 148
Huang-Lao 5, 146–149; antithetical concepts 131–133; definition of

126–128; interpretation of the Way or the *Dao* (道) 128–129; on right conduct after war 143–145; on right conduct in war 141–143; on right to war 137–140; on war 134–137; political philosophy of 128–131; *wuwei* (無為) in ruling 128–129; *see also* punitive expedition
Humanitarian rescue 67–68, 70, 75–76, 78–80, 83, 86

International law 6, 47–49, 52, 54, 57, 102–105, 284, 300–301, 316; of armed conflict (LOAC) 99; of war 91; post-Westphalian 73
Iraq, Invasion of 30, 247
Isidore of Seville 72
Islamic narrative 55–56

Japanese pacifist tradition 5–6, 297–298; and Confucianism 303–306; and Contingent pacifism 306–309; and Kant 299–303
Jiang, Guozhu 84
Justice: retributive 67, 70, 72–74; rectifying 67, 69, 70, 72, 74–75, 84

Kant 298, 304, 306–307, 309; on democracy and republic 299–303; on the abolition of war and the military 299–300
Korean war 192, 246
Kuomintang (KMT) 171–174, 180–191, 200–201, 244, 246
Kuwait, Invasion of 247, 271

Laozi 121, 126–128, 130–138, 141–142, 144, 146–148, 150–152, 227; *see also* *Daodejing*
Last resort 18, 21–22, 35, 93, 135, 148, 170, 172, 174, 181, 186, 188, 203–204, 280–281, 285
Lau, D. C. 66, 69–71, 74, 86
Legal warfare (or Lawfare) 228, 234, 245, 252
Legitimate authority 4, 39, 148, 171–172, 181, 203, 257, 265, 267, 305–306
Liberal international order 49, 50–56
Luther, Martin 258, 261–265, 274

Mao Zedong (or Mao Tse-tung) 24, 43, 157, 170–196; command virtues 176–177, 191–192; guerrilla war 178–180; just war and unjust war 175; on protracted war 182–183

Marxist 223–227, 231, 236
Media warfare 228, 244–245, 247
Meiji 297–299, 304
Mengzi (or Mencius) 2–5, 39–41, 65–71, 74–86, 161, 191–192, 201, 211–217; benevolent governance 66–70, 76–77, 79–80, 82–83; just cause 65–66, 74–80; moral self-cultivation 67, 69, 70, 85; way of the hegemon (*Badao* 霸道) 81, 305; way of the True King (*Wangdao* 王道) 81, 85, 305; *see also* punitive expedition
Mohacs, Battle of 262
Moral warfare 40, 234, 240–252
Muye, Battle of (牧野之戰) 82

Nakae Chōmin 298–300, 304–305
Noncombatant immunity 23, 26, 35–36, 170, 172, 212, 214,

Orend, Brian 77–78, 86–87

People's Liberation Army (or PLA) 2, 4, 24, 27, 36, 84, 170, 186–188, 223–230, 233–236, 240, 243–252, 315
Prisoners of war 97, 179, 183, 185, 190, 209
Probability of success 101, 203–204, 206
Proportionality 35, 141, 148, 190, 209, 214, 281, 284, 287
Psychological warfare 191, 228, 242, 245, 247, 249
Pufendorf, Samuel 65, 73–78
Punitive expedition 50, 67, 77–78, 92, 153, 167; Huang-Lao 133, 136, 138–140, 142, 144, 146; Mengzi (or Mencius) 39, 66, 69, 71, 78, 86, 305; *Zheng* (征) 66, 69, 74–75
Putin, Vladimir *see* Russia's narrative

Qing dynasty 53, 158–160, 162, 201

Realism 5, 105, 114, 298, 307, 310
Ren 161, 163, 166, 211, 305
Responsibility to protect (or R2P) 76, 85, 149, 287–288
Retributive justice 67, 70, 72–74
Revolutionary war 86, 170, 175, 181
Roman law 19
Rousseau 299, 304
Rule of law 98, 280, 283–284, 286
Russia's narrative 54–55

Science and technology ethics 225–226, 229, 231–235, 252

Scobell, Andrew 75
Self-defense 6, 24, 39, 66, 104–105, 133,
  136, 146, 163, 172, 184, 203, 266,
  269, 299, 302–303; as a just cause 18,
  138, 171, 178, 204, 207; against
  tyranny and oppression 280; and
  natural law 18; collective right of 4,
  285–287; human right of 283–284;
  individual 5, 306–309; of the state 3;
  right of 264, 280–288
Sino-Japanese War 171, 175, 180,
  184–186, 188, 200–201, 203–205,
  210–211, 218–220
Sino-Vietnamese War 75
Suarez, Francisco 65, 73, 77–78
Sunzi (or Sun Tzu) *see The Art of War*
Supreme emergency 189, 207, 219

Tao (討) 66
The Seven Military Classics 2, 7, 37, 39,
  101; and jus ad bellum 98–101; and
  jus in bello 96–97; legitimate cause for
  war 92–93; violence as last resort
  93–94; winning without fighting
  94–95; Yizhan 20–24
Three Warfares 228, 244, 252
Thucydides Trap 51, 57

United Nations 6, 48–49, 51, 54, 57, 76,
  86, 91, 247, 278, 305, 316; Charter
  53, 91, 284–287
Universal Declaration of Human Rights
  (or UDHR) 205, 278–280, 284–292
Unrestricted Warfare 30–32

Ukraine, Invasion of 48, 54–55

Van Norden, Bryan W. 66
Virtue ethics 42, 92, 98, 100, 105, 165–166

Wang Yang-ming 20, 24–26, 201–202,
  211, 214, 217
War crimes 149, 210–211
Warring States Period 1, 20, 39, 65, 70,
  121, 123, 126, 128, 143, 146, 148,
  192, 239
Weapon restrictions 19, 35
Westphalian 3, 49, 52, 55, 57, 75–76,
  99, 306
World War I 49
World War II 4–6, 11, 43, 48–49, 51,
  54–55, 97, 200, 245, 280–282, 285,
  296, 316

Xi Jinping 52, 226
Xunzi 39–41, 115, 120, 137, 161, 217

Yokoi Shōnan 298, 303–305

Zeng Guofan 2–3, 157–166; moral
  leadership 162; on Ren 161; ruling the
  army according to rituals 161; solider's
  virtues 162–163
Zhao, Qi 67, 68, 69, 81
Zheng (征) 66–78, 86, 129, 145; *see also*
  punitive expedition
Zhu (誅) – 66–68, 74, 78, 81; *baojun* 暴君
  139
Zhu Xi 65, 67–69, 74, 85, 158, 304